ENCYCLOPEDIA OF PHYSICS

EDITED BY

S. FLÜGGE

VOLUME XI/2

ACOUSTICS II

WITH 264 FIGURES

SPRINGER-VERLAG

BERLIN · GÖTTINGEN · HEIDELBERG

1962

HANDBUCH DER PHYSIK

HERAUSGEGEBEN VON
S. FLÜGGE

BAND XI/2

AKUSTIK II

MIT 264 FIGUREN

SPRINGER-VERLAG
BERLIN · GÖTTINGEN · HEIDELBERG
1962

ISBN 978-3-642-45978-8   ISBN 978-3-642-45976-4 (eBook)
DOI 10.1007/978-3-642-45976-4

Contents.

Generation and Measurement of Sound in Gases.

By

R. W. LEONARD.

With 96 Figures.

A. Generation of sound.

I. General ideas and definitions.

1. Classification of sources. It is convenient to classify sound sources in terms of three basic types: Monopole, dipole, and quadrupole. Although sources of higher order are realizable, most sources of practical interest will reduce to one of these three types as the wavelength of the radiated sound becomes very large compared to the dimensions of the source.

α) The monopole source. The monopole or simple source radiates uniformly in all directions. The classic example is the pulsating sphere, the surface of which moves radially with a velocity which is uniform in both magnitude and phase.

The specific radiation impedance (cf. Sect. 2) for a pulsating sphere is given by

$$\frac{p}{u_r} = z_r = \varrho c \, \frac{k^2 a^2 + jka}{1 + k^2 a^2} \quad (1.1)$$

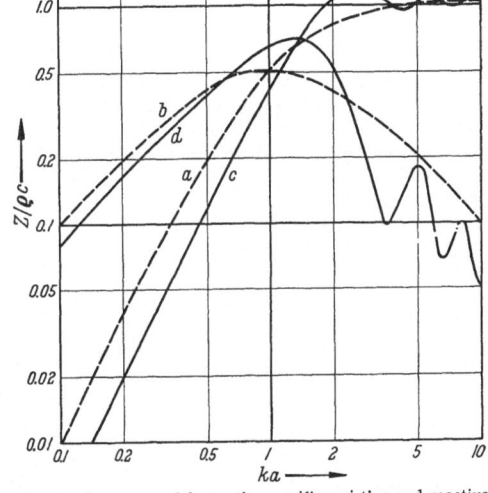

Fig. 1. Curves a and b are the specific resistive and reactive components of the radiation impedance of a pulsating sphere. Curves c and d are the average specific resistive and reactive components of the radiation impedance of a circular piston in an infinite plane baffle. The parameter k is $2\pi/\lambda$. The specific resistance and reactance presented here would appear as series elements in an equivalent circuit.

where $k = 2\pi/\lambda$, a is the radius of the sphere, ϱ is the density of the gas, c is the velocity of sound, p is the acoustic pressure, u_r is the acoustic particle velocity, and λ is the acoustic wavelength. Fig. 1, curves a and b, show plots of the real and imaginary parts of the specific radiation impedance as a function of ka.

It is interesting to compare this with the average specific radiation impedance for a rigid circular piston of radius a vibrating in an infinte rigid baffle and radiating into a half space. The average specific radiation impedance for the piston is given by

$$z_r = \varrho c \left[1 - \frac{J_1(2ka)}{ka} + j \, \frac{K_1(2ka)}{2(ka)^2} \right] \quad (1.2)$$

where

$$K_1(2ka) = \frac{2}{\pi} \left[\frac{(2ka)^3}{3} - \frac{(2ka)^5}{3^2 \cdot 5} + \frac{(2ka)^7}{3^2 \cdot 5^2 \cdot 7} \cdots \right] \quad (1.2b)$$

The real and imaginary parts appear as curves c and d in Fig. 1.

This comparison shows two things. First, that piston radiator essentially reduces to a monopole type source when the radiated wavelength becomes large compared to the radius of the piston ($ka < 1$). Second, that the radiation impedance for a pulsating sphere gives a reasonable smoothed approximation to the specific radiation impedance of a piston in an infinite plane baffle if the radius of the equivalent sphere is taken as 0.7 times the radius of the piston. This approximation is useful in setting up an equivalent circuit representation for direct radiator loud-

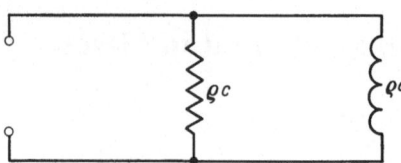

Fig. 2. The equivalent circuit representing the specific radiation impedance of a pulsating sphere as a parallel combination of inductance and resistance.

speaker since the radiation impedance for the pulsating sphere may be represented by a resistance and inductance in parallel. This is clearly the case if Eq. (1.1) is rewritten as

$$z_r = \cfrac{1}{\cfrac{1}{\varrho c} + \cfrac{1}{j\omega \varrho a}} \qquad (1.3)$$

where ω is $2\pi f$ and the other symbols are defined as previously. Fig. 2 gives the equivalent circuit for the specific radiation impedance of a pulsating sphere. It is interesting to note that although the pulsating sphere radiates uniformly and the piston in an infinite plane baffle has the marked directivity in the far field radiated intensity shown in Eq. (1.4)

$$I = \frac{1}{2} U^2 \varrho c k^2 \frac{a^4}{r^2} \left[\frac{2 J_1 (ka \sin \vartheta)}{ka \sin \vartheta} \right]^2 \qquad (1.4)$$

their radiation impedance functions are so similar. U is the velocity amplitude of the piston. This indicates that the effect of source geometry and environment modify the directional characteristic of a single source to a much greater extent than its radiation impedance.

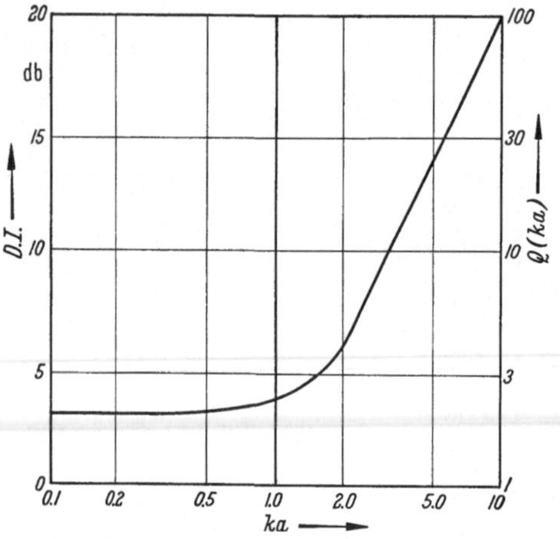

Source directivity may be expressed in terms of a directivity factor $Q(ka)$ where the directivity factor is the ratio of the intensity on axis of a source to the intensity of a non-directional source

Fig. 3. The directivity index in decibels as a function of ka for a circular piston in an infinite plane baffle. The corresponding directivity factor is indicated by the right-hand scale.

radiating the same total power. The directivity index D.I. in decibels (db) is defined by

$$\mathrm{D.I.} = 10 \, \mathrm{Log_{10}} \, [Q(ka)] \, \mathrm{db}. \qquad (1.5)$$

An excellent treatment of the calculation of directivity factors has been published by MOLLOY[1]. Fig. 3 shows the directivity index as a function of ka for the circular piston in an infinite baffle. Special note is to be taken of the linear region for $ka = 2$ and above. Here, the directivity is increasing with frequency at such a rate that the axial intensity is quadrupled for each doubling of frequency. Expressed in decibels, the curve is said to have a positive slope of 6 db per octave.

[1] C. T. MOLLOY: Directivity index for various types of radiators. J. Acoust. Soc. Amer. 20, 387 (1948).

The compensating effect of directivity on axial intensity will be referred to in connection with many of the practical radiators to be discussed in the following sections. As might be expected, the directivity index of a line source has a maximal slope of 3 db per octave.

β) The dipole source. The dipole or doublet source may be constructed of two monopoles of equal strength A and of opposite phase placed a distance d apart which is small compared to the radiated wavelength. The acoustic pressure field around the doublet is given by

$$p = A\, d\varrho c\, \frac{(k^2 r - jk)}{4\pi r^2}\, \cos\vartheta\; e^{j(\omega t - kr)} \tag{1.6}$$

and the intensity at large distances compared to the wavelength by

$$I_d = \frac{A^2\, d^2\, \varrho\, c\, k^4 \cos^2\vartheta}{16\pi^2 r^2} \tag{1.7}$$

where ϑ is the angle between the radial distance r and the axis of the doublet. The directional pattern is a figure of revolution about the doublet axis. Comparing this expression with the intensity radiated by a single monopole

$$I_m = \frac{A^2\, \varrho\, c\, k^2}{16\pi^2 r^2}$$

it is apparent that the two expressions differ by a factor $(kd)^2 \ll 1$. Thus, the dipole is a weak radiator compared to the monopole. The radiated intensity from a dipole varies as the f^4 whereas that from a monopole varies as f^2.

The classical example of the dipole is the oscillating sphere, a source which has been used for calibration purposes. The radiation from a vibrating string may be considered as due to a linear array of colinear dipoles with their axes in the direction of motion of the string. The low-frequency radiation of an unbaffled direct-radiator loudspeaker is that of a dipole.

γ) The quadrupole source. Quadrupole sources are of two types: lateral and longitudinal. The lateral quadrupole results from the out-of-phase radial motion of alternate quadrants of a spherical surface. The radiated pressure field is symmetrical about the equatorial plane and has a four-lobed angular dependence about the polar axis. The lateral quadrupole may be synthesized from four monopoles in a square array, the phases changing by π from corner to corner around the square. It is clear that this array may be synthesized out of two antiparallel dipoles placed side by side.

The longitudinal quadrupole consists of two antiparallel dipoles placed end to end in a linear array. It may be considered as a linear array of four monopoles with the inner pair 180° out of phase with the two at the ends of the array. The motion of the tines of a tuning fork results in a longitudinal quadrupole type of radiation when the radiated wavelength is very large compared to the separation of the tines. The directivity pattern of this type of quadrupole radiator is characterized by symmetry about the axis of the array and a two-lobed dependence on the polar angle.

2. Equivalent circuits for mechanical and acoustical systems. It is convenient to construct equivalent or analogous electrical networks when considering the performance of sound sources even in cases where the transduction does not involve electrical quantities directly. Since physicists are, as a rule, more familiar with electrical circuits than with the combinations of acoustic and mechanical elements that make up sound generating and detecting systems, there is considerable to be

gained by an equivalent network description. This is particularly true in understanding the role of the elements of a system in determining its frequency response
characteristics. It will be necessary to use three types of impedance in many of
the systems to be considered. These are mechanical impedance Z_m, acoustic
impedance Z_a, and specific acoustic impedance z. These are defined by the
equations

$$Z_m = \frac{\text{force}}{\text{velocity}} = \frac{f}{u},$$

$$Z_a = \frac{\text{pressure}}{\text{volume current}} = \frac{p}{Su},$$

$$z = \frac{\text{pressure}}{\text{particle velocity}} = \frac{p}{u}$$

since force is the product of pressure and area, $f = pS$, the impedances are related
by

$$Z_m = Sz = S^2 Z_a$$

where S is the area over which the force is developed. In general, all three types
of impedances will be complex numbers; the resistances, reactances, and phase
angles depending on the details of the system.

In constructing equivalent circuits for mechanical systems, we will make use
of the direct analogy in which voltage and force correspond, and electrical current
and velocity correspond.

In constructing the equivalent circuits for acoustic systems, voltage and
pressure will correspond, and electrical current and volume current will correspond.
Since the interaction of the acoustic elements of a system with an electrical
transducer will involve a mechanical element such as a diaphragm, the acoustic
impedances forming the system will be converted to mechanical impedances in
setting up the equivalent circuits.

3. Electrodynamic transduction. An electrodynamic transducer is one in which
the mechanical force is produced by a current flowing in a conductor immersed
in a constant magnetic field. Since the force per unit length of conductor depends
on the vector product $\boldsymbol{I} \times \boldsymbol{B}$, it is maximized if the conductor carrying the current
is perpendicular to the lines of induction. For this case, the force F in newtons
is given by

$$F = BlI \tag{3.1}$$

where B is in webers per square meter and I in amperes and l in meters. If the
conductor moves in the magnetic field, an electromotive force is generated. This
generated voltage is given by

$$E_m = -BlU \tag{3.2}$$

where U is the magnitude of the velocity of the conductor in meters per second.
The velocity and the force are related by the *mechanical* impedance of the system
Z_m as follows

$$U = \frac{F}{Z_m}. \tag{3.3}$$

Combining Eqs. (3.1), (3.2), and (3.3), we have

$$E_m = -\frac{B^2 l^2 I}{Z_m} \tag{3.4}$$

and we may define a *motional* impedance Z_M as

$$Z_M = -\frac{E_m}{I} = \frac{B^2 l^2}{Z_m} \tag{3.5}$$

or

$$Z_M = B^2 l^2 Y_m \tag{3.6}$$

where Y_m is the mechanical admittance of the system including the conductor. This transformation of a mechanical admittance into an electrical impedance is common to all types of magnetic transducers since the motional electromotive force can be related to the velocity of some part of the transducer. Since the *total electrical* impedance is given by

$$Z_e = Z'_e + Z_M \qquad (3.7)$$

where Z'_e is the *blocked or clamped electrical* impedance of the transducer, it is clear that the efficiency of transduction will be greatest when Z_M is maximized and a real number. This is attained by making the product Bl as large as possible and Y_m both large and real. The second condition is satisfied by series resonance (velocity resonance) in the mechanical system if the mechanical resistance is small. If an efficient sound source is to be obtained, the mechanical resistance should result from the radiation of sound.

Since electrodynamic transduction is a reversible process, one may expect that when the transducer is connected to an electrical source of impedance Z_s and a force applied to the mechanical side of the transducer, a current will flow in the electrical side. The current so produced is given by

$$I = \frac{BlU}{Z'_e + Z_s} \qquad (3.8)$$

where BlU is the voltage generated and $Z'_e + Z_s$ impedance of the electrical side. This current in interacting with the magnetic flux results in a force given by

$$F_b = -BlI = -\frac{B^2 l^2 U}{Z'_e + Z_s} \qquad (3.9)$$

and we may define a *mechano-electric* impedance Z_{me} as

$$Z_{me} = -\frac{F_b}{U} = \frac{B^2 l^2}{Z'_e + Z_s} = B^2 l^2 Y_e \qquad (3.10)$$

where Y_e is the admittance of the electrical side of the transducer. We are now in a position to set down the mechanical equivalent circuit of the electrodynamic transducer when acted upon by an external force F_m of mechanical origin as shown in Fig. 4. It is clear that if Y_e is real (conductive), Z_{me} will be a pure resistance and provide additional damping to the mechanical system. A classic example of this is the critical damping of a moving coil galvanometer by the proper value of resistance connected across the electrical terminals. This mechanoelectric impedance does not appear

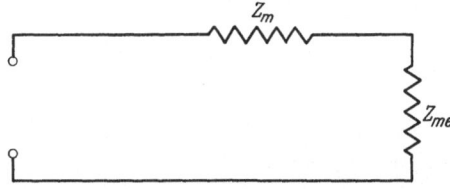

Fig. 4. The mechanical equivalent circuit for an electrodynamic transducer driven by an external force F of mechanical origin.

in the equation for the velocity due to a current in the conductor but may be introduced if the force is calculated from the blocked current as shown in Eq. (3.11) where I is the actual current in the conductor and I' is the blocked current.

$$U = \frac{BlI}{Z_m} = \frac{BlI'}{Z_m + Z_{me}}. \qquad (3.11)$$

The currents I and I' are given by

$$I = \frac{E}{Z'_e + Z_m}, \qquad I' = \frac{E}{Z'_e}. \qquad (3.12)$$

Although the formulation in terms of the blocked current is sometimes used in electrodynamic source calculations, the mechano-electric impedance is of most importance in electrodynamic microphones where the force acting is of mechanical origin.

4. Electrostatic transduction. Electrostatic transduction is probably the earliest known form of electro-mechanical transduction. An exhaustive study of the theory of electrostatic transduction and its historical background has been given by Hunt[1]. Only a linearized first order theory will be considered here.

The force between two plane parallel conducting surfaces is given by

$$f = \frac{\varepsilon_0 A V^2}{2 x_0^2} \tag{4.1}$$

where x_0 is the separation of the surfaces in meters, A the area of one surface in square meters, V the potential difference in volts and ε_0 is the permittivity of free space ($\varepsilon_0 = 8.85 \times 10^{-12}$ coulombs² per newtonmeter²)*. If we choose V as follows

$$V = V_0 + V_1 e^{j \omega t} \tag{4.2}$$

the instantaneous force f is given by

$$f = \frac{\varepsilon_0 A}{2 x_0^2} [V_0^2 + 2 V_0 V_1 e^{j \omega t} + V_1^2 e^{2 j \omega t}] \tag{4.3}$$

where the second term in the brackets represents the fundamental component of the driving force. The third term represents a second order term resulting in a second harmonic in the driving force f. The fundamental component of the velocity of the movable plate is given by

$$u = \frac{\varepsilon_0 A V_0 V_1 e^{j \omega}}{x_0^2 Z'_m} \tag{4.4}$$

where Z'_m is the effective mechanical impedance of the system including the negative compliance resulting from charge variation on the plates as the spacing changes. To relate velocity to current, we write for the charge on the plates

$$q = \frac{\varepsilon_0 A V_0}{x_0} \tag{4.5}$$

and for the current due to plate motion

$$i_M = \frac{dq}{dt} = -\frac{\varepsilon_0 A V_0}{x_0^2} \cdot \frac{dx}{dt} = -\frac{\varepsilon_0 A V_0 u}{x_0^2}. \tag{4.6}$$

Combining Eqs. (4.4) and (4.6) we have

$$i_M = \frac{\varepsilon_0^2 A^2 V_0^2 V_1 e^{j \omega t}}{x_0^4 Z'_m}, \tag{4.7}$$

and adding to this the current which would flow if the plates were immobile, we have for the total current

$$i = V_1 e^{j \omega t} \left(j \omega C_0 + \frac{\varepsilon_0^2 A^2 V_0^2}{x_0^4 Z'_m} \right). \tag{4.8}$$

The electrical circuit including the motional impedance is made up of two parallel branches as indicated in Fig. 5. In the electrostatic case, the mechanical impedance

---

[1] F. V. Hunt: Electroacoustics. New York: Harvard University Press; John Wiley & Sons, Inc. 1954.
* The rationalized M.K.S. system is used here.

is transformed directly and appears in parallel with the blocked impedance in contrast to the electrodynamic case where it is transformed reciprocally and appear in series with the blocked impedance.

Next, let us consider the effect of maintaining a constant voltage difference between the plates as this separation is varied. The force required to overcome the attractive force between the plates is given by

$$f = \frac{\varepsilon_0 V_0^2 A}{2 x^2} \tag{4.9}$$

and differentiating with respect to time, we have

$$\frac{df}{dt} = \frac{- \varepsilon_0 A V_0^2 u}{x_0^3}, \tag{4.10}$$

and writing

$$u = C_{me} \frac{df}{dt} \tag{4.11}$$

we see that the mechanical compliance C_{me} due to the change in electrical charge required to maintain a constant V_0 is given by

$$C_{me} = \frac{- x_0^3}{\varepsilon_0 A V_0^2} = \frac{- x_0^2}{C_0 V_0^2} \tag{4.12}$$

where C_0 is the electrical capacitance of the transducer and x_0 is the equilibrium spacing. With the evaluation of this additional term, we may write for the effective mechanical impedance of the transducer Z'_m

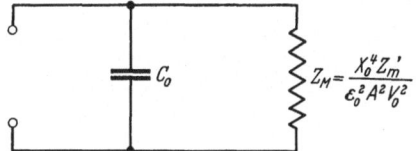

Fig. 5. The electrical circuit for an electrostatic transducer including the motional impedance.

$$Z'_m = R_m + j \left[\omega M - \left(\frac{1}{\omega C_m} - \frac{C_0 V_0^2}{\omega x_0^2} \right) \right] \tag{4.13}$$

where R_m is the resistance, M the mass, and C_m the mechanical compliance of the uncharged ($V_0 = 0$) transducer. If the transducer is coupled to a fluid, the effective stiffness of the transducer K'_m is given by

$$K'_m = \frac{1}{C_m} - \frac{C_0 V_0^2}{x_0^2}. \tag{4.14}$$

Since the restoring force Δf for a small displacement Δx is given by

$$\Delta f = - K'_m \Delta x \tag{4.15}$$

we observe that the transducer will not have a stable region unless

$$\frac{1}{C_m} > \frac{C_0 V_0^2}{x_0^2}. \tag{4.16}$$

The extent of the stable region for constant V_0 is indicated in Fig. 6 where the negative of the attractive force and the mechanical restoring force are plotted. The straight line is the mechanical restoring force vs. displacement from the unpolarized equilibrium position x_0'. The solid curve is the negative of the attractive force for polarization $V = V_0$, and the intersection with the straight line determines the polarized equilibrium position x_0. The dashed curve is the negative of the attractive force for a polarization $V = \sqrt{2} V_0$. It is clear that the diaphragm will be stable and return to x_0 if for $V = V_0$ it does not get closer to the fixed plate than the position indicated by the second intersection of the two solid curves at S. For displacements from x_0 less than that at S, the restoring force exceeds that of electrostatic attraction and return to x_0 is to be expected. However, if the polarization is increased ($V = \sqrt{2}\,V_0$) the attractive force will exceed the restoring force everywhere except at S' where the intersection is an unstable one, and the diaphragm will collapse against the fixed plate. $V = \sqrt{2}\,V_0$ represents the peak voltage (a.c. + d. c.) that may be applied in the low-frequency range where the diaphragm impedance is determined by the mechanical stiffness of the diaphragm. In a practical design, the margin of safety would be considerably greater than in this example.

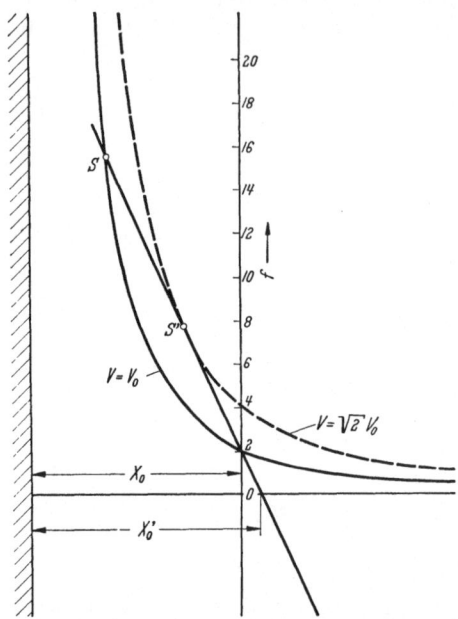

Fig. 6. The straight line plots the mechanical restoring force vs. displacement acting on the movable plate of a parallel plate electrostatic transducer. The curves represent the negative of the attractive forces between the two electrodes for polarization voltages of V_0 and $\sqrt{2}\,V_0$. Point S limits the stable region for polarization V_0. No stable position exists for polarization $\sqrt{2}\,V_0$ indicated by the dashed curve.

Harmonic distortion. The nonlinear attraction between the movable and fixed plate of the electrostatic transducer combined with the second harmonic in the driving force may result in serious distortion. Bobb[1] has shown that the percent second harmonic distortion is given by

$$d_2 = 50\,\frac{V_1}{V_0} \qquad (4.17)$$

where V_1 is the peak value of the a.c. driving voltage. This is just twice the percent second harmonic present in the driving force as can be seen by expanding the real part of Eq. (4.3)

$$f = \frac{\varepsilon_0 A}{2\,x_0^2}\left[V_0^2 + 2\,V_0\,V_1\cos\omega t + \frac{1}{2}\,V_1^2\cos 2\omega t + \frac{1}{2}\,V_1^2\right] \qquad (4.18)$$

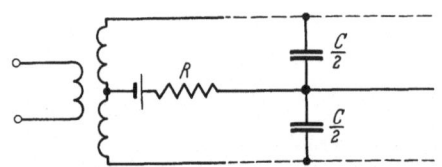

Fig. 7. The electrical circuit for a push-pull electrostatic transducer. The polarization potential is supplied by the battery through R. The capacitors represent the capacity between the driving electrodes and the movable plate. The push-pull transformer appears at the left.

and taking the percent ratio of the amplitudes of the third term to the second, we have

$$100\,\frac{|f_2|}{|f_1|} = 25\,\frac{V_1}{V_0}. \qquad (4.19)$$

The nonlinear drive and nonlinear mechanical compliance of the single-sided electrostatic transducer can largely be overcome by push-pull operation as indicated in Fig. 7. Hunt[2] has pointed out the fact that if the charge on the diaphragm remains essentially constant during an oscillation of the diaphragm, the electrostatic forces will be independent of the position of the diaphragm in the

    [1] L. Bobb, R. G. Goldman and R. W. Roop: Design and performance of a high-frequency electrostatic speaker. J. Acoust. Soc. Amer. **27**, 1128—1133 (1955).
    [2] F. V. Hunt: Electroacoustics, p. 189. New York: Harvard University Press; John Wiley & Sons, Inc. 1954.

gap between the fixed electrodes. This demands that the electrical time constant of the diaphragm-plate combination (RC) must be large compared to the period of oscillation of the diaphragm. This condition can be met in the push-pull case but not in the single-sided transducer. The other effect of push-pull is to reduce the second harmonic component in the driving force. In the push-pull case, the driving force is given by

$$f = \frac{\varepsilon_0 A}{2 x_0^2} \left[(V_0 + V_1 e^{j\omega t})^2 - (V_0 - V_1 e^{j\omega t})^2 \right] \qquad (4.20)$$

which on expansion yields no second harmonic terms.

The combination of constant charge operation and distortionless driving force makes the push-pull electrostatic transducer a very linear device.

II. Direct-radiator loudspeakers.

5. Electrodynamic loudspeakers. Electrodynamic transduction has proven to be the most effective means of converting an electrical signal into the motion required to radiate low-frequency airborne sound. Because of the low intrinsic impedance (ϱc) of air, large amplitudes are required to radiate appreciable power. This is especially true at low frequencies where the specific radiation resistance is further reduced by the small size of the source compared to a wavelength. The large amplitude requirements rule out the use of electrostatic, magnetostrictive, and piezoelectric transduction for frequencies below 1000 c.p.s. Fig. 8 shows the constructional details of a typical electrodynamic loudspeaker of modern design. The magnetic field intensity B (≈ 12000 gauss or 1.2 webers/m²) in the airgap containing the voice coil is supplied by a cylindrical slug of permanently magnetized material. The voice coil is of edge-wound copper ribbon and fills about 80% of the air gap. The electrical resistance of the voice coil is usually about 10 ohm. The

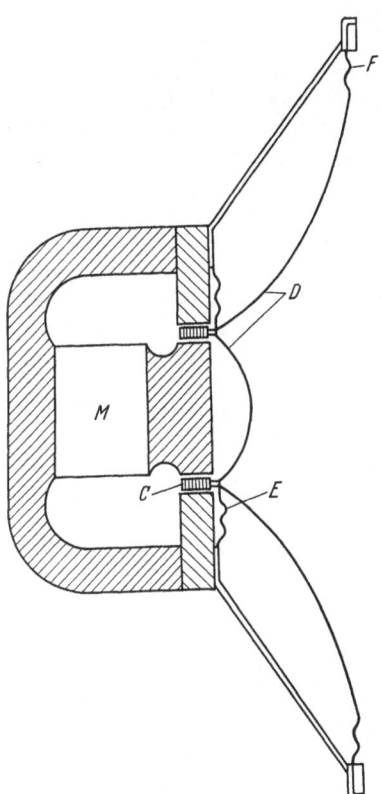

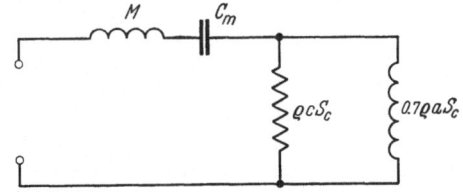

Fig. 8. A modern electrodynamic loudspeaker. The voice coil C, the diaphragm D, the centering suspension E, the edge compliance F, and the permanent magnet M are indicated in the sectional view.

Fig. 9. Mechanical equivalent circuit for direct-radiator electrodynamic loudspeaker. M is the mass of the moving system, C_m the compliance of the cone, a the radius of the cone base, ϱ the density of air, c the velocity of sound in air, and S_c the area of the cone base.

length l of the voice coil wire is about 20 meter. The combined mass of the paper cone and voice coil is about 40 gm. The effective diameter of the cone as a piston is 35 cm.

α) *Equivalent circuits of direct-radiator electrodynamic loudspeaker in an infinite baffle.* The total mechanical impedance Z_m may be obtained from the circuit in Fig. 9. Since the mechanical impedance Z_m is transformed reciprocally by the electrodynamic transduction, it is necessary to obtain the mechanical admittance. This may be done in terms of admittances of the elements making up the mechanical circuits. The total mechanical admittance is given by

$$Y_m = \frac{1}{\frac{1}{Y_1} + \frac{1}{Y_2} + \frac{1}{Y_3 + Y_4}} \tag{5.1}$$

where

$$\frac{1}{Y_1} = j\omega M, \quad \frac{1}{Y_2} = \frac{1}{j\omega C_m}, \quad \frac{1}{Y_3} = \varrho c S_c, \quad \frac{1}{Y_4} = j\omega\,(0.7\varrho\,a_c\,S_c).$$

The motional impedance is then

$$Z_M = B^2\,l^2\,Y_m = \frac{1}{\frac{1}{B^2 l^2 Y_1} + \frac{1}{B^2 l^2 Y_2} + \frac{1}{B^2 l^2 (Y_3 + Y_4)}} \tag{5.2}$$

and may be derived from the circuit in Fig. 10 by taking the reciprocal of the sum of the reciprocals of the three branch electrical impedances $B^2 l^2 Y_1$, $B^2 l^2 Y_2$, and $B^2 l^2 (Y_3 + Y_4)$. Inserting the mechanical parameters, we have the complete equivalent circuit including the resistance R_e' and inductance L_e' of the voice coil shown in Fig. 11. It is to be noted that the mechanical inductances in Fig. 9 appear as capacitors and the mechanical compliance as an inductor. And that elements formerly in series are now in parallel, and those formerly in parallel are now in series. We have, in essence, constructed the dual of the mechanical circuit of Fig. 9 and connected it in series with the electrical side of the transducer introducing the proper dimensional changes through the parameter $B^2 l^2$. The inductance is in henrys, the capacitors in farads, and the resistance in ohms. The acoustic power radiated is given by

$$P_A = I_3^2\,\frac{B^2\,l^2}{\varrho c S_c} \tag{5.3}$$

where I_3 is the root mean square current in the transformed mechanical radiation resistance. The fractional efficiency is given by

$$\eta = \frac{I_3^2\,\dfrac{B^2\,l^2}{\varrho c S_s}}{I_1^2\,R_e' + I_3^2\,\dfrac{B^2\,l^2}{\varrho c S_c}}. \tag{5.4}$$

The evaluation of η as a function of frequency is a rather lengthy task. The fractional efficiency at mechanical resonance is readily obtained since the mesh in which I_2 flows presents an anti-resonant impedance making I_1 essentially equal to I_3 and η_r is given by

$$\eta_r = \frac{\dfrac{B^2\,l^2}{\varrho c S_c}}{R_e' + \dfrac{B^2\,l^2}{\varrho c S_c}}. \tag{5.5}$$

The resonant efficiency may exceed 0.6 when a carefully designed magnetic structure is used. Efficiencies as high as this can be obtained only over a very limited band width.

For the reproduction of music and speech, a uniform radiated power vs. frequency characteristic is desirable over a large band of frequencies. This is possible above the frequency of mechanical resonance if the electrical network is fed by a constant current source. The circuit for this condition appears in Fig. 12. It is clear that as long as the impedance through which I_3 flows is controlled by the reactance of the capacitor C_2 that the current I_3 will remain constant, and constant power will be radiated. This will be the case nearly up to the frequency at which the reactance of C_2 is equal to R_2 for which

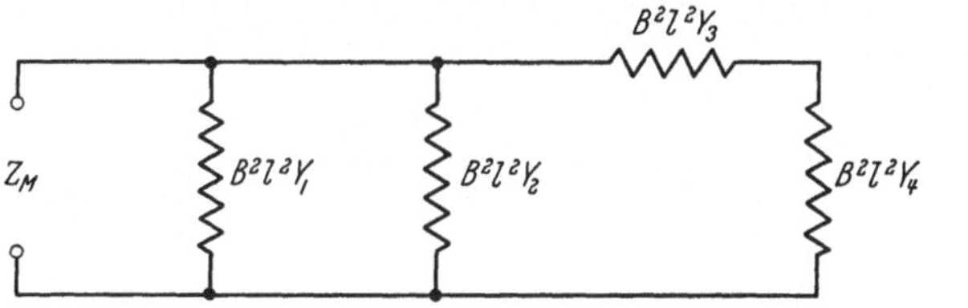

$$0.7 \omega \varrho \, a_c \, S_c = \varrho \, c \, S_c \quad (5.6)$$

or

$$\frac{\omega a_c}{c} = \frac{2\pi a_c}{\lambda} = 1.4. \quad (5.7)$$

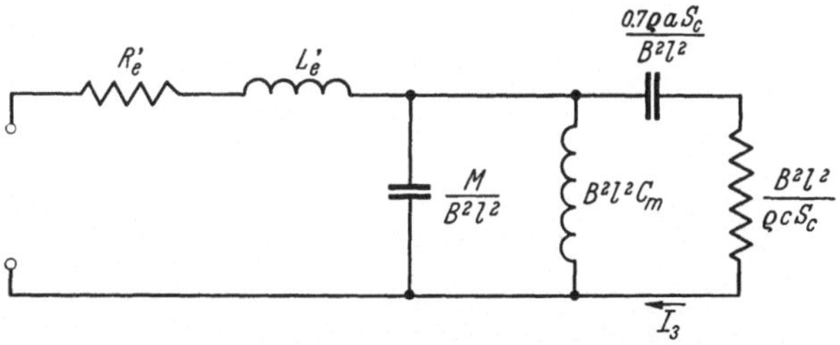

Above this frequency, the power radiated will fall off as $1/f^2$ or 6 db per octave.

At frequencies below mechanical resonance, the equivalent circuit for the constant current fed system is given in Fig. 13. It is clear that the current I_3 will increase as f^2 or that the power will be increased by 12 db per octave as the frequency is increased[1].

Fig. 14 shows the relative power radiated vs. frequency for different source impedances. These curves were derived from an electrical analogue in which both frequency and impedance were scaled to permit the use of available circuit elements. The equivalent circuits discussed above are useful in predicting total power radiated vs. frequency but do not include the effects of directivity. Directivity must be taken into account in predicting the sound intensity at any point in front of the loudspeaker. The directivity of a cone-type radiator is similar to that of a flat piston as shown in Fig. 3. Measurements of the axial intensity radiated by a cone loudspeaker follow the prediction of the equivalent circuit up to frequencies where the directivity begins to be appreciable $(\lambda = \pi a_c)$, and the radiated power is beginning to drop off. Above this frequency, the decrease in power radiated is compensated for by the decrease in solid angle of the central lobe into which most of the sound is radiated, with the result that the axial intensity is nearly independent of frequency.

Fig. 10. Equivalent circuit for Eq. (5.2).

Fig. 11. The electrical circuit for the electrodynamic loudspeaker including the transformed mechanical quantities entering into the motional impedance.

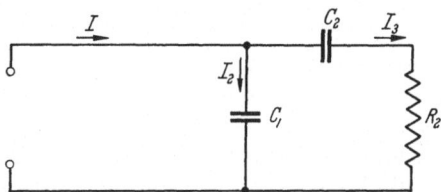

Fig. 12. The equivalent electrical circuit for the direct-radiator loudspeaker above mechanical resonance.

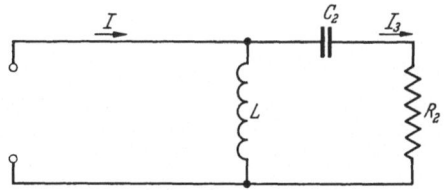

Fig. 13. The equivalent circuit for the direct-radiator electrodynamic loudspeaker at frequencies below mechanical resonance.

---

[1] The number of decibels (db) = 10 Log_{10} (Power Ratio) or 20 Log_{10} (Current Ratio).

The preceding discussion assumes that the loudspeaker cone moves as a rigid unit. This assumption is reasonably good up to about 500 c.p.s. but not valid above this range for large cones. The paper cone has normal modes of vibration having radial nodal lines and/or nodal circles. The presence of the standing flextural waves on the cone has a marked effect on the radiated sound. These standing waves may result in peaks and dips in the axial pressure frequency curve separated by as much as 10 db. CORRINGTON[1] has made a systematic study of these effects.

β) *Amplitude distortion.* There are two important sources of non-linear distortion in the directradiator electrodynamic loudspeaker. The first is amplitude

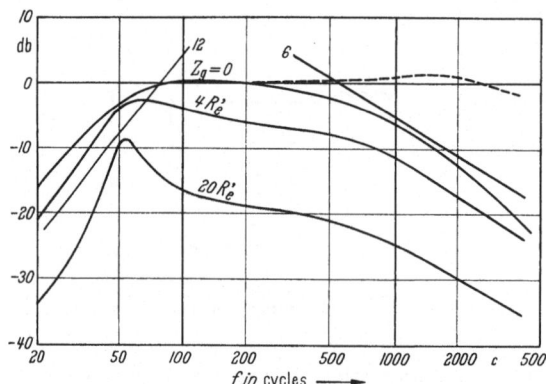

Fig. 14. The solid curves are relative power radiated in decibles vs. frequency in cycles per second for different volumes of source impedance. The straight lines have slopes of 12 and 6 db per octave. The dashed curve indicates the expected acoustic pressure on the axis of the speaker cone.

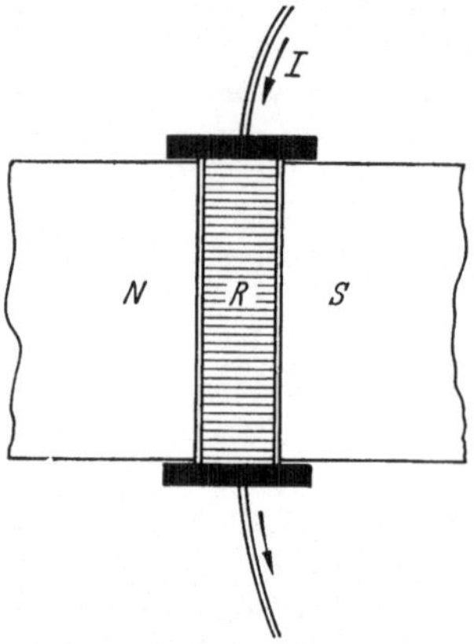

Fig. 15. The ribbon loudspeaker. The ribbon marked R is stretched between the poles of a magnetic structure marked N and S.

limiting by the compliance at the edge of the cone. In principle, the solution to this is simple, but in practice, it is complicated by resonances in the compliance, itself.

The second source of distortion is associated with voice coil travel. If the amplitudes are large enough, the voice coil will move out into the fringing field of the gap and amplitude distortion introduced. This distortion may be reduced either by making the voice coil long compared to the gap length or the gap length long compared to the voice coil travel[2].

γ) *The ribbon loudspeaker.* This device consists of a thin metal ribbon stretched between the poles of a magnetic structure with the magnetic field in the plane of the ribbon. A current is then passed through the ribbon at right angles to the magnetic field resulting in a force normal to the surface of the ribbon. The arrangement of parts is shown in Fig. 15.

Unless the ribbon source is to be used at resonance where the mass can be tuned out, it is necessary to use very thin ribbon of the order of 5.0 micron thick. Aluminum alloys are suitable due to their relatively low density and good electrical conductivity. The width of the ribbon is usually limited to about 1 cm in order to maintain a high field. The force in newtons per ampere is just the product of the field B, and l, the length of the ribbon. The efficiency is proportional to B^2, and fields of the order of 1 weber per square meter are desirable. Efficiencies of the order of 10% are possible with the ribbon source coupled to a tube or horn,

[1] M. S. CORRINGTON: Amplitude and phase measurements on loudspeaker cones. Proc. Inst. Radio Engrs. **39**, 1021 (1951).
[2] H. F. OLSON: RCA Review **2**, 265 (1937).

but the low impedance and limited power output make this source useful as a research tool only[1].

6. Electrodynamic loudspeaker enclosures. *α) Simple enclosure.* An enclosure covering the back surface of the loudspeaker cone is essential to good low-frequency radiation. If no baffle or no enclosure is used, the wave from the back of the cone will interfere destructively with that from the front surface at frequencies below that for which the wavelength is approximately equal to the diameter of the cone. Below this rough limit, the unbaffled loudspeaker will radiate as a dipole source. As indicated in Eq. (1.7) this results in a radiated intensity which is lower than that of a simple source by a factor $k^2 d^2$ for $kd < 1$

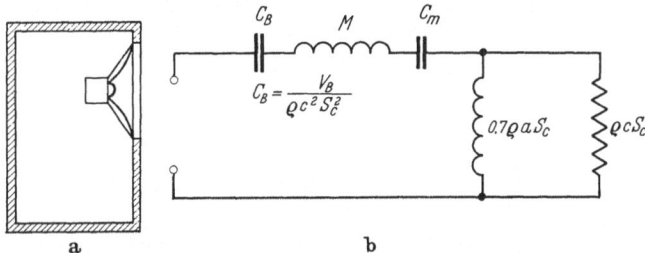

Fig. 16 a and b. The simple enclosure and equivalent circuit for a direct-radiator cone speaker. C_B is the mechanical compliance of the enclosure air volume V_B referred to the cone area S_c. M' includes the inertia of the air moved by the back of the cone.

where d is of the order of magnitude of the diameter of the loudspeaker cone. For a diameter of 30 cm and a frequency of 50 c.p.s., the factor $k^2 d^2$ is approximately one-tenth.

Since the infinite plane baffle is impractical, the simplest solution is the closed box shown in Fig. 16a. If the volume of the box is large enough, it will have only a small effect on the mechanical impedance of the loudspeaker. The box acts as an added stiffness to the loudspeaker cone. The equivalent mechanical circuit in Fig. 16b. The reciprocal of the added stiffness appears as a series compliance C_B in the equivalent circuit. The only effect of C_B is to raise the frequency of the mechanical resonance. If the volume V is of the order of $0.1\,M^3$ for a cone 30 cm in diameter, the change in resonant frequency will be unimportant The dimensions of a simple enclosure are not critical, but their ratios should not be more extreme than $\sqrt{2}$. It is necessary to treat the enclosure with sound absorbing material to damp the resonances of the enclosure due to standing waves. A standing wave in the enclosure having a pressure antinode at the center of the rear surface of the loudspeaker cone will tend to block the motion of the cone producing a dip in the radiated intensity. The magnitude of this dip will be greater the larger the Q of the standing wave. It will depend on the magnitude of the driving point impedance presented to the back of the cone. Sound absorbing material 1 cm thick can be fastened diaginally across opposite corners of a rectangular enclosure in the form of equilateral triangles. Material spaced out in this manner is much more effective than many times the same area fastened directly in contact with the enclosure walls. Small leaks or cracks at the joints in the enclosure have no serious effect as long as the vibration of the walls excited by the pressure variations inside the enclosure does not result in rattling.

Mounting the loudspeaker in a simple enclosure reduces its directivity over that obtained in a large flat baffle such as the wall of a room. The diffraction

---

[1] F. A. ANGONA: Absorption of sound in gases by a tube method. J. Acoust. Soc. Amer. **25**, 1111 (1953).

around the edges of the face of a small enclosure increases the angular spread of the radiated sound field. Placement of the enclosure in the corner of a rectangular room reduces the angular spread required for reasonable uniformity of radiated intensity to all parts of the room.

β) *Bass-reflex enclosure.* The bass-reflex enclosure or acoustic phase inverter[1,2] is a ported enclosure designed to improve the low-frequency radiation of a direct-radiator loudspeaker. Fig. 17 shows the enclosure and simple equivalent circuit

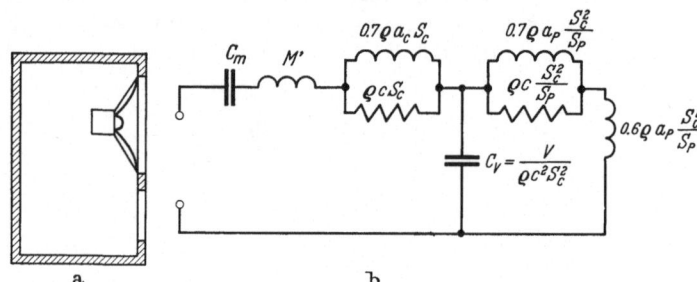

Fig. 17 a and b. A sectional view of a bass-reflex enclosure is shown in (a). The electrical analogue of the mechano-acoustic system is shown in (b). M' is the mass of the cone plus the effective mass of the air moved by the back of the cone. The inductance in series with the radiation impedance of the port represents inertia of the air just inside the port.

of the mechanical system. The additional parameters are a_p the radius of the circular port, and S_p the port area. This equivalent circuit does not take into account the fact that the volume of the bass-reflex enclosure is driven from the back of the cone while the radiation impedance presented to the cone is driven by the front. It is evident that at frequencies below the resonant frequency of the port-volume combination where the reactance of C_v becomes very large, that the particle velocity in the port will be in phase with the back of the cone and

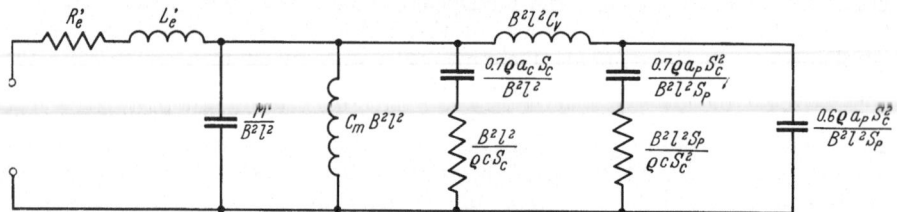

Fig. 18. The equivalent electrical circuit including the motional impedance of the mechano-acoustic system.

hence out of phase with the front of the cone. This results in a dipole type radiation similar to that of an unbaffled loudspeaker. The radiated power falls off as f^4 as the frequency decreases if the cone velocity is held constant. Since the cone velocity decreases as f below the cone resonance, the combination of these two effects leads to an 18 db per octave slope for the radiated power. Above the port-volume resonance, phase inversion occurs, and the port velocity is approximately out of phase with the back of the cone or in phase with the front of the cone, and the cone and port radiate in phase as two simple sources. This in-phase condition continues as the frequency increases, but the port velocity falls at 12 db

[1] H. F. Olson: Elements of Acoustical Engineering, 2nd ed., pp. 154—156. New York: D. Van Nostrand Company, Inc. 1947.

[2] F. V. Hunt: Electroacoustics, pp. 159—164. New York: Harvard University Press; John Wiley & Sons, Inc. 1954.

per octave and is only important near the port-volume resonance. Thus, the effect of the enclosure is to extend the low-frequency range about an octave and then cut it off at 18 db per octave.

The equivalent mechanical circuit if Fig. 17b for the bass-reflex loudspeaker can be transformed into the equivalent electrical circuit presented to the terminals of the voice coil by constructing the dual of the mechanical circuit, representing the mechanical admittance and multiplying by $B^2 l^2$ as before. The result of this transformation appears in Fig. 18. In making use of this circuit to compute the total power radiated at low frequencies, where the separation of the cone and port is very small compared to the radiated wavelength, it is necessary to take the square of the weighted vector difference of the currents in the two resistive branches multiplied by the resistance of the first branch. The total power radiated is given by

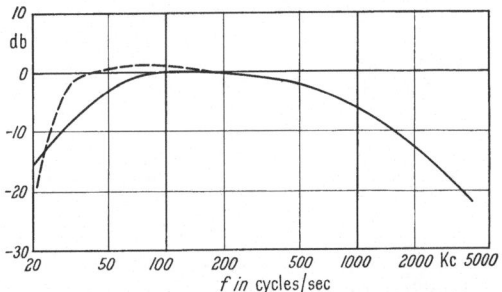

$$P = \left(I_c - I_P \frac{a_P}{a_c}\right)^2 \frac{B^2 l^2}{\varrho c S_c} \quad (6.1)$$

where I_c and I_P are the currents in the branches representing the transformed mechanical radiation admittances of the cone and port, respectively. The difference in the two currents is used because the port is driven by the back of the cone, and the direct radiation from the cone is derived from its front surface.

Fig. 19. Shows the increase in low-frequency radiation achieved by a bass-reflex enclosure where the port area is made equal to the cone area. The curves are the result of an analogue computation using the equivalent circuit of Fig. 18.

For the case of equal areas for cone and port, it is clear that the dipole nature of this double source would be lost below the port resonance if the currents were added vectorially rather than subtracted. The weighting factor a_P/a_c weights the contribution of the port to the total free field pressure. Eq. (6.1) was derived on the assumption that pressure at any point in space is the vector sum of the pressure contributions of the cone and port. Thus, for two simple sources, one sums the radiated pressures rather than the radiated powers. Thus, for two equal monopole sources in phase and separated by a distance small compared to a wavelength, the total power radiated is four times that of one of the sources alone. The increase in radiation impedance necessary to account for the radiation of double the power by each of the sources is not taken into account in the simple analogue shown in Fig. 18. However, in most practical cases where the $a_P < a_c$ the mutual impedance resulting from the interaction of the two sources has a small effect on the loading of the electrical generator and the approximation involved in using the simple analogue is justified. LOCANTHI[1] has computed the required mutual impedance and published a paper on the more accurate analogue.

Fig. 19 shows the increase in low-frequency radiation achieved by a bass-reflex enclosure where the port area is made equal to the cone area. The curves are the result of an analogue computation using the equivalent circuit of Fig. 18.

7. Electrostatic loudspeaker. The electrostatic loudspeaker has received considerable attention in the past six years, both in the United States of America and in Germany. The work of KUHL at the Third Physical Institute in Göttingen and that of JANSZEN at the Acoustics Research Laboratory, Harvard University has resulted in marked advances in the development of electrostatic loudspeakers.

---

[1] B. N. LOCANTHI: Trans. Inst. Radio Engrs., PGA-6, March, 1952.

The work of Janszen[1] has been largely along the lines of developing a highvoltage high-impedance push-pull type of electrostatic loudspeaker. The work of Kuhl[2,3] has resulted in the development of low-voltage low-impedance solid dielectric type loudspeakers. Both developments were assisted very materially by the availability of new plastic materials having the proper mechanical and electrical properties.

α) *The high-voltage electrostatic loudspeaker.* This type may require polarization voltages of 1 to 3 kilovolts. The insulating dielectric is air although the electrodes which drive the membrane are usually covered by an insulating material to avoid a short-circuit in case the membrane touches one of them. Fig. 20 shows a section of a push-pull loudspeaker. The membrane is a thin film of plastic (rubber hydrochloride 0.0015 gm/cm^2) with a thin layer of conducting material on the surface. Fig. 21 shows the relative axial pressure vs. frequency for a single pushpull unit developed by Janszen[1]. The dashed curve is that predicted from the

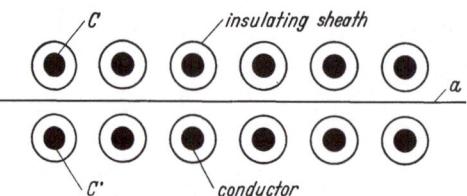

Fig. 20. A sectional view of a push-pull electrostatic loudspeaker showing the membrane *a*, and the driving electrodes *c—c'* which are insulated wires stretched between the frame members.

electrical analogue and corrected for directivity. While the analogue is clearly somewhat oversimplified, it predicts the general shape of the curve over most of the useful frequency range. The absence of a resonant peak in the dashed curve is due to the damping effect of the enclosure which may be arranged to present a mechanical impedance of $\varrho c S_d$ to the back of the diaphragm. This high degree of damping may be achieved by plane wave radiation from the back of the diaphragm into an absorbing medium such as glass wool. Plane waves even at the lower frequencies may be assured by subdividing the back cavity into cells containing absorbing material. A 12 cm square push-pull unit may radiate of the order of 0.1 watt of acoustic power in the range between 1 and 3 kc. At higher frequencies, the maximal power radiated will decrease due to the mass reactance of the diaphragm membrane.

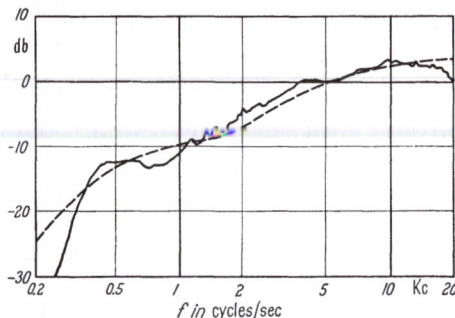

Fig. 21. The relative axial pressure vs. frequency for a single push-pull unit (after Janszen).

β) *The low-voltage electrostatic loudspeaker.* This type of loudspeaker requires only a few hundred volts d.c. for polarization and a few tens of volts a.c. from the driver. The insulating dielectric is the diaphragm itself which is in contact with and supported by a perforated or grooved backplate. The front surface of the diaphragm is metalized (by evaporation) with a very thin film of gold, silver, or aluminum. The plastic diaphragm is under only slight tension to prevent wrinkling. Fig. 22 shows a portion of such a unit in section. The fringing electric field is shown at *C*. It is the alternating component of this fringing field acting

[1] A. A. Janszen, R. L. Pritchard and F. V. Hunt: Electrostatic loudspeakers. Technical Memorandum No. 17, 1. April, 1950.

[2] Privately communicated by Dr. Erwin Meyer, Drittes Physikalisches Institut, Göttingen, Germany.

[3] W. Kuhl: Acustica 4, 519 (1954).

on the unsupported part of the membrane that results in the radiation of sound. While the supported part of the membrane may make a small contribution to the radiated sound, most of the radiation comes from the unsupported sections. This is verified by the fact that closing the grooves or perforations at the bottom end has a marked effect on the resonant frequency of the unit as a radiator. For use in the range 5 to 20 kc, open grooves or perforations are used. For the low ultrasonic range 30 to 100 kc, grooves machined in solid metal are used as a backing surface. For still higher frequencies, a sand-blasted backplate may be used. The two plastic foils

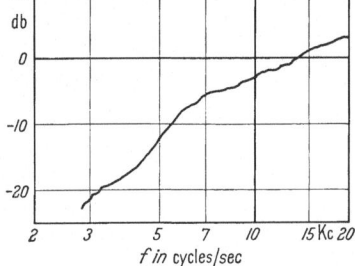

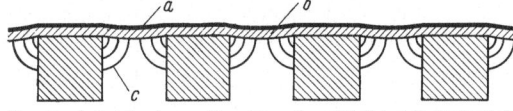

Fig. 22. A sketch of a small section of a solid dielectric electrostatic loudspeaker. The metal film is shown at a, the plastic membrane at b, and the fringing electric field at c.

Fig. 23. The relative axial acoustic pressure of a solid dielectric electrostatic loudspeaker (after Hass).

found to be most satisfactory by KUHL were 10 micron polystyrol and 12 micron poly-ethylene.

A change in radiating efficiency with time was noted and attributed to an accumulation of charges on the surface of the foil next to the backplate. This decrease in efficiency with time at constant polarization voltage does not occur when the back of the foil is cleaned with alcohol before assembling the unit.

Fig. 23 shows the relative axial acoustic pressure vs. frequency for a design by Hass[1] based on the work of KUHL. This loudspeaker has a greater acoustic output above 5 kc than an electrodynamic direct-radiator loudspeaker when the two are driven by equal voltages at the power tube of the driving amplifier

III. Horn loudspeakers.

The low efficiency and limited power handling capacity of the wide band direct-radiator loudspeaker has resulted in the development of the horn loudspeaker. The horn and associated acoustic elements can function as a mechano-acoustic transformer. It can be used to match a low-impedance medium such as air to the relatively high mechanical impedance of the electrodynamic driving unit. This sort of matching is partially accomplished by the use of a large area cone in the case of a direct-radiator loudspeaker, but at the expense of increase cone mass and the added complication of cone resonances in the useful frequency range of the loudspeaker. The use of a horn makes it possible to use a smaller diaphragm with a resultant reduction in mass and a considerable increase in the lowest frequency at which the diaphragm resonates as a distributed system exhibiting diametral and circular nodes. By coupling the driver unit more tightly to the medium, the horn reduces the diaphragm amplitudes required for a given radiated power with a corresponding reduction in harmonic distortion due to excessive voice coil travel. The horn loudspeaker is more efficient than the direct-radiator loudspeaker by about a factor of ten when the band width is maximized for both types. It is not to be assumed from the foregoing discussion that the horn loudspeaker does not have inherent limitations and that it presents no problems of its own.

---

[1] H. HASS: Funkschau **24**, H. 2, 23 (1952).

8. Horn driver units. α) *Low-frequency units.* For the low-frequency range (20 to 500 c.p.s.), conventional paper cone electrodynamic loudspeakers are used. Since the radiation load is more closely coupled to the driver unit, a larger $(Bl)^2$ factor is desirable than required for the direct-radiator loudspeaker. The arrangement of components for coupling a large paper cone loudspeaker to the throat of a horn is shown in Fig. 24a.

The equivalent circuit in Fig. 24b shows the increased loading of the cone by making $S_t < S_c$. If $S_t = S_c$ the mechanical impedance presented to the cone would be $Z_t S_c$. This impedance has been transformed to a higher value $Z_t \dfrac{S_c^2}{S_t}$ by just the area ratio S_c/S_t. This increased loading has been achieved at the expense of introducing the shunt capacitance C_f due to V_f. This shunt capacitance will limit the high-frequency radiated power of the horn loudspeaker. Here, a compromise between efficiency and band width is required. V_f may be reduced by changing the throat structure to conform to the shape of the cone, but this

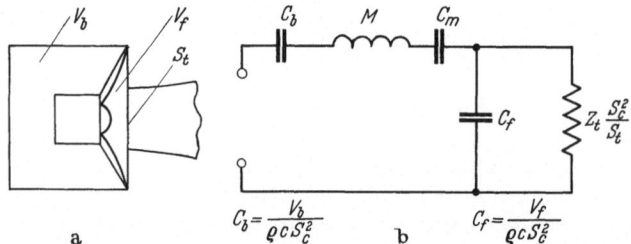

Fig. 24 a and b. The arrangement of components for coupling a large paper cone loudspeaker to the throat of a horn are shown in (a). V_f is the volume between the front of the cone and the plane of the horn throat. V_b is the closed volume back of the cone. The mechanical equivalent circuit is shown in (b). z_t is the specific throat impedance of the horn and S_t the throat area.

change will be limited by the amplitudes required at the low-frequency end of the band. V_f must be large compared to the volume displacement of the cone if C_f is to remain essentially constant in the circuit. If C_f varies with time, the low frequencies will modulate the high frequencies with the production of intermodulation distortion[1].

It is not profitable to carry the discussion of the equivalent circuit further in the case of the low-frequency horn loudspeaker since it is seldom possible to use a horn for which the throat impedance can be represented by a sufficiently simple analytic form to make computations worth considering. Locanthi[2] has set up the complete electrical analogue of the loudspeaker-horn combination and obtained remarkably good agreement between electrical measurements made on the analogue and those made on the actual loudspeaker-horn combination simulated by the analogue.

β) *High-frequency horn driver units.* The high-frequency driver units usually have aluminum diaphragms domed to increase their rigidity. Since the wavelength of the sound radiated may be smaller than the radius of the diaphragm, some care is required in combining the radiation from different annular zones of the diaphragm in such a way as to minimize destructive interference. Fig. 25 shows a sectional view of the important features of the high-frequency horn driver unit. The aluminum alloy diaphragm d, the voice coil C, the fronting volume V_f, and the back volume V_b are labeled in the diagram. Special attention is called

[1] H. F. Olson: Elements of Acoustical Engineering, 2nd ed., p. 199. New York: D. Van Nostrand Co., Inc. 1947.

[2] B. N. Locanthi: Trans. Inst. Radio Engrs., PGA-6, March, 1952.

to the annular throat structure which provides nearly equal sound paths from annular zones on the diaphragm to the horn throat at t. SMITH[1] has made a comprehensive study of the design of such throat structures taking into account the normal modes (standing waves) of the fronting volume between the diaphragm and the throat structure and their effect on the radiated power. The radial flux across the gap in which the voice coil moves is supplied by the cylindrical permanent magnet. The voice coil is usually of edge-wound aluminum ribbon bonded to-

gether with a thermo-setting cement that insulates the turns from each other electrically. The factor Bl is approximatelyc 15 newtons per ampere in a well-designed unit. The mass M of the diaphragm and voice coil assembly will be close to 0.003 kg and the diaphragm area S_d about 0.008 m². Such a driver provided with a suitable horn will radiate, conservatively, about 5 watt of sound in its mid-frequency range with an efficiency of about 25 %.

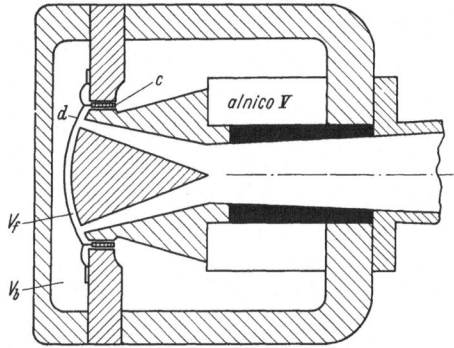

Fig. 25. Sectional view of high-frequency horn driver unit. The unit is a figure of revolution about the indicated center line.

The frequency vs. radiated power characteristics of the high-frequency horn driver unit can be predicted reasonably well from the equivalent circuit. This is largely due to the fact that there is little difficulty in providing an adequate horn in the high-frequency range. Here, the wavelength is short enough so that the mouth of the horn can be large enough to present, essentially, a real impedance at its throat ($Z_t \approx \varrho c$). If this condition is realized, the equivalent mechanical circuit is given in Fig. 26.

This circuit is transformed by taking its dual and multiplying by $B^2 l^2$ as before with the resulting circuit including the voice coil impedance shown in Fig. 27.

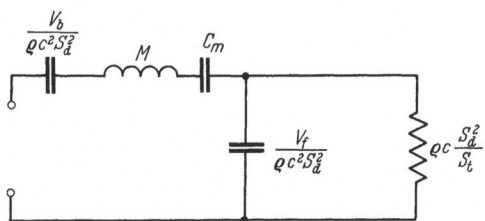

Fig. 26. The mechanical equivalent circuit for the high-frequency horn driver unit. V_b and V_f are the back and fronting volumes, respectively; S_d and S_t the diaphragm and throat areas, respectively; and C_m the mechanical compliance of the diaphragm.

As MASON[2] has pointed out, the circuit in Fig. 27 is, essentially, a band pass filter section. The lower limit of the pass band is the frequency of mechanical resonance of diaphragm and voice coil mass M with the combined compliance of the diaphragm and back volume C'_m. The upper limit of the pass band is determined by the voice coil inductance and the shunt capacitance $M/B^2 l^2$ if the series inductance at the right is made equal to L'_e by adjusting V_f to the proper value. Thus, the upper limit can be increased only by decreasing the product $M L'_e/B^2 l^2$. It is clear that a reduction of the number of turns will reduce both L'_e and $B^2 l^2$ by the same factor and is thus ineffectual. A decrease in M is effective in raising the upper limit of the pass band provided C'_m is increased correspondingly to hold the lower band limit constant. It is clear that the terminating resistance should be adjusted to be equal to the midband characteristic impedance of the

[1] B. H. SMITH: J. Acoust. Soc. Amer. **25**, 305 (1953).
[2] W. P. MASON: Electromechanical Transducers and Wave Filters, p. 228. New York: D. Van Nostrand Company, Inc. 1942.

filter section by a proper choice of the ratio of the throat to diaphragm area. It is also evident that the resistance of the voice coil R'_e plus the electrical source impedance should equal the characteristic impedance of the filter. The design equations derived from MASON's book (ref. 2, p. 19) are

$$f_1 = \frac{1}{2\pi\sqrt{C'_m M}}, \tag{8.1}$$

$$f_2 = f + \frac{B^2 l^2}{2\pi^2 M L'_e}, \tag{8.2}$$

$$Z_0 = 2\pi f_2 L'_e \tag{8.3}$$

where f_2 and f_1 are the frequencies of the upper and lower limits of the pass band respectively and Z_0 is the midband characteristic impedance.

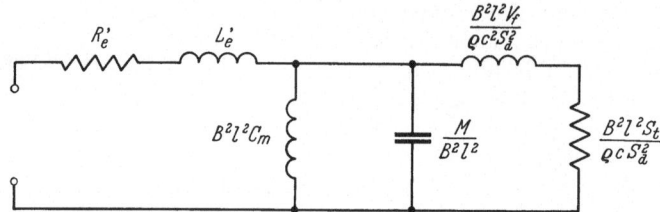

Fig. 27. The equivalent electrical circuit for the high-frequency horn driver unit including the motional impedance of the transformed mechano-acoustic system. C'_m is the mechanical compliance due to C_m and $V_b/\varrho c^2$ in series as shown in Fig. 26.

Since all the power not dissipated in the voice coil resistance is radiated, the midband fractional efficiency is given by

$$n = \frac{\dfrac{B^2 l^2 S_t}{\varrho c S_d^2}}{R'_e + \dfrac{B^2 l^2 S_t}{\varrho c S_d^2}}. \tag{8.4}$$

With the constants given earlier for a good high-frequency horn driver unit with a voice coil resistance of 8 ohm and inductance of 0.3 mh, the fractional efficiency is 0.51.

9. Horns. α) *Infinite horns.* The pioneer work on the theory of infinite horns was carried out by WEBSTER[1] who pointed out the importance of the throat impedance as a measure of the performance of the horn. The exponential horn appeared to be the only horn at that time exhibiting a sharp cut-off in its transmission characteristics at low frequencies. This is indicated in its throat impedance by the vanishing of the real part of the throat impedance indicating the inability of the horn to accept power from a driver unit. Thus, until the work of SALMON[2], little progress in the development of horns was recorded.

Starting with the wave equation for plane waves in a conduit of varying area

$$\frac{1}{S}\frac{d}{dx}\left(S\frac{dp}{dx}\right) + k^2 p = 0 \tag{9.1}$$

where p is the pressure in the plane wave, S is the conduit area, and x the axial distance along the conduit, SALMON obtained solutions for a new family of horn contours. This family presented in dimensionless form is given as

$$\varrho = \mathrm{Cos}\,\alpha + T\,\mathrm{Sin}\,\alpha \tag{9.2}$$

[1] A. G. WEBSTER: Proc. Nat. Acad. Sci. U.S.A. **5**, 275 (1919).
[2] VINCENT SALMON: A new family of horns. J. Acoust. Soc. Amer. **17**, 212 (1946).

where the symbols are defined as follows

$$\varrho = \frac{\text{diameter}}{\text{diameter of throat}},$$

$$\alpha = \frac{x}{x_0},$$

$$x_0 = \frac{c}{2\pi f_0},$$

$f_0 = $ cut off frequency,

$c = $ velocity of sound,

$T = $ horn parameter. $\qquad (9.3)$

As the parameter T is varied from zero to infinity, the horn shape changes from catenoidal ($\varrho = \mathrm{Cos}\,\alpha$) to conical as indicated in Fig. 28. The exponential is obtained for $T=1$. The throat admittance ratio y_t is given by

$$y_t = \frac{\varrho c}{z_t} = \sqrt{1 - \frac{f_0^2}{f^2} - \frac{jTf_0}{f}}. \qquad (9.4)$$

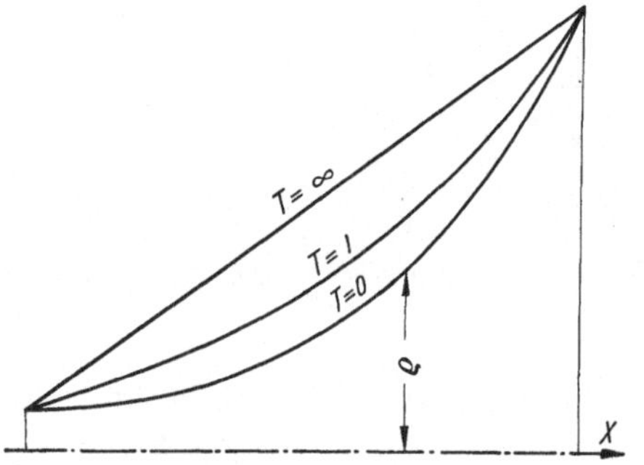

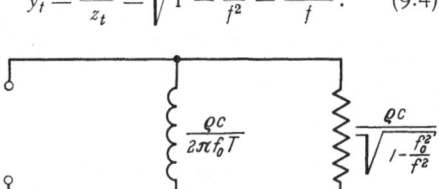

Fig. 28. Horn contours for different values of T. They are catenoidal, exponential, and conical starting at $T=0$ and ending with $T=\infty$.

Fig. 29. The equivalent circuit for the specific acoustic impedance at the throat of the new family of horns (after SALMON).

This expression can be represented by an equivalent circuit in terms of specific acoustic resistance and reactance as shown in Fig. 29. This circuit is more useful than a series form where both the inductance and resistance would have very complicated frequency dependences. The complexity of the series representation is indicated in the curves showing the specific resistance ratio as a function of the frequency ratio in Fig. 30. The sharp cut-off for $T < 1$ is clearly evident in the resistance ratio curves of Fig. 30 whereas there is no obvious indication of it in Eq. (9.4) or in the circuit in Fig. 29. However, the equivalent circuit is very convenient when transformed along with the mechanical impedances of the driver unit to give the electrical equivalent circuit including the motional impedance and voice coil impedance as shown in Fig. 31. Since the horn characteristics are represented by a fixed capacitor in series with a monotonicly frequency-dependent resistor, an analogue type calculation is quite convenient. The results of such a calculation for fixed f_0 and various values of T are shown in Fig. 32. The relative power radiated is expressed in decibels and plotted against the ratio of the frequency f to f_0, the cut-off frequency of the horn. The resonant frequency of the driver is 0.7 f_0, and the other constants of the driver are essentially those given in Sect. 8.

β) *Finite horns.* The analysis of a horn of finite length is complicated by reflections occuring at the mouth of the horn. OLSON[1] has carried out the analysis

---

[1] H. F. OLSON: Elements of Acoustical Engineering, 2nd ed., p. 109. New York: D. Van Nostrand Co., Inc. 1947.

for both the conical and exponential horns. Fig. 33 shows the resistance and reactance ratios at the throat of a finite exponential horn having a mouth diameter of 0.225 λ_0 and a length 0.55 λ_0 where λ_0 is the cut-off wavelength. The fluctuations shown are not sufficiently great to affect the power radiated appreciably (less than 1 db) if the driver unit is at least 50% efficient at mechanical resonance and driven by an amplifier of proper internal impedance. The amplifier impedance R_g plus the voice coil resistance R_e' should be equal to the trans-

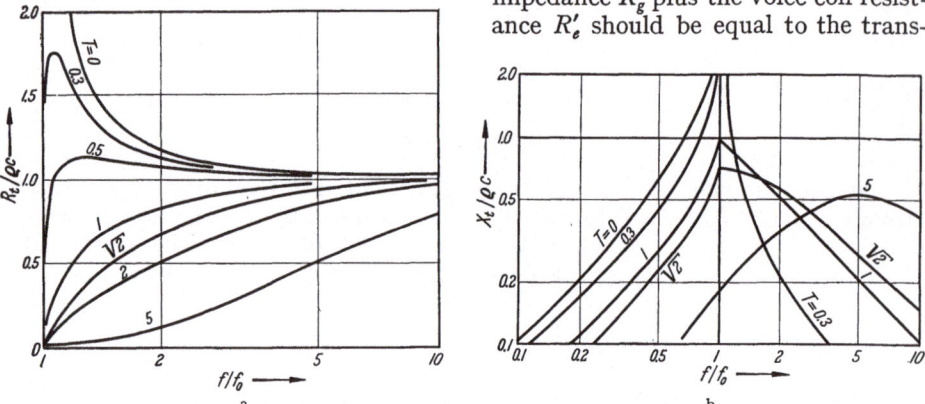

Fig. 30 a and b. The specific acoustic resistance and reactance ratios vs. the frequency ratio f/f_0. The reactance ratio is zero above f_0 for the $T=0$ horn.

formed mean throat conductance which will be, essentially, that for an infinte horn. Stated, concisely, a calculation based on the infinite horn throat admittance is amply accurate if the mouth of the horn is at least 0.225 λ_0 in diameter and if an efficient driver is properly matched by the driving amplifier. The details of this type of calculation are presented by Wente[1] for a case where even larger fluctuations are encountered. The reason the efficiency of the transducer must be high is to compensate for peaks in the transformed admittance by a reduction in current delivered by the amplifier with the result that the power delivered to the transformed acoustic is essentially constant. In other words, the transducer must be efficient enough to permit the amplifier to feel the transformed acoustic load. A transducer with an efficiency of 10% would reproduce the fluctuations in transformed acoustic load quite closely and would therefore require a smoother horn. A mouth diameter of 0.3 λ_0 would probably be satisfactory in this case.

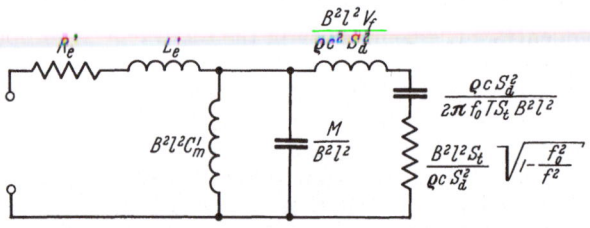

Fig. 31. The equivalent electrical circuit for a horn driver unit coupled to the throat of an infinte horn of the family desribed by Salmon.

γ) *Folded horns.* For the efficient radiation of low frequencies (40 c.p.s.), rather large horns are required. The length may be reduced by folding the horn back on itself maintaining the proper rate of growth of area with mean length. The bends in the horn channel will result in serious reflections if the dimensions of the

---

[1] E. C. Wente and A. L. Thuras: Bell. Syst. Tech. J. **13**, 265 (1934).

channel in the plane of the bends are not small compared to the half-wavelength of the transmitted sound. An interesting folded horn is described by KLIPSCH[1].

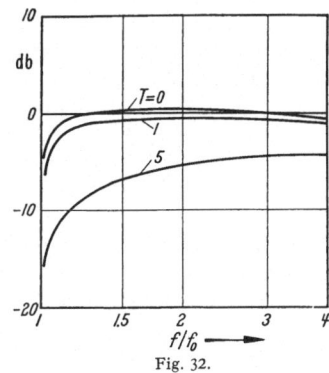

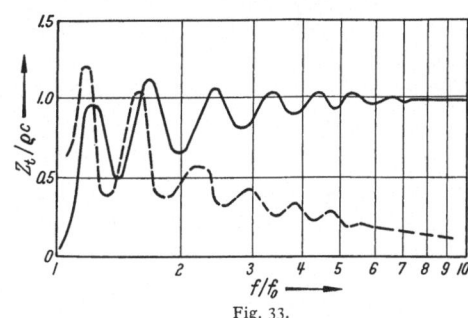

Fig. 32.

Fig. 33.

Fig. 32. The relative acoustic power in db radiated by a high efficiency horn driver unit coupled to the throat of individual members of the new family of horns. The curves for various values of T are plotted against the ratio of the frequency f to the cut-off frequency f_0.

Fig. 33. The throat resistance and reactance ratios for a finite exponential horn of mouth diameter 0.225 λ_0 and length 0.55 λ_0 (after OLSON).

IV. Resonant bar source.

10. The St. Clair generator. A resonant bar transducer for fixed frequencies between 10 and 20 kc has been described by ST. CLAIR[2]. The essential elements of the apparatus are shown in Fig. 34. The bar is driven at its lower end by a circular ring which is an integral part of the bar. A fixed coil C wound on the center pole piece induces circulating currents in the ring R, and it is the interaction of these currents with the radial magnetic field that provides the driving force for the resonent bar. It is necessary to machine the ring from the metal of the bar since peak accelerations of the order of 40000 g may be obtained. The vibrating bar used was made of duraluminum which was probably the best material vailable at the time. Fifty-four turns were used on the fixed coil, and the field in the gap was 1.25 webers per square meter. The length and diameter of the bar were chosen so that the center of gravity of the bar remained fixed during vibration, and mechanical losses due to the center supporting rod were minimized. The Q of the rod as mounted was 2.3×10^4 at a frequency of 17 kc. With a Q of this magnitude, it was necessary to use a self-excited system. The necessary feedback was supplied by an insulated polarized electrostatic pickup electrode place near the lower surface of the bar around the supporting rod. This was amplified

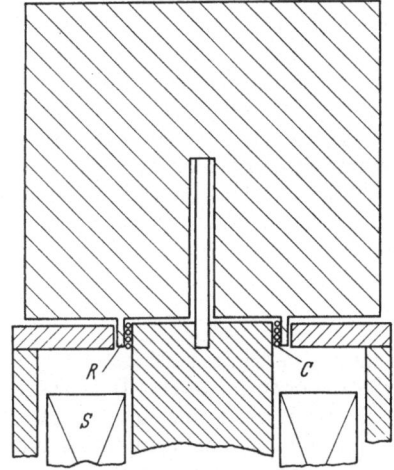

Fig. 34. A sectional view of St. Clair Generator showing the fixed coil C, the driving ring R, and the magnetizing coil S.

---

[1] P. W. KLIPSCH: J. Acoust. Soc. Amer. **13**, 137 (1941).
[2] HILLARY W. ST. CLAIR: Rev. Sci. Instrum. **12**, 250 (1941).

by a preamplifier having automatic gain control. The power amplifier was capable of delivering 200 watt.

The electrical efficiency of the transducer was 69.5%, but the mechanical losses in the bar reduced the efficiency as a sound generator to 30%. With this efficiency, sound pressure levels of the order of 150 db close to the vibrating surface were possible. ST. CLAIR mentions the floating of lead shot and a small coin in the standing wave field above the bar when a reflecting plate was placed parallel to and a few wavelengths above the upper end of the bar.

V. Air driven sources.

11. The siren. The siren, which dates back to 1801, was at first used only as a frequency meter. It consisted of a rotating disc with a ring of evenlyspaced holes driven by a single air jet and in this form is found today among the lecture demonstration equipment of most beginning Physics courses. In this form its efficiency is extremely small due to the dipole nature of the radiation field. The siren owes its name to Baron CHARLES CAGNIARD DE LA TOUR who made the discovery that it would sing under water. SEEBECK[1] brought the direct-radiator siren essentially to its present form by placing the perforated disc in front of a chamber, the front of which was perforated to match the disc. Thus, as is the case with the modern design, all ports opened and closed together. This increased the efficiency to the order of one percent. The next and final stride in the development of the siren was to couple it properly to a horn radiator. This was done in 1941 by a group of researchers at the Bell Telephone Laboratories and is reported in a paper by JONES[2]. The result of this work was a siren with an efficiency of over 70%.

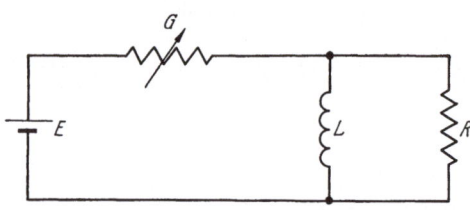

Fig. 35. The equivalent circuit for a siren (after JONES).

The analysis of the high efficiency siren presented here is that of JONES. To start with, consider the equivalent circuit for the siren and acoustic load shown in Fig. 35. The modulating port resistance is indicated as variable resistor G. L and R form a parallel representation for the acoustic impedance of the horn throat. They are given by

$$L = \frac{c}{2\pi f_0 S T}, \quad R = \frac{\varrho c}{S}\sqrt{1 - \frac{f_0^2}{f^2}} \tag{11.1}$$

where f_0 is the cut-off frequency of the horn, S is the throat area, and $T = 1$ for the exponential horn. It is assumed that the radiation looking upstream from the siren is negligible compared to that presented by the horn throat. For maximal efficiency, the ports should be fully open half the time and fully closed the other half. That is, G should be either zero or infinite. Unfortunately, G has a minimal value G_0 which is not zero. If the pressure drop through G_0 is small compared to the absolute pressure, then Δp is given by

$$\Delta p = \tfrac{1}{2}\varrho u^2 \tag{11.2}$$

and G_0 by

$$G_0 = \frac{\Delta p}{S u} = \frac{1}{2}\varrho\frac{V}{S^2} \tag{11.3}$$

[1] SEEBECK: Pogg. Ann. **35**, 417 (1841).
[2] R. CLARK JONES: A fifty horsepower siren. J. Acoust. Soc. Amer. **18**, 371 (1946).

where $V = Su$ is the volume velocity through the open port. To simplify the analysis for the square wave case, L will be made infinite so that no alternating current can flow through it. This corresponds to letting $T=0$ or the case of the catenoidal horn.

Let I_0 be the steady current through L. When the port is closed, the current in L flows up through R, that is a current $-I_0$ flows in R. When the port is open, it is necessary that a current $2I_0$ flow through the port to provide the average current I_0 in L. Thus, while the port is open, a current I_0 must flow down through R. The electromotive force E which is the excess pressure p above atmospheric delivered by the compressor is equal to the potential drop through G_0 and R when the port is open:

$$E = 2I_0G_0 + I_0R. \tag{11.4}$$

Setting $V = 2I_0$ in Eq. (11.3), we have for G_0

$$G_0 = \varrho \frac{I_0}{S^2}. \tag{11.5}$$

Table 1.	
p in atmospheres	$n = F(y)$
0.0	1.0
0.035	0.954
0.07	0.916
0.175	0.828
0.35	0.732
0.70	0.618
1.75	0.463
3.50	0.358

Combining Eqs. (11.4) and (11.5) and solving the resulting quadratic equation for I_0, we may write

$$I_0 = \frac{E}{R} \cdot F(y) \tag{11.6}$$

where y is defined as

$$y = \frac{4\varrho E}{R^2 S^2} \tag{11.7}$$

and

$$F(y) = (\sqrt{1 + 2y} - 1) \frac{1}{y}. \tag{11.8}$$

If we insert for E the excess pressure p and $\frac{\varrho c}{S}$ for R and γP_0 for ϱc^2, we have for y

$$y = \frac{4p}{\gamma P_0} = 2.85 \frac{p}{P_0}. \tag{11.9}$$

The power delivered to the acoustic load is $I_0^2 R$ and the power supplied by the compressor is $E I_0$. The efficiency n is then

$$n = \frac{I_0 R}{E} = F(y). \tag{11.10}$$

We may now tabulate the efficiency $F(y)$ in terms of y or better still in terms of the excess pressure delivered by the compressor (in atmospheres) as shown in Table 1.

This efficiency is the total efficiency for square wave generation. The fraction of the power in the fundamental is $8/\pi^2$ of the total power. The efficiency for generation of the fundamental may be written

$$n_f = \frac{8}{\pi^2} F(y). \tag{11.11}$$

Thus, as p becomes very small compared to P_0, n_f is approximately 0.81. A more elaborate analysis by JONES gives as the limiting efficiency for a sine wave siren 0.50. This comparison indicates that if power in the fundamental and not wave form is of primary importance, a square wave siren is most efficient. It must be said that the generation of square waves is only approached as a limit as the port angle becomes vanishingly small with a corresponding reduction in total power output.

A brief description of the 50 horsepower siren seems appropriate here. Fig. 36 shows a longitudinal section of the air chamber rotor and cluster of six horns.

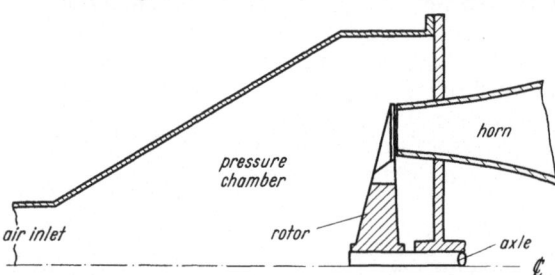

This view shows the careful matching of the horn throat to the siren port which is necessary for high efficiency. Fig. 37 shows an axial view of the rotor and ports. The ports were open 38% of the time and closed 38% of the time; 24% of the time being used for opening and closing. The horns were exponential with cut-off frequency of 125 c.p. s. The mouth dia-

Fig. 36. A longitudinal section of the 50 horsepower siren showing a section of the port designed as part of the exponential horn contour. The rotor and conical pressure chamber are also shown (after Jones).

meter of the cluster was about a wavelength at 500 c.p.s. insuring a resistive throat impedance for the individual horns. 36.6 kw or air at 0.33 atm was supplied by the compressor. 15 kw were required to turn the rotor at full air flow. The rotor was a duraluminum casting 25 inches or 0.635 meter in diameter.

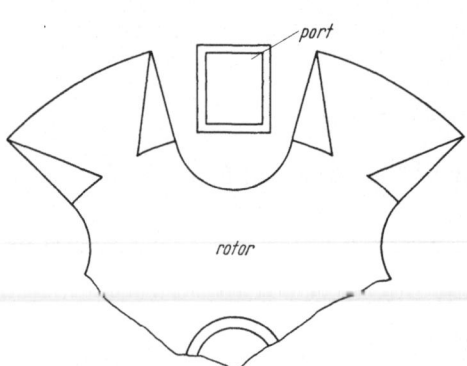

α) **High-frequency sirens.** Sirens of moderate power and efficiency have been constructed for the frequency range between 1 kc and 35 kc. To reach 35 kc, relatively small rotors with a large number of teeth are required. The rotors are given a gaussian section to increase the bursting strength at the large angular velocities required. With many small ports, the total port area is small, and the clearance between the rotor teeth and the ported surface must be correspond-

Fig. 37. An axial view of the rotor port combination used in the 50 horsepower siren (after Jones).

ingly small to prevent excessive leakage when the ports are covered by the rotor teeth.

ALLEN and RUDNICK[1] describe a siren for the frequency range between 2.5 and 35 kc. The essential features of this siren are shown in section in Fig. 38. It had 100 circular ports (0.094 inch diameter) spaced equally on a 6 inch circle. The total port area was 4.43 cm². The rotor had 100 teeth slightly larger than the ports. The clearance was slightly less than 0.001 inches between the rotor and the ported surface of the stator. The rotor was made of stainless steel and the stator of brass. This siren operated with two atmospheres pressure in the air chamber and delivered 2 kw of acoustic power over the 2.5 to 35 kc frequency range at an efficiency of 20%. An annular exponential horn was required from 2.5 to 8 kc. Above 8 kc the ring of conical ports radiated more energy without

---

[1] C. H. ALLEN and I. RUDNICK: A powerful high-frequency siren. J. Acoust. Soc. Amer. **19**, 857 (1947). — See also p. 98 in this Volume.

the horn than with it. This was explained in terms of the attenuation of waves of finite amplitude (weak shock waves)[1].

The lower efficiency of this siren compared to the 50 horsepower siren may be attributed to the high pressure (2 atm) and the departure from square wave generation due to the large port angles used. However, the intensities produced were sufficient to cause cotton to burst into flame by the absorption of acoustic energy and to support glass spheres in a progressive wave field.

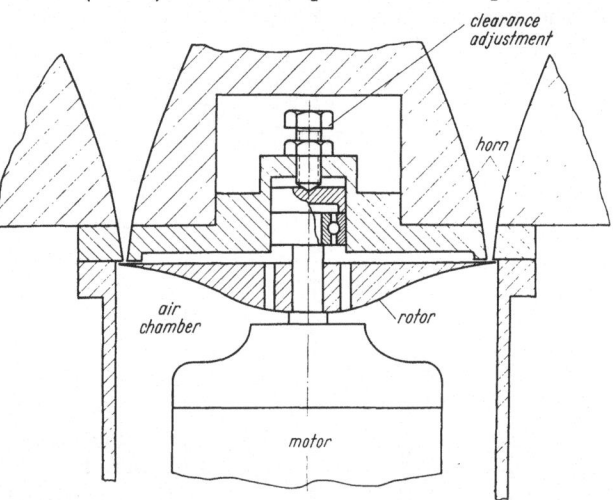

β) *Low-frequency sirens.* A low-frequency siren designed by LEONARD and RUDNICK[2] operating in the frequency range 20 to 200 c.p.s. delivers 100 kw of sound to a 25.4 cm diameter circular tube which is terminated to minimize reflections.

Fig. 38. A sectional view of a high-frequency siren (100 ports) showing the rotor profile and the exponential horn used from 2.5 to 8 kc. The pressure chamber surrounds the electric motor (after ALLEN and RUDNICK).

Sound levels in excess of 180 db are produced in the tube. The siren is operated at a chamber pressure of about one atmosphere and an efficiency of about 30%.

12. Air jet generators. The generation of sound by the instability of hydrodynamic flow certainly antidates all other sources. The study of the frequencies generated by subsonic jets impinging on obstacles has received considerable attention in the last one hundred years. Recently, with the advent of the jet engine, considerable attention has been given to sound generation by supersonic jets.

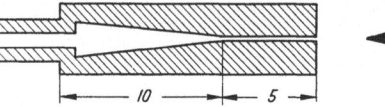

α) *Subsonic jets.* Small area subsonic jets are not effective sound sources unless they inpinge on an obstacle. One of the basic systems

Fig. 39. A jet-edge system in section (after NYBORG).

is the jet edge combination where a thin, essentially two-dimensional jet of air strikes a sharp edge. The edge is coplanar with the jet as indicated in section in Fig. 39. The jet edge system was studied by SONDHAUS[3] over a century ago. Since that time, many investigators have been interested in "Schneidentöne". BROWN[4] has reviewed the literature up to 1937, and in addition has presented an empirical formula for the frequencies of the jet edge system for its various stages of operation (regions of instability in flow). The expression for the frequency f_m is

$$f_m = 0.466\, j_m (u_0 - 40) \left(\frac{1}{h} - 0.07 \right) \tag{12.1}$$

[1] I. RUDNICK: J. Acoust. Soc. Amer. **25**, 1012, (LE) (1953).

[2] R. W. LEONARD and I. RUDNICK: Acoustic instrumentation for high intensities. J. Acoust. Soc. Amer. **24**, 451 (A) (1952).

[3] C. SONDHAUS: Pogg. Ann. **91**, 128, 214 (1854).

[4] G. B. BROWN: Proc. Phys. Soc. Lond. **49**, 493 (1937).

where u_0 is the average jet velocity in cm/sec at the orifice and j_m has the values

$$j_m = 1.0, 2.3, 3.8, 5.4 \qquad (12.2)$$

for the first four stages of operation. h is the distance from the orifice slit to the edge.

Nyborg[1] has made a detailed study of frequency and sound pressure for a jet edge system in stage 1. These results are presented in Fig. 40. Both Nyborg and von Gierke[2] have studied the intensity and directivity of the jet edge system as a source. For frequencies where the width of the jet is small compared to the wavelength, the angular dependence of the acoustic pressure around the axis of the jet is given by

$$p(\vartheta) = p_0 \cos \vartheta \qquad (12.3)$$

indicating the dipole nature of the source. Von Gierke has observed the same angular dependence about an axis parallel to the edge.

Nyborg[1] has observed the acoustic pressure to be a linear function of the jet velocity when the frequency is held constant. This is indicated by the pressure values in dynes per cm² recorded on the isofrequency lines of Fig. 40. Von Gierke[2] has observed the sound pressure to be proportional to f^3 for a fixed jet edge combination with variable jet velocity in agreement with the dipole nature of the source. Bouyoucos[3] has photographed a liquid model of the jet edge system showing the dynamic trajectories of the liquid jet as it oscillates in a ribbon-like manner. Although vortex motion is evident beyond the edge, it seems probable that the rate of formation of vortices is the result rather than the controlling factor in the jet oscillations. Nyborg[4] has developed a dynamical theory for the instability of the jet based on the assumption that the lateral force on the jet between the orifice and the edge is proportional to the lateral displacement of the jet at position of the edge. This theory, while involving some over-simplification, predicts rather well the form of the trajectories and the ratios of the frequencies in the various stages. Basic to this dynamical description is the idea of a feedback mechanism making the lateral forces on the jet depend on the motion of the jet near the edge. It is assumed that the alternation of the jet

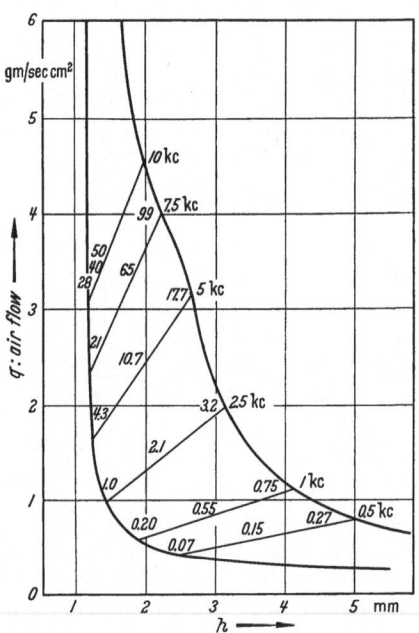

Fig. 40. The domain of stage one operation showing isofrequency lines and the acoustic pressures for different flows in dynes/cm². The jet cross-section was 0.38 mm × 25.4 mm. The pressures were measured at a perpendicular distance of 20 cm from the jet-edge system which was surrounded by a 30 cm square baffle about 0.3 cm thick to increase the radiation efficiency (after Nyborg).

[1] W. L. Nyborg, M. D. Burkhard and H. K. Schilling: J. Acoust. Soc. Amer. 24, 293 (1952).
[2] H. von Gierke: Z. angew. Phys. 2, 97 (1950).
[3] J. V. Bouyoucos and W. L. Nyborg: J. Acoust. Soc. Amer. 26, 511 (1954).
[4] W. L. Nyborg: J. Acoust. Soc. Amer. 26, 174 (1954).

to either side of the edge results in the radiation of sound waves which interact
with the jet causing welf-maintained oscillations. Strong evidence for this interacting sound field is provided by the excellent schlieren photograph of Powell[1]
showing the nature of the waves radiated from the edge back along the jet toward
and beyond the orifice. This photograph was taken with a choked jet which
generated sound waves at the edge of sufficiently high amplitude to make them
visible by the schlieren technique. Fig. 41 shows a reproduction of Powell's
schlieren photograph. Powell also reports the presence of pressure waves for
the subsonic case, the observations being made with a pair of microphones to
show the 180° phase difference above and below the jet. The variation of pressure
with distance indicated that the sources
of these waves were downstream from the
edge splitting the flow.

β) Subsonic whistles. The combination
of a jet, an edge, and a resonator is basic
to the design of whistles and organ pipes.
Nyborg *et. al.* have studied the combination of a jet edge system loosely coupled
to a resonator of variable length as shown
in Fig. 42a. Fig. 42b—e shows graphs of
the interesting interactions obtained for
three different flow rates. Graph b shows
the two stages A and B and their extrapolated behavior as dashed extensions. Graph c
shows the participation of stage B at a flow
too low to produce it in the absence of the
resonator. As is to be expected, stage A is
present. Graph e shows the participation of

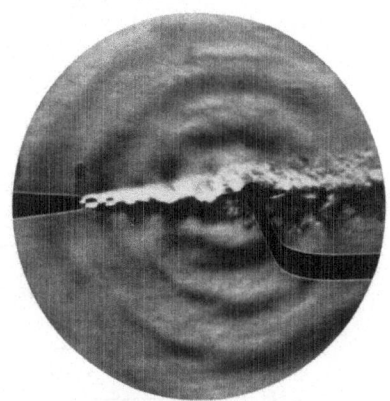

Fig. 41. A schlieren photograph of a supersonic jet-
edge system showing the sound waves radiated
upstream from the edge (after Powell).

stage A at a flow too high to produce it in the absence of the resonator. As is
to be expected, stage B is present. Graph d shows an intermediate case. The
point of interest lies in the fact that stages of the jet edge system are permitted
by the presence of the resonator at flows where they would not be excited without
the resonator. This seems to strengthen the position of the acoustic feedback
mechanism and weaken that of the vortex-controlled mechanism as an explanation
of the oscillations of the jet edge system.

In conventional whistles and organ pipes, the jet edge system may be very
tightly coupled acoustically to the resonator with the result that continuous
operation in a single stage is possible over an octave frequency range with a
variable-length resonator. One might say that the acoustic coupling between
the jet and the resonator is so great compared to that between the edge and the
jet that the former almost completely dominates the system. Mercer[2] has
concluded that in the presence of this strong coupling, the edge tone does not seem
to explain tone production in the organ flue pipe.

γ) Supersonic jets. When a jet of air leaves a nozzle at a speed greater than
sound, in addition to the wide-band noise created, there is a screeching which
may be described as a line spectrum containing several components. Powell[1]
has studied both three-dimensional and two-dimensional flows by the two-mirror
Toepler schlieren method. Fig. 43 shows a two-dimensional jet with the characteristic cellular structure close to the jet followed by a region of instability in

[1] Alan Powell: The noise of choked jets, J. Acoust. Soc. Amer. **25**, 385 (1953).
[2] Derwent M. A. Mercer: J. Acoust. Soc. Amer. **27**, 208 (A) (1955).

which the jet has a sinuous motion which is in turn followed by a turbulent wake. The screeching appears to come from this region of sinuous oscillation. The source region is indicated more accurately by the center of curvature of the wave fronts travelling upstream. POWELL gives the following expression for the upstream wavelength

$$\lambda = 0.63 \sqrt{R - R_c} \qquad (12.4)$$

where R is the pressure ratio across the orifice, and R_c is the critical pressure ratio for sonic exit velocity.

A careful study of the upstream radiation shows the waves on the two sides of the jet to be 180° out of phase.

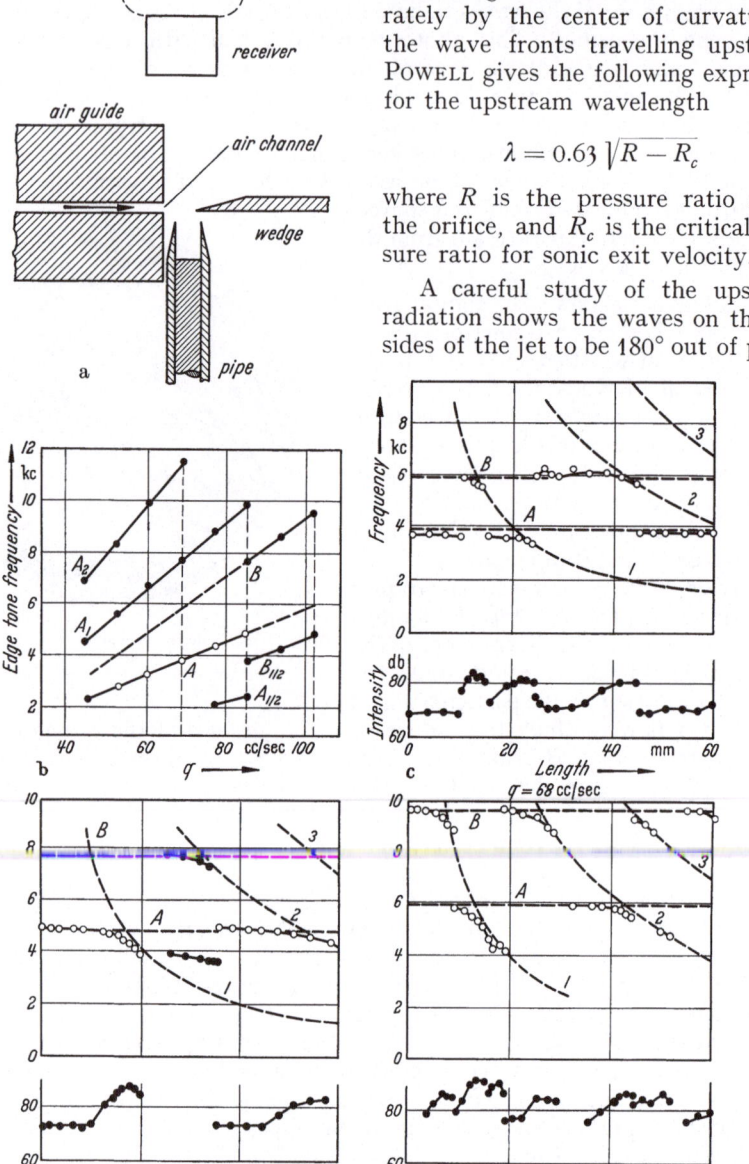

Fig. 42 a—e. The interaction of the jet-edge system with a resonator of variable length L. (a) shows the arrangement of the system in section, (b) the frequency flow dependence of stages A and B, and (c), (d), and (e) the frequency dependence on resonator length L for three flow rates (after NYBORG).

The resulting pressure gradient transverse to the jet is just that required to produce the instability resulting in the sinuous oscillations further downstream. POWELL has suggested that this acoustic feedback mechanism is responsible for the screeching of supersonic jets as well as the production of tones by a jet edge system.

δ) *The Hartmann generator.* HARTMANN[1], in making measurements with a Pitot tube in the supersonic jet resulting from flow through a converging nozzle, observed a periodic curve for the pressure indicated by the Pitot tube. The nozzle

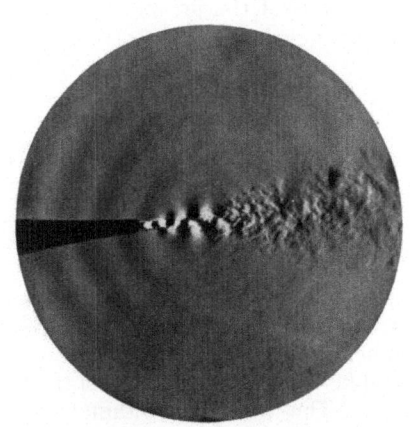

Fig. 43. A schlieren photograph of a two-dimensional supersonic jet showing the characteristic cellular structure and sound waves propagating upstream in the still air surrounding the jet (after POWELL).

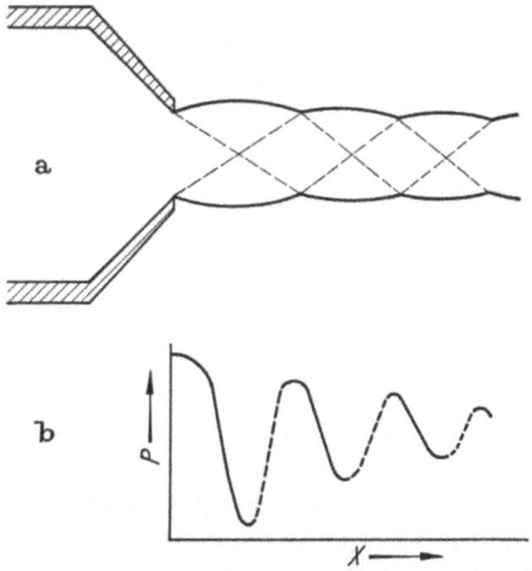

Fig. 44 a and b. The cellular pattern of supersonic flow from a circular nozzle (a) and the accompanying stagnation pressure vs. distance x from nozzle as measured by a PITOT tube (after HARTMANN).

is shown in (a) and the pressure distribution in (b) of Fig. 44. The dotted portions of the pressure curve indicate regions of instability where Pitot tube measurements were erratic for large bore tubes. HARTMANN observed that an acoustic resonator was excited if its mouth was placed in any of these unstable regions. A systematic study of the acoustic output of the jet resonator led to the development of the configuration shown in Fig. 45. An efficiency of about 4% is obtained by making $d_0 = d = l$. With $d = 1.2$ mm, an output of 6 watt at 50 kc was obtained. With $d = 6$ mm, an output of 150 watt at 10 kc was obtained. The wavelength is observed to be dependent on excess pressure and the distance

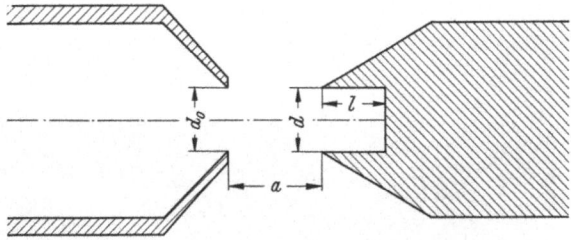

Fig. 45. The essential features of the Hartmann generator (after HARTMANN).

a between the orifice and the resonator mouth. For the case $d_0 = d = l$, the wavelength is given by HARTMANN[2] as

$$\frac{\lambda}{d} = 5.8 + 2.5\left[\frac{a}{d} - \{1 + 0.041\,(p - 0.9)^2\}\right] \tag{12.5}$$

[1] JUL. HARTMANN and BIRGIT TROLLE: Kgl. danske Vid. Selsk., math.-fys. Medd. **7**, 6 (1926). — J. Sci. Instrum. **4**, 101 (1927).
[2] JUL. HARTMANN: J. Sci. Instrum. **16**, 140 (1939).

where p is in atmospheres. This may be approximated,

$$\frac{\lambda}{d} = 5.2.\tag{12.6}$$

The deviation of l from $\lambda/4$ is due to the large end correction required by a resonator for which $d = l$. The optimal value of a appears to be approximately $1.5\,d$ for $p = 3$ atmospheres. Precise alignment of the axis of the jet and that of the resonator is necessary for maximal power output.

VI. Explosive sound sources.

Sounds associated with explosions, their great range of audibility and anomalously high velocity of propagation have interested physicists since the middle of the 19th century. The theoretical work of REIMANN[1] on the propagation of waves of finite amplitude represents one of the most complete discussions of that era. Most of the early experiments involved the measurement of the velocity of propagation near and far from the explosion. One of the early attempts to measure the wave form of explosive sound waves was that of MILLER[2] in 1919 using the "phonodeik". With the development of modern microphones and oscillographs, wave form recordings are possible which show the discontinuous form of the shock front in an explosive wave. The work of FURRER[3,4] from which most of the following discussion is taken, is particularly outstanding.

Attention, here, will be centered on the wave form and spectral distribution of two types of explosive sources: the free explosion of trinitrotoluol and the explosive sounds of guns.

13. Free detonation of trinitrotoluol. Measurements were made by FURRER on charges ranging from 20 gm to 1 kg. Fig. 46 shows the wave form of a 1 kg charge at a distance of 1 meter. The rise time of the leading edge of the wave is less than 10^{-5} sec. This probably represents the rise time of the quartz microphone used, since the rise time for a shock front is of the order of 10^{-8} sec. The peak pressure p_0 and the time to return to the ambient pressure t_0 were studied as a function of the quantity of explosive Q and the following empirical relationships obtained

$$p_0 = a\,Q^{0.44},\tag{13.1}$$

$$t_0 = b\,Q^{0.12},\tag{13.2}$$

where a and b are given by

$$a = \frac{0.34}{r},\tag{13.3}$$

$$b = 2.85,\tag{13.4}$$

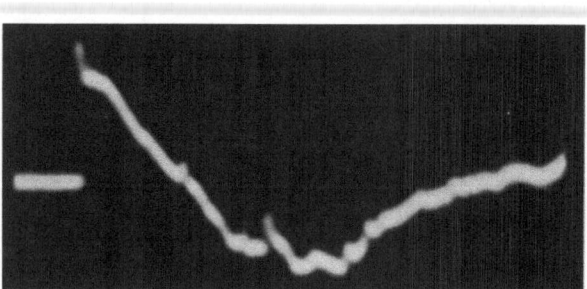

Fig. 46. The pressure wave produced by 1 kg of trinitrotoluol at a distance of one meter (after FURRER).

where p_0 is in kg/cm², t_0 in milliseconds, r in meters, and Q in kg of explosive. The coefficient a was found to follow the $1/r$ relation for pressure peaks less than

[1] REIMANN: Nachr. Ges. Wiss. Göttingen **8** (1860).
[2] D. C. MILLER: Sound Waves, Their Shape and Speed. New York: Macmillan 1937.
[3] W. FURRER: Schweizer Arch. angew. Wiss. Techn. **12**, 213 (1946).
[4] W. FURRER: Techn. Mitt. schweiz. Telegr. u. Teleph.-Verw. **1946**, No. 6.

25×10^{-3} kg/cm². At pressures equal to and below this, the velocity head or stagnation pressure of the explosive gases was observed to be neglibible compared to p_0 since the measured value of p_0 was found to be independent of the orientation of the sensitive surface of the microphone. It is interesting to note that at a peak pressure of 0.1 kg/cm², the stagnation pressure was 0.01 kg/cm², and the velocity of propagation was only 4% higher than the normal velocity of sound.

14. Sound waves from guns. Although the process of the generation of sound by guns is a more complicated one than that due to the free explosion, the resulting sound wave has a very similar shape. The jetting of gases from the gun has an effect on the directivity of the source as shown in Fig. 47 where two curves are presented. Curve A is for a bare muzzle, and curve B is for the muzzle covered by a muzzle brake. Fig. 48 shows the wave form from a 10.5 cm gun at a distance of 100 m. An analysis of the wave forms from 12 guns ranging from 2 to 15 cm in bore results in the following empirical equations for p_0 and t_0

$$p_0 = a' Q^{0.44}, \quad (14.1)$$
$$t_0 = b' Q^{0.24} \quad (14.2)$$

where a' and b' are given by

$$a' = \frac{0.43}{r}, \quad (14.3)$$
$$b' = 1.8 \quad (14.4)$$

where p_0 is in kg/cm², t_0 in milliseconds, r in meters, and Q in kg of explosive.

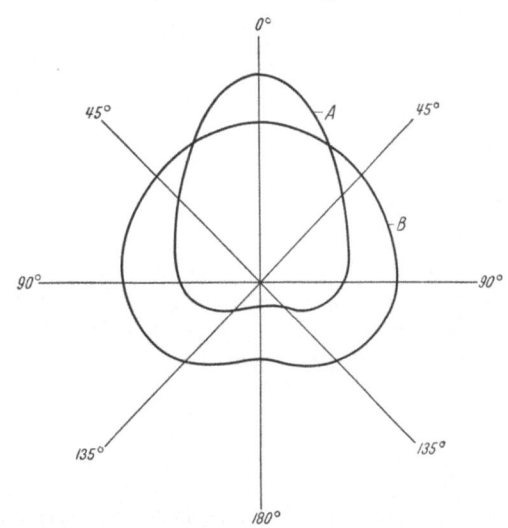

Fig. 47. The directivity of the acoustic radiation from a gun. Curve A is for a bare muzzle and B for a muzzle brake (after FURRER).

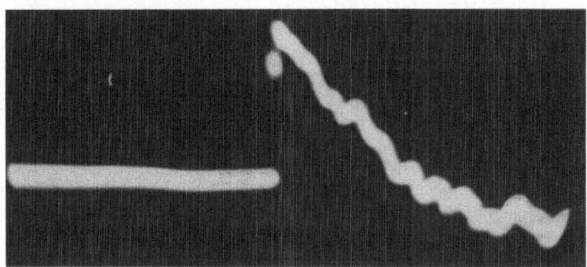

Fig. 48. The pressure wave produced by a 10.5 cm gun at 100 meters (after FURRER).

15. The acoustic energy from explosions. The acoustic energy per cm² may be obtained by integrating the intensity over the duration of the sound pulse. The small amplitude expression for the intensity may be used for $p_0 < 0.1$ kg/cm² as a reasonable approximation. The total energy per cm² is given by

$$W = \frac{1}{\varrho c} \int_0^\infty p(t)^2 \, dt. \quad (15.1)$$

This integral is readily evaluated by replacing the actual $p(t)$ by the function

$$p(t) = \frac{p_0}{\cos \varphi} e^{-bt} \cos(\Omega t + \varphi) \quad (15.2)$$

subject to the conditions that

$$\int_0^\infty e^{-bt} \cos(\Omega t + \varphi)\, dt = 0 \tag{15.3}$$

and

$$\frac{p_0}{\cos\varphi} e^{-bt_0} \cos(\Omega t_0 + \varphi) = 0 \tag{15.4}$$

giving

$$\Omega t_0 + \varphi = \frac{\pi}{2}, \tag{15.5}$$

$$\left(\frac{dp}{dt}\right)_{t=0} \approx -\frac{p_0}{t_0} = -p_0(b + \Omega \tan\varphi). \tag{15.6}$$

The result of these three conditions is

$$d = 23°, \tag{15.7}$$

$$\Omega = \frac{1}{t_0}\left(\frac{\pi}{2} - \varphi\right), \tag{15.8}$$

$$b = \frac{1}{2t_0}, \tag{15.9}$$

also

$$\tan\varphi = \frac{b}{\Omega}. \tag{15.10}$$

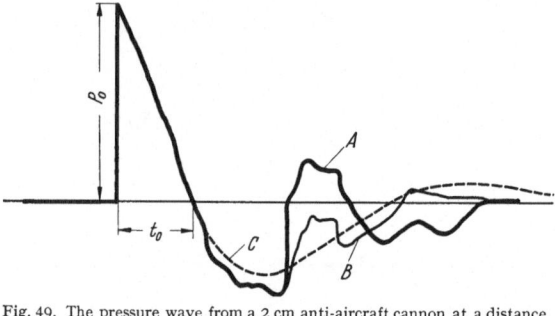

Fig. 49. The pressure wave from a 2 cm anti-aircraft cannon at a distance of 2 meters (after Furrer).

Fig. 49 shows the accuracy of the curve fitting applied to the sound pulse from a 2 cm antiaircraft gun. Curve A is the recorded wave at a distance of 2 m. Curve B is the wave form corrected for reflections from the ground. Curve C is the expression given by Eq. (15.2). Using the expression in Eq. (15.2), the integration in (15.1) can be carried out with the result that

Table 2. *The acoustic efficiency of various explosions (after Furrer).*

Explosion	Efficiency
Free Trinitrotoluol	1.85%
Curbine	0.85%
2 cm anti-aircraft gun	1.5%
10.5 cm gun	1.8%
15 cm gun	2.85%

$$W = \frac{p_0^2}{4\varrho c b}. \tag{15.11}$$

The integration of W over a spherical surface yields the total acoustical energy radiated by the explosive source. This permits the evaluation of the source efficiency as shown in Table 2.

16. The frequency spectrum of explosions. The frequency spectrum of the sound pulse from an explosion may be calculated approximately from the Fourier integral of the function in Eq. (15.2). The pressure amplitude per 1 c.p.s. band $P_0(\omega)$ is given by

$$P_0(\omega) = \frac{P_0}{\sqrt{\left(\dfrac{\Omega^2 + b^2 - \omega^2}{\omega}\right)^2 + 4b^2}} \tag{16.1}$$

and the phase by

$$\psi = \arctan\left(\frac{\Omega^2 + b^2 - \omega^2}{2\omega b}\right). \tag{16.2}$$

$P_0(\omega)$ has a maximal value for $\omega^2 = \Omega^2 + b^2$ given by

$$P_0(\omega)_{\max} = \frac{p_0}{2b} \tag{16.3}$$

and falls to zero at zero frequency. Fig. 50 shows the frequency spectrum of the pressure amplitude per c.p.s. for three different sources. The peak in the spectrum of a gun wave bears no simple relation to the lowest resonance of the barrel of the gun, the latter being lower in the case of a pistol. This is explained by the very high amplitudes involved. The damped pressure oscillations of the barrel are too small compared to the explosive pressures to be observed in the radiated wave. This is confirmed by an experience of the author with an oxiacetylene source where the mixture was ignited in a cylindrical cavity 50 cm long and 15 cm in diameter. With this explosive source, the strength of the explosion could be controlled. For very weak explosions, the fundamental longitudinal mode of the cylinder was observable in the radiated wave, but on increasing the strength of the explosion, the damped oscillations were unobservable in the wave form. The pressure amplitudes due to the oxyacetylene combustion were considerably smaller than those from military guns.

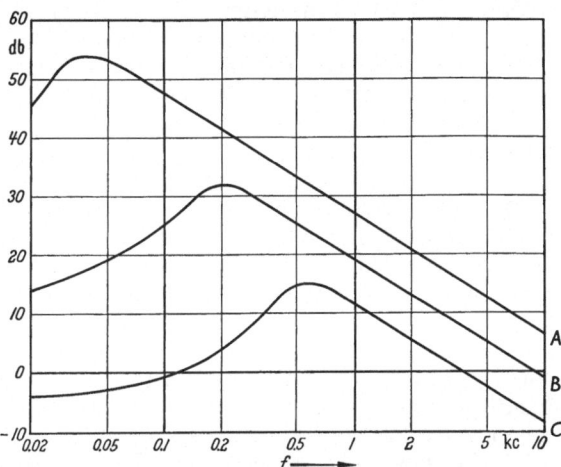

Fig. 50. The frequency spectra of guns; A 15-cm gun at 10 meters. B 2-cm anti-aircraft gun at 2 meters. C Pistol at 0.85 meters. (0 db = 1 dyne/cm²/c.p.s.) (after FURRER).

B. Measurement of sound.

α) *Measurable quantities.* The measurable quantities associated with a freely propagating sound wave may be divided into first order and second order quantities.

The *first order* quantities are pressure variations, temperature variations, particle velocities, and particle displacements. Although of the same order, these small quantities are not equally measurable. In a disturbingly loud sound, the acoustic pressure amplitude may be 10 dyne/cm² or about 10^{-5} atmospheres. The corresponding temperature amplitude will be 10^{-3} °C, and the particle velocity 0.24 cm/sec. The particle displacement amplitude depends on the frequency, and if we choose 1000 c.p.s., it will have a value of about 4×10^{-5} cm or 4000 Å.

The *second order* quantities are intensity and radiation pressure. The values of these quantities corresponding to an acoustic pressure amplitude of 10 dyne/cm² are 1.2 ergs sec^{-1} cm^{-2} (1.2×10^{-7} watt/cm²) and of the order of 4×10^{-5} dyne/cm², respectively. The radiation pressure cannot be specified within better than the correct order of magnitude until acoustic field and boundaries involved are taken into account.

The magnitudes indicated emphasize the extreme sensitivity necessary in measuring even the first order quantities in a relatively strong sound wave. It may be mentioned in passing that the human ear has little difficulty in detecting acoustic pressure amplitudes of 10^{-3} dyne/cm² (10^{-9} atmospheres). The measurement of the second order quantities is even more exacting. The measurement of the intensity by a calorimetric method involves the measurement of heat accumulated at a rate of about one calorie per year.

3*

β) *Measurement of acoustic pressures.* Acoustic pressure is the most easily measured quantity associated with a sound field. This is, in part, due to its scalar nature and, in part, due to technological progress along existing lines.

It is appropriate to describe the characteristics of an ideal pressure-measuring instrument before considering the details of existing microphones.

1. The ideal microphone should disturb the measured pressure field to a negligible extent. This requires it to be small compared to a wavelength to avoid appreciable scattering and diffraction effects. A microphone small compared to the wavelength of the sound measured is equally sensitive to waves incident from any direction[1].

2. The ideal microphone should have constant pressure sensitivity (volt/dyne/cm^2) at all frequencies in the desired frequency range.

3. The ideal microphone should have an acoustic impedance which is large compared to that of the sound field in which the pressure is measured. This criterion is sometimes overlooked in measuring the standing wave field in a tube with the result that the pressure maxima are not correctly determined. This will occur if the volume current required to actuate the microphone is not very small compared to the volume current in the standing wave system in the tube at the plane (perpendicular to the particle velocity) in the tube. Thus, the ideal pressure microphone may be considered as the acoustic analogue of the ideal voltmeter.

4. The ideal microphone should be a linear reversible transducer. Such a microphone may be calibrated by the reciprocity technique.

5. The ideal microphone should have a low thermal noise output. Of course, the importance of this criterion depends on the magnitude of acoustic pressures to be measured. In any case, low noise output means extended dynamic range.

None of the microphones to be discussed in the following sections will meet all the criteria for an ideal pressure microphone, but each has its place in pressure measurement.

I. Pressure sensitive microphones.

17. Electrodynamic pressure microphones. The electrodynamic or moving coil microphone is characterized by a low electrical impedance of the order of 20 ohm an a correspondingly low thermal noise output of about 6×10^{-8} volt at 300° K over a frequency band width of 20 kc. The pressure sensitivity in the mid-frequency range for this type of microphone is of the order of 30 microvolt (open-circuited) per dyne/cm^2, measured at the terminals of the 20 ohm moving coil. The pressure sensitivity of a moving coil microphone in volts per dyne/cm^2 is given by

$$S_p = \frac{B l S_d}{10 Z_m} \tag{17.1}$$

where the flux density B, the coil length l, the diaphragm area S_d, and the mechanical impedance Z_m are all given in M.K.S. units. The low electrical impedance makes it possible to use long cables between the microphone and amplifier without appreciable loss of signal. This characteristic makes the dynamic microphone convenient for measurements in the open.

Fig. 51 shows a sectional view of a dynamic microphone. Since the voltage generated by the moving coil is proportional to the coil velocity, the mechanical impedance of the moving system must be resistive throughout the frequency range where the pressure sensitivity remains constant. The mechanical resonance

---

[1] For special applications, directional properties may be desirable.

of the coil diaphragm assembly is usually placed in the middle of the useful frequency range and sufficient mechanical resistance added to completely suppress the resonant peak. One complication results from the fact that pressure equalization must be provided on the two sides of the diaphragm for slow changes in atmospheric pressure. This permits the acoustic pressure to reach the back as well as the front surface of the diaphragm. The pressure on the back of the diaphragm must be small compared to or out of phase with that on the front surface of the diaphragm in the useful frequency range. This necessitates the introduction of an acoustic filter between the vent in the microphone case and the back surface of the diaphragm. The mechanical equi-

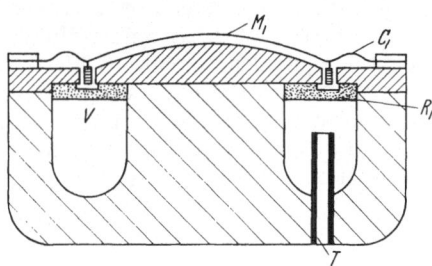

Fig. 51. A sectional view of a dynamic microphone.

valent circuit is shown in Fig. 52. M_1 is the mass of the coil and diaphragm, C_1 is the mechanical compliance of the diaphragm, R_1 is the mechanical resistance referred to the diaphragm area resulting from the acoustic resistance of the porous material just below the moving coil. R_1 is given by

$$R_1 = R_{A1} S_d^2 \tag{17.2}$$

where R_{A1} is the acoustic resistance of the porous material. M_2 and R_2 are the effective mechanical mass and resistance of the vent tube T.

$$M_2 = \frac{4}{3} \varrho \left(l_T + 0.7 \sqrt{S_T} \right) \frac{S_d^2}{S_T}, \tag{17.3}$$

$$R_2 = \frac{8 \mu S_d^2}{r_0^2 S_T} \tag{17.4}$$

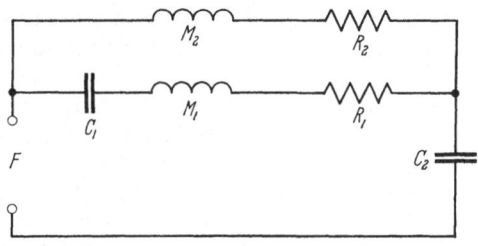

Fig. 52. The mechanical equivalent circuit for a dynamic microphone.

where S_d and S_T are the areas of the diaphragm and vent tube respectively, μ ist the viscosity of air, r_0 and l_T the radius and length of the vent tube and ϱ the density of air. The expressions for M_2 and R_2 assume a parabolic velocity distribution in the vent tube. CRANDALL[1] presents more general expressions covering the transition from the narrow tube to the wide tube case. C_2 the equivalent mechanical compliance of the volume V referred to the diaphragm area given by

$$C_2 = \frac{V}{\varrho c^2 S_d^2}. \tag{17.5}$$

The effect of the venting branch containing M_2 and R_2 at very low frequencies equalizes the forces on the back and front of the diaphragm since the impedance of this branch becomes very small compared to the branch containing C_1. At frequencies just above the frequancy where M_2 and C_2 are series resonant, the force developed across C_2 will be out of phase with the driving force F, and the velocity of M_1 will be increased. Thus, the vented volume may be designed to extend the low-frequency response of the moving coil microphone. The similarity between this and the bass reflex loudspeaker enclosure discussed in Sect. 6

---

[1] IRVING B. CRANDALL: Vibrating Systems and Sound. New York: D. Van Nostrand Co., Inc. 1926.

is quite evident. Fig. 53 shows relative free-field pressure response in decibels vs. frequency in cycles per second for an early electrodynamic microphone designed by WENTE[1]. The dashed curve indicates the decrease in sensitivity to be expected when the back cavity is vented through a very high flow resistance vent. The high-frequency end of the curve has some residual irregularities due to resonance in the air space under the diaphragm which are not taken into account by the equivalent circuit of Fig. 52. The gradual decrease in sensitivity to be expected from the equivalent circuit does not appear in the free-field pressure calibration due to the scattering of sound by the microphone as the wavelength becomes comparable with the diameter of the microphone. This scattering or diffraction effect may result in an increase in sensitivity of as much as 10 db for perpendicular incidence on a diaphragm in a cylindrical housing. WIENER[2] has published an interesting paper on this subject which contains a large number of reference to the earlier work. With large microphones like the early electro-dynamic microphones, the diffraction resulted in sensitivity differences as great as 15 db between perpendicular and grazing incidences at 5000 c.p.s. and above. A reduction of the diameter of the housing from 8 cm to 3 cm has increased the frequency range over which diffraction effects are negligible.

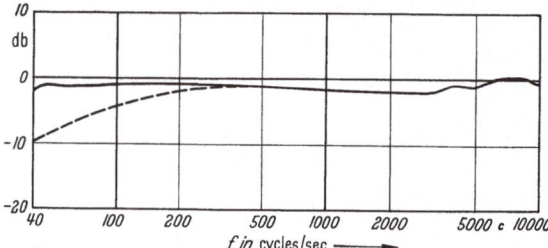

Fig. 53. Free-field pressure response for an early dynamic microphone (after WENTE).

There are two precautions to be considered in the use of moving coil microphones. First, they are quite sensitive to alternating magnetic fields and hence may not be useful measuring the noise fields around large transformers or alternating current machines. Second, they are limited by the coil amplitudes required in the measurement of large acoustic pressures at low frequencies. This limitation should not be overlooked in the measurement of the noise from jet aircraft. The limit of linear operation at low frequencies can best be checked in a closed coupler type of calibrator.

The moving coil microphone is rather sensitive to acceleration of its housing due to the mass of the diaphragm and voice coil assembly. The pressure equivalent of an acceleration of 100 cm/sec^2 acting on the housing of a moving coil microphone having a diaphragm-coil mass of 0.1 gm and an area of 10 cm^2 is 1 dyne/cm^2 which corresponds to a sound pressure level of 74 db reference 0.0002 dyne/cm^2.

18. The condenser microphone. The condenser microphone merits careful consideration since it approaches an ideal microphone more closely than any other type of pressure-sensitive transducer of similar sensitivity. This type of microphone can be made sufficiently small to make it omnidirection over most of the audible frequency range. It lends itself to a variety of methods of calibration and shows excellent stability of sensitivity when not abused. Its one disadvantage is the requirement that the first tube of the amplifier must be within less than a meter of the microphone if a serious loss in sensitivity is to be avoided.

[1] E. C. WENTE and A. L. THURAS: J. Acoust. Soc. Amer. **3**, 44 (1931).
[2] FRANCIS M. WIENER: J. Acoust. Soc. Amer. **19**, 444 (1947).

The first type of condenser microphone to be considered will be the stretched diaphragm air dielectric type. This type has received more study and, as a result, is better understood than the solid dielectric microphone which will be considered later. The essential features are shown in Fig. 54. The cavity behind the diaphragm D is vented to the atmosphere through a leak L. The fixed elec-trode E behind the diaphragm is perfo-rated or grooved to reduce the effect of viscous damping on the motion of the dia-phragm. The fixed electrode is supported by an insulator I. The spacing between the diaphragm and the electrode is about 25 micron or 0.001″. The diaphragm may be either a stretched membrane of plastic or metal foil. The microphone shown in Fig. 54 has the general appearance of a modern stretched diaphragm microphone but differs very little from the microphone described by WENTE[1,2] in 1922. His analy-sis was sufficiently complete so that only minor changes in the structure of the microphone have been made. The analy-sis of the distortion by RADEMAKERS[3] is an important clarification. Most of the changes in the condenser microphone have been in the way of miniaturization and improved circuitry.

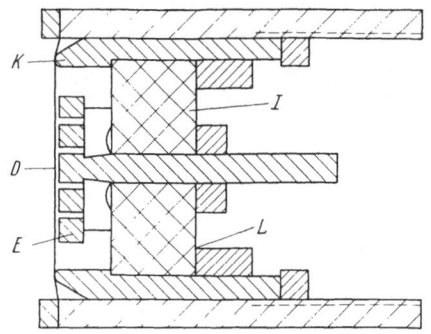

Fig. 54. Condenser microphone showing diaphragm D, fixed electrode E, insulator I, stretching ring K, venting leak L, and backing volume V_B.

Fig. 55 shows the condenser microphone and associated circuit. The non-linearity of the condenser microphone for diaphragm displacements which are not small compared to the equilibrium spacing results from variations in charge on the fixed electrode. For plane parallel plates the force of attraction is inde-pendent of the separation if the charge on the plates is constant. Furthermore, the potential difference between the plates is a linear function of their separation for constant charge. In as far as the condenser microphone approximates the behavior of a pair of parallel pla-tes with constant charge, it ap-proximates a linear system pro-vided, of course, that the mechani-cal restoring force on the dia-phragm is a linear function of the displacement from the equi-

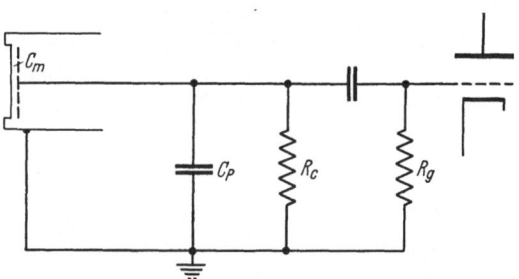

Fig. 55. Condenser microphone and associated circuit.

librium spacing. The real problem in linearizing the pressure amplitude response of the condenser microphone as a transducer lies in maintaining constant charge on the diaphragm. Examination of the circuit in Fig. 55 indicates two causes of charge variation with spacing. As the spacing changes, currents flow in R_c and R_g resulting in a transfer of charge to or from the capacitor C_m. A portion of this charge is supplied by C_p the total fixed shunt capacitance including the input capacitance of the vacuum tube at the right. If the time constant of the circuit τ is very large compared to the period of motion of the diaphragm,

[1] E. C. WENTE: Phys. Rev. **10**, 39—69 (1917).
[2] E. C. WENTE: Phys. Rev. **19**, 498—503 (1922).

the charge variation due to the currents in the resistors R_c and R_g will be negligible

$$\tau = (C_m + C_p)\left(\frac{R_c R_g}{R_c + R_g}\right). \tag{18.1}$$

This is not sufficient to linearize the transducer, for the shunt capacitance C_p requires a transfer of charge as its potential difference changes with the variation in spacing of C_m. The computation of the output voltage including the distortion term given below is taken from a publication of Rademakers[1].

If C_m is the capacitance of the microphone for the equilibrium spacing x_0, then the instantaneous capacitance for an instantaneous displacement Δx may be written

$$C(\Delta x) = \frac{C_m}{1 + \alpha} \tag{18.2}$$

where

$$\alpha = \frac{\Delta x}{x_0} \tag{18.3}$$

and the instantaneous voltage V is given by

$$V = \frac{Q}{C_p + \dfrac{C_m}{1 + \alpha}} = \frac{Q}{C_m + C_p} \cdot \frac{1 + \alpha}{1 + \dfrac{\alpha C_p}{C_p + C_m}} = V_0 \frac{1 + \alpha}{1 + \dfrac{\alpha C_p}{C_p + C_m}} \tag{18.4}$$

where Q is the total charge which is assumed to be fixed for the open-circuited condition $(R_c = R_g = \infty)$. V_0 is voltage at the equilibrium spacing x_0. The variation in voltage ΔV is given by

$$\Delta V = V - V_0 = \frac{V_0 C_m}{C_m + C_p} \cdot \frac{\alpha}{1 + \dfrac{\alpha C_p}{C_m + C_p}} \tag{18.5}$$

which may be approximated by

$$\Delta V \approx \frac{V_0 C_m}{C_m + C_p} \cdot \alpha\left(1 - \alpha \cdot \frac{C_p}{C_m + C_p} + \cdots\right) \tag{18.6}$$

and if α is a harmonic time function

$$\alpha = \alpha_0 \sin \omega t \tag{18.7}$$

$$\Delta V \approx \frac{V_0 C_m}{C_m + C_p}\left(\alpha_0 \sin \omega t - \frac{\alpha_0^2}{2}(1 - \cos 2\omega t) + \cdots\right) \tag{18.8}$$

and the percent second harmonic distortion is

$$\text{percent second harmonic} \approx \frac{100\,\alpha_0 C_p}{2(C_p + C_m)} \tag{18.9}$$

the quality of the approximation depending on the condition that

$$\frac{\alpha_0 C_p}{C_p + C_m} < 1. \tag{18.10}$$

This condition is not difficult to satisfy since $C_p/(C_p + C_m)$ can be of the order o 0,2, and α_0 is usually less than 0.1.

---

[1] A. Rademakers: Philips Techn. Rev. 9, 330—338 (1947).

Thus, the analysis of RADEMAKER's clarifies the role played by the inactive capacity C_p in reducing the fundamental component and increasing the percentage of second harmonic distortion. The author is not aware of any exact treatment of the distortion introduced by both the parallel capacity C_p and finite resistance R_c and R_g.

The linearized circuit in which the condenser microphone is replaced by an equivalent generator is shown in Fig. 56. The generator has an open-circuited voltage $\alpha_0 V_0 \sin \omega t$ and an internal impedance $1/j\omega C_m$. The form shown in Fig. 56b is obtained from Fig. 56a by the use of THEVENIN's theorem. It is clear that at the low frequencies, the voltage across R will be less than the generated voltage in Fig. 56b. The voltage across R will be down 3 db when

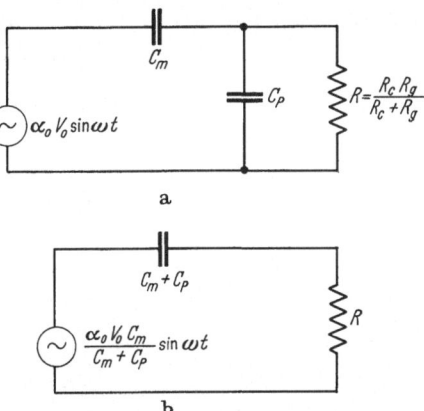

Fig. 56a and b. Two forms of electrical circuit where the condenser microphone has been replaced by a voltage generator and series impedance.

$$\omega R (C_p + C_m) = 1. \qquad (18.11)$$

This limits the low-frequency sensitivity of the condenser microphone. For a microphone having a capacitance $C_m = 40 \,\mu\mu\text{fd}$, and a shunt capacitance $C_p = 10 \,\mu\mu\text{fd}$, and $R = 10^8$ ohm, the sensitivity of microphone and circuit will be down 3 db at 32 c.p.s.

α) *Noise threshold for condenser microphones.* SCHOTTKY[1] and JOHNSON[2] have discussed the generation of thermal noise in conductors, and WEBER[3] has considered the noise in microphone circuits. The spectrum voltage level E_n in volt per c.p.s. is given by

$$E_n = \sqrt{4\,k\,T\,R\,(f)} \qquad (18.12)$$

where k is the Boltzmann constant (k $= 1.37 \times 10^{-23}$ joules per °K), T is the absolute temperature, and $R(f)$ is the series equivalent resistance between the terminals where E_n is measured. In the case of the condenser microphone, $R(f)$ is given by

$$R(f) = \frac{R}{1 + (\omega C R)^2} \qquad (18.13)$$

and the total r.m.s. noise voltage E_T is given by

$$E_T = \sqrt{4\,k\,T \int_{f_1}^{f_2} R(f)\,df} = \sqrt{\frac{2kT}{\pi C}} \sqrt{\arctan (\omega_2 C_R) - \arctan (\omega_1 C_R)} \qquad (18.14)$$

where $\omega_1 = 2\pi f_1$ and $\omega_2 = 2\pi f_2$ and f_1 and f_2 are the frequency band limits. In most cases $\omega_2 C R$ is sufficiently large that setting it equal to infinity does not change the value of the first term under the second square root sign. E_T is now given by

$$E_T = \sqrt{\frac{2kT}{\pi C}} \sqrt{\frac{\pi}{2} - \arctan (\omega_1 C R)}. \qquad (18.15)$$

[1] W. SCHOTTKY: Ann. Physik **57**, 541—567 (1918).
[2] J. B. JOHNSON and F. B. LLEWELLYN: Bell Syst. Techn. J. **14**, 85—96 (1935).
[3] W. WEBER: Akust. Z. **8**, 121—127 (1943).

The minimal value of $\omega_1 C R$ is unity if f_1 is chosen as the frequency of the lower half power point where the sensitivity is down 3db. Examination of the behavior of Eq. (18.15) shows that for each doubling of R ($\omega_1 C =$ const) reduces the total noise voltage E_T by about 3db. This reduction in noise requires that the lower band limit be determined at some other point in the chain of amplification where noise generation is unimportant. Eq. (18.15) also indicates that a doubling of C for $\omega_1 R =$ const, will decrease the noise E_T by 6db. It must also be noted for $C_m =$ const. that doubling C decreases the signal voltage by 6db so that no change in signal to noise ratio is to be gained by a change in C alone. Since high resistances may be difficult to maintain, one might consider letting $\omega_1 C R =$ const

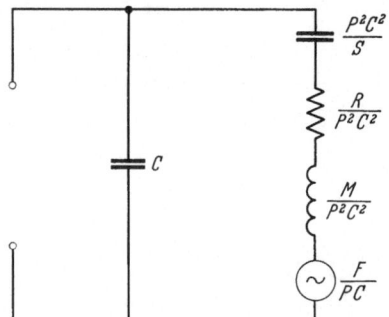

Fig. 57. The equivalent electrical circuit for a condenser microphone including the transformed mechanical impedances. P represents the electric field between the diaphragm and the fixed electrode. S is the mechanical stiffness of the diaphragm (after BECKING [2]).

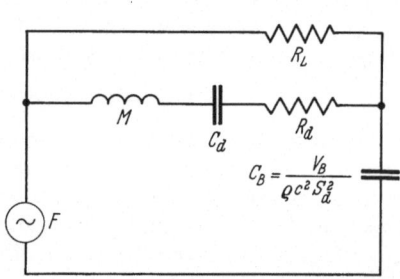

Fig. 58. The mechanical equivalent circuit of the condenser microphone shown in Fig. 54. R_L is the mechanical resistance of the leak L referred to the diaphragm area S_d. M_d, C_d, and R_d are the mass, compliance, and resistance of the diaphragm. C_B is the mechanical compliance of the backing volume V_B.

and increasing C. Each doubling of C will then reduce the noise 3db and the signal 6db with a net loss in the signal to noise ratio of 3db. Finally, it should be clear that for $RC =$ const, each doubling of f_1 will result in a reduction of the noise voltage E_T by 3db and an increase in the signal to noise ratio of 3db.

As an example, consider a microphone with a sensitivity of 3×10^{-3} volt per dyne/cm² having a capacity of 50×10^{-12} farad. If $\omega_1 R C = 1$, E_T will be 0.4×10^{-6} volt at 20° C which is equivalent to 21.3×10^{-4} dyne/cm² or a sound pressure level of 21db reference 0.0002 dyne/cm².

A reduction of the noise threshold of a condenser microphone through the use of a carrier-frequency system in which the microphone is placed in a low-resistance circuit has been studied by ZAALBERG VAN ZELST[1]. The bridge circuit described in his paper yield an equivalent noise generating resistance of 400 ohm and an equivalent noise pressure spectrum which is independent of frequency. The total noise pressure over a 14 kc band is computed to be 1×10^{-4} dyne/cm² or a sound level of -6 db reference 0.0002 dyne/cm². Direct measurements showed the actual noise to be somewhat higher than this. BECKING[2] has explained the discrepancy in terms of a fluctuating force generated in the viscous damping resistance between the diaphragm and the fixed electrode. Fig. 57 shows the electrical circuit including the mechanical impedances as transformed into their electrical equivalents. S, R, and M are the mechanical stiffness, resistance, and mass, respectively. C is the capacity of the microphone and P the electric field between the diaphragm and fixed electrode. A computation of the

[1] J. J. ZAALBERG VAN ZELST: Philips Techn. Rev. 9, 357 (1948).
[2] A. G. TH. BECKING and A. RADEMAKERS: Acustica 4, 96 (1954).

pressure equivalent of the fluctuation noise generated in the mechanical resistance of a high quality condenser microphone for a 14 kc band gives a value of 10^{-3} dyne/cm² or an equivalent sound pressure level of 14 db reference 2×10^{-4} dyne/cm². The results of computations from the theory were confirmed by exceptionally fine experiments.

BECKING points out that the r.m.s. amplitude of the diaphragms due to its Brownian motion over a 14 kc band, is only 2×10^{-3} Å which is a thousand times smaller than the diameter of an atom and a hundred times smaller than the amplitude of the temperature vibration of separate atoms in the diaphragm.

β) *Frequency response.* The frequency response of a condenser microphone is determined by its mechanical constants. Fig. 58 shows the equivalent mechanical circuit for a stretched diaphragm condenser microphone. The leak R is required to equalize slow changes in atmospheric pressure. The output voltage is proportional to the displacement of the diaphragm which is represented by a drop in force across the compliance of the diaphragm C_d. At very low frequencies where M_d and R_d may be neglected, the equivalent circuit indicates that the response will be down 3 db when

$$\omega R_L (C_d + C_B) = 1. \tag{18.16}$$

C_B is usually large compared to C_d. The resistance R_L can be made large enough to satisfy Eq. (18.16) at 10 c.p.s. The circuit of Fig. 58 predicts the pressure response only and does not take into account dffraction effects. Fig. 59 shows the normal incidence free field pressure calibration of a Western Electric 640 AA condenser microphone without grid as measured by RUDNICK[1]. The dotted curve shows the pressure calibration supplied by the Bell Telephone Laboratories (ROMANOW and HAWLEY) for the same microphone. The diffraction effecis clearly evident as the difference between the two curves. The pressure calibration indicates that the microphone was slightly over-damped as there is no evidence of a diaphragm resonance between 8 and 9 kc. The 12 db per octave decrease in sensitivity in the pressure calibration at frequencies above diaphragm resonance is to be expected from the equivalent circuit of Fig. 58.

γ) *Directional characteristics.* The condenser microphone has a directivity which depends on its size and the frequency measured. Since the size of the microphone and associated preamplifier vary considerably and both are important in determining the directivity, a single polar pattern will be presented to indicate approximately what is to be expected. Fig. 60 shows the relative sensitivity vs. angle for a Western Electric 640 AA microphone at 18.7 kc. This microphone has a diameter of 2.4 cm and was used with a cylindrical support of diameter 2.4 cm and length about 15 cm.

δ) *Recent developments in condenser microphones.* Two interesting types of microphones have been developed since 1946. The first is the Solid Dielectric Microphone developed by KUHL. This type shows great promise for relatively high sensitivity over extremely broad frequency bands.

The second is the development by HILLIARD of a miniature clamped plate type condenser microphone of moderate band width which can be constructed to withstand very high sound levels by the insertion of different diaphragms.

ε) *Solid dielectric microphones.* Paralleling the development of solid dielectric electrostatic loudspeakers by KUHL is his development of solid dielectric condenser

---

[1] I. RUDNICK and M. N. STEIN: J. Acoust. Soc. Amer. **20**, 818 (1948).

microphones[1]. This work has produced microphones with a remarkably uniform sensitivity over an exceptionally large frequency range. Fig. 61 shows details of such a microphone. The metalized plastic diaphragm D (styroflex) is

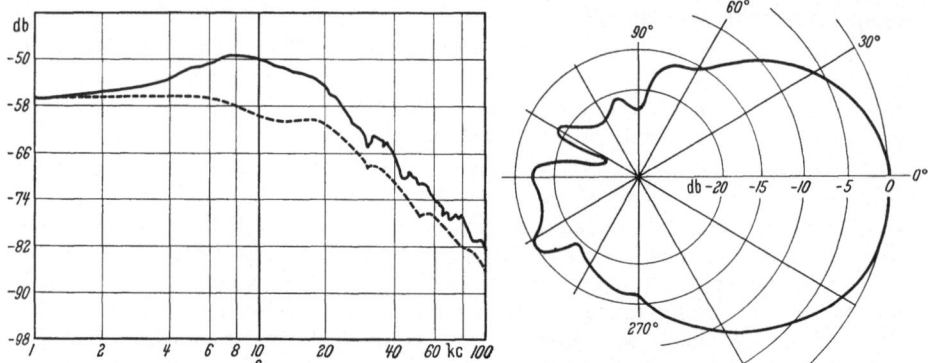

Fig. 59. Free-field and pressure calibrations of a Western Electric 640 AA microphone without protective grid. The difference in the two curves represents the diffraction correction for normal incidence (after RUDNICK et al.).

Fig. 60. The pressure sensitivity of a Western Electric 640 AA microphone without protective grid at a frequency of 18.7 kc (after RUDNICK et al.).

stretched over the grooved plate E with the metalized surface turned away from the back plate. The back plate is insulated from the spherical housing by plexiglas insulators. The tension in the diaphragm is obtained by pushing the back plate against it by means of the threaded stem. The two plastic materials found to be useful for diaphragms were 10 micron styroflex and 12 micron polyethylene. KUHL has made a careful systematic study of the effect of diaphragm tension and back plate configuration on the frequency response of this type on condenser microphone. Fig. 62 shows the free field pressure sensitivity of two such microphones of identical construction. The sensitivity is in db reference 1 volt/dyne/cm². The circular grooves in the back plate were 0.1 mm deep, 0.25 mm wide, and separated by 0.25 mm flat rails. The polarization voltage was 150 volt. The diaphragm was silver-coated styroflex 10 micron thick. KUHL remarks that it is almost impossible to produce two microphones with the same frequency response since the pre-strain in the plastic diaphragm cannot be completely controlled. The deviation in the two curves in Fig. 62 shows about the variation to be expected when considerable care is exercised.

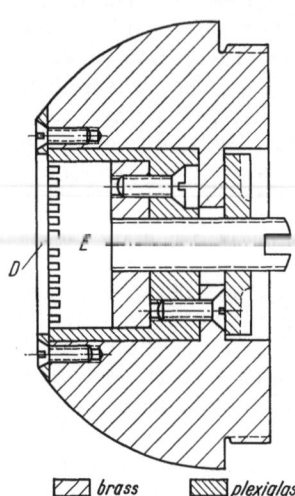

▨ brass   ▨ plexiglas

Fig. 61. A solid dielectric microphone in section with the rear half of the spherical housing removed. The metalized diaphragm D is stretched over the grooved back plate E with the metalized surface turned away from the back plate (after KUHL[1]).

The stability of these microphones seems quite good. Variations of the order of 0.5 db were observed over a period of seven months. To obtain this short of consistency, the diaphragm must be glued to the retaining ring, and the unit must remain assembled in the same housing since the pre-strain will not be the same in different housings.

[1] W. KUHL, G. R. SCHODDER and F. K. SCHRÖDER: Acustica 4, 519 (1954).

Because of the spherical housing, these microphones show remarkably smooth directivity patterns. The directivity of pressure microphones in spherical housings is covered in several papers[1-3].

Noise measurements on solid dielectric microphones indicated the self-noise of this type of microphone is well below the circuit noise. This was demonstrated by no change in noise level when the microphone was replaced by a fixed condenser.

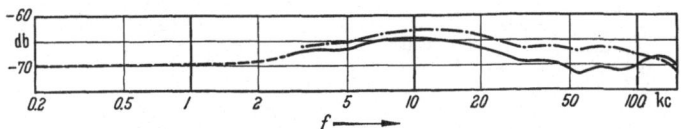

Fig. 62. The free-field normal incidence pressure sensitivity of two identical solid dielectric condenser microphones. The ordinate is in db reference 2×10^{-4} dynes/cm² (after KUHL et al.).

ζ) *The clamped plate condenser microphone.* This miniature condenser microphone is shown in section in Fig. 63. The diaphragm is a glass disc a few thousandths of an inch thick clamped at the edge between flat surfaces. The glass diaphragm is coated with gold on the side away from the back plate. The capacity between the back plate and the diaphragm is about 6 $\mu\mu$. Such a small capacity requires a very high input resistance at the first tube of the amplifier as well as a very high value for the resistance of the insulation in the microphone. Electrical leakage in the microphone is reduced by the presence of a guard ring which is

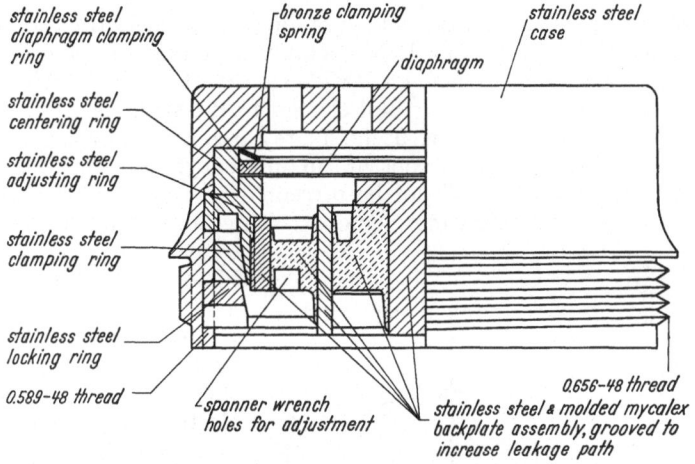

Fig. 63. Miniature plate microphone construction (Altec Lansing 21).

at a d. c. potential essentially equal to that of the back electrode. The floating grid cathode-follower circuit used with this microphone is shown in Fig. 64. The guard ring is connected to the cathode which is about 200 volt above the grounded shell of the microphone. The input resistance of the floating grid cathode-follower is about 3×10^{10} ohm. A circuit of this type is essential with a microphone of such low capacitance. The circuit has some disadvantages: one of which is the fact that the first tube contributes no amplification. The second disadvantage

---

[1] See footnote 1, p. 44.

[2] W. KUHL: Acustica **2**, 226 (1952).

[3] R. L. PRITCHARD: Acustica **3**, 359 (1953).

is the greater noise output of the cathode-follower circuit which may be of the order of 20 db noisier than indicated by a calculation based on the input re-

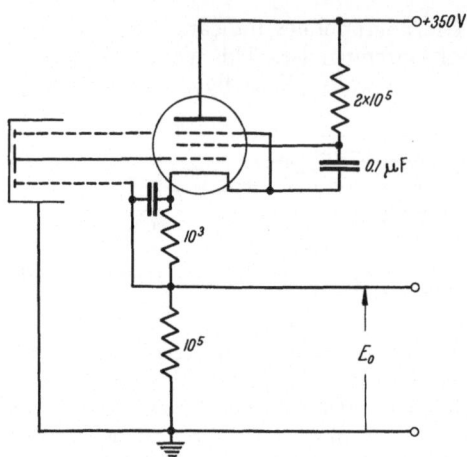

Fig. 64. Circuitry of floating grid cathode follower used with Altec Lansing 21 microphone.

sistance of the cathode-follower. This results from the fact that the cathode-follower voltage feedback is ineffective in reducing the noise generated by the resistance between cathode and grid. Thus, while the voltage gain of the follower is unity as far as the signal generated by the microphone is concerned, it may be much greater than unity for the noise generated in the grid to cathode leakage resistance.

The clamped plate type condenser microphone is well adapted to use in high intensity sound fields where circuit noise is of little importance, and the inherent ruggedness of this type of microphone makes it difficult to damage. Mercer[1] has studied its stability and sensitivity at temperatures up to $500° F$ and Goff[2] its linearity up to 176 db sound pressure level.

19. Piezoelectric microphones. The use of the piezoelectric properties of crystalline materials has developed considerably over the past 20 years. Piezoelectric microphones are constructed in a wide variety of forms involving different combinations of mechanical and crystal elements. We shall consider, here, only a few examples of configurations useful in research type measurements. The details of the orientation of the various cuts in different crystals has been discussed by Cady[3] and Mason[4] and will not be presented here. An excellent summary of this material is to be found in "Sonics"[5]. Three types of pressure sensitive elements will be considered: bimorph benders, cylinders, and the volume sensitive stack.

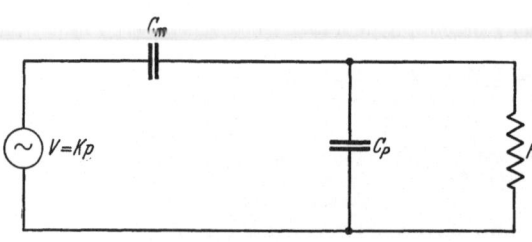

Fig. 65. The equivalent circuit for a piezoelectric microphone.

Before considering specific types of microphones, some of the general properties of piezoelectric microphones will be discussed. The equivalent electrical circuit is given in Fig. 65. Since the piezoelectric element behaves like a constant voltage generator with a capacitative internal impedance determined by C_m, the shunt capacitance of the cable lowers the voltage delivered to the amplifier but does not reduce the high-frequency components of the signal selectively. However, it does increase the temperature dependence of the over-all sensitivity of the system

[1] D. M. A. Mercer: J. Acoust. Soc. Amer. **26**, 936 (A) (1954).
[2] K. W. Goff and D. M. A. Mercer: J. Acoust. Soc. Amer. **27**, 1133 (1955).
[3] W. G. Cady: Piezoelectricity. New York: McGraw-Hill 1946.
[4] W. P. Mason: Piezoelectric Crystals and their Application to Ultrasonics. New York: D. Van Nostrand Co. 1950.
[5] T. F. Hueter and R. H. Bolt: Sonics. New York: John Wiley & Sons, Inc. 1955.

if the dielectric constant of the piezoelectric element depends markedly on temperature. The ratio of the voltage across C_p to the generated voltage for $R(C_p + C_m) > 1$ is given by

$$\frac{V_0}{V_g} = \frac{C_m}{C_p + C_m} \quad (19.1)$$

and it is clear that the voltage ratio will be proportional to C_m for $C_p > C_m$ and exhibit the temperature dependence of C_m, whereas for $C_p < C_m$ this effect is negligible. Fig. 66 shows the fractional change in dielectric constant vs. temperature for a number of piezoelectric materials in common use. There is, also, a smaller effect due to the temperature dependence of the piezoelectric constants, but this is

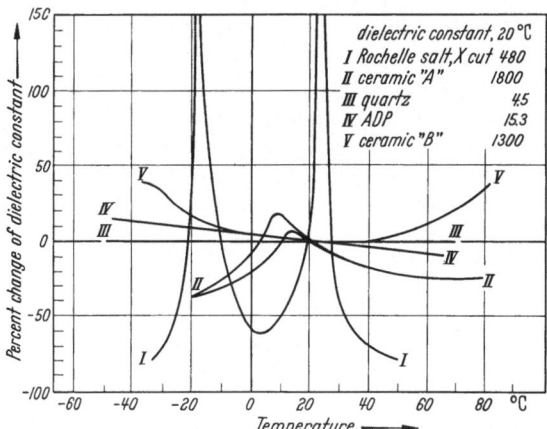

Fig. 66. The variation of dielectric constant with temperature for several piezoelectric materials. (Courtesy of Brush Electronics Co.)

essentially negligible in most materials compared to the effect of the variation in dielectric constant.

The value of the product $R(C_p + C_m)$ determines the low-frequency cut-off $(-3$ db point). R may be seriously reduced by leakage through the piezoelectric material. This effect is quite marked in ammonium dihydrogen phosphate as shown in Fig. 67.

α) *Bimorph type microphones.* The term "bimorph" applies to a laminated structure composed of two layers of crystal slabs cut in such a way as to be sensitive to elongation. Fig. 68 shows the bimorph configuration[1]. The two crystal slabs are oriented so that when the bimorph is bent and the upper slab shortened while the lower slab elongates, the voltages generated add. The two outer foils are connected together to the grounded side of the electrical system, and the inner foil Q is connected to the high side. The bimorph element must be supported so as to bend when pressure is applied to a single surface, and the other surface must be shielded from the pressure field. The bimorph microphone

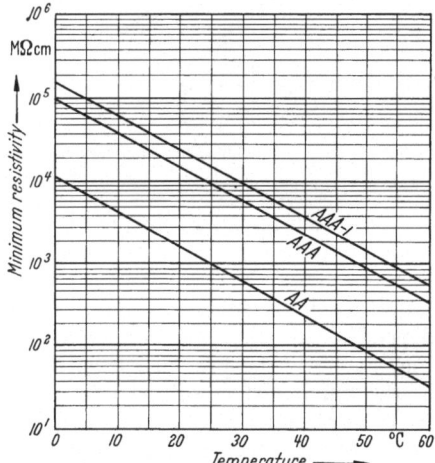

Fig. 67. The resistivity of ADP (ammonium dihydrogen phosphate) as a function of temperature. (Courtesy of Brush Electronics Co.)

in Fig. 68 is made of two X-cut quartz slabs cemented together and supported by knife edges. The pressure seal was obtained with a nondrying lacquer. The quartz slabs were 30 mm long, 10 mm wide, and 1.5 mm thick. The resonant frequency of the cemented pair was 19 kc. The sensitivity at frequencies well

---

[1] W. FURRER and H. WEBER: Techn. Mitt. schweiz. Telegr. u. Teleph.-Verw. **1946**, No. 6.

below resonance was 6 microvolt per dyne/cm². The electrical capacitance of the microphone and cable is 90 μμfd. Quartz was used since the microphone was employed to measure explosive sounds where the peak pressures were of the order of one atmosphere, and temperature and chemical stability were more important than sensitivity.

β) *Cylindrical ceramic microphones.* The use of barium titanate cylinders as pressure sensitive elements for small microphones and hydrophones for the measurement of high intensity acoustic fields has proved highly satisfactory. Langevin[1] has developed expressions for the pressure sensitivity of cylindrical elements for a number of different polarizations, electrode configurations, and types of exposure to the acoustic pressure field. We shall consider only the case of radial polarization with the electrodes covering the entire cylindrical surfaces both inside and out and rigid caps which cover the total area of the ends.

The cylinder is subjected to three stresses when placed in a uniform pressure field. The axial and circumferential

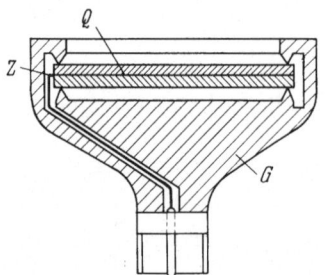

Fig. 68. A quartz bimorph microphone for the measurement of intense sound waves (after Furrer).

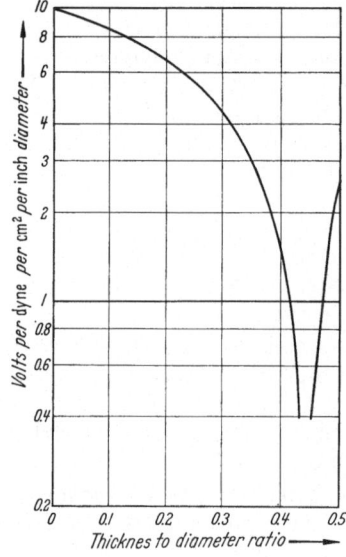

Fig. 69. The voltage sensitivity of barium titanate as a function of the thickness to diameter ratio (after Langevin).

stresses produce radial electric fields of the same sign which contribute most of the potential difference between the cylindrical electrodes. The radial stress produces a radial electrical field of opposite sign and thus reduces the potential difference between the electrodes. The relative strength of these two effects depends on the ratio a/b where a and b are the inner and outer radii, respectively. Fig. 69 shows the pressure sensitivity in volts per dyne/cm² per inch of diameter of the outer surface of the cylinder. Complete cancellation yielding zero sensitivity is obtained for a/b equal to 0.43. Since the sensitivity is proportional to the diameter, very small microphones of this type are only useful in high-intensity sound fields.

Allen[2] has described a microphone made of a barium titanate cylinder $\frac{1}{16}$ inch in diameter. Fig. 70 shows a sectional view of this tiny microphone. The frequency response was flat with ±1 db from 10 to 100 kc. with a sensitivity of −143 db reference 1 volt per dyne/cm².

γ) *High impedance piezoelectric microphones.* Microphones of very high acoustic impedance are required for the measurement of acoustic pressures in

[1] R. A. Langevin: J. Acoust. Soc. Amer. **26**, 421 (1954).
[2] H. Schilling et. al.: Final Report on Atmospheric Physics (1950), Pennsylvania State College.

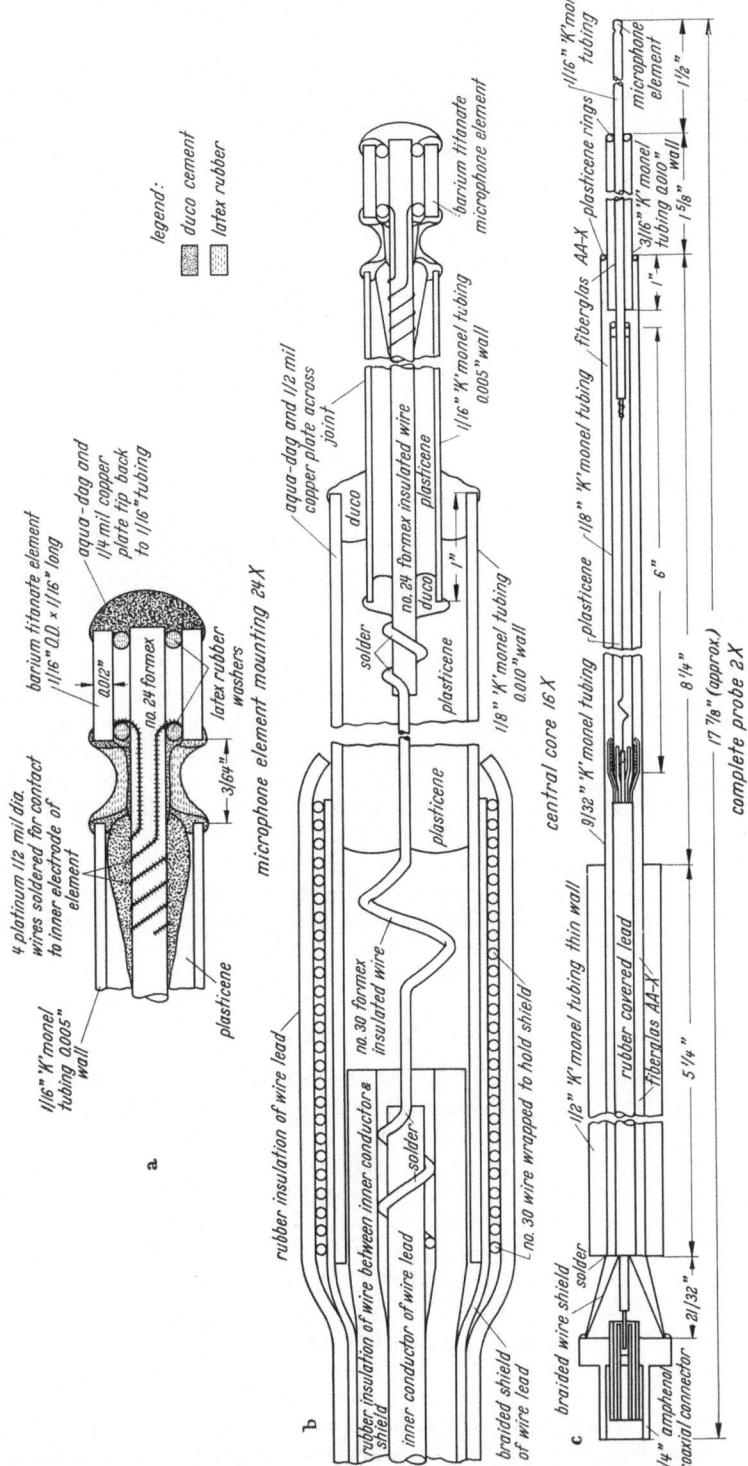

Fig. 70a–c. A cylindrical barium titanate microphone for intense sound fields (after ALLEN).

liquid-filled resonant systems where a compliant microphone may produce a serious detuning of the resonant system. A microphone for this type of measurement must have an effective compressibility no larger than that of the liquid in which the measurements are made. This condition is satisfied by piezoelectric materials, but only a few such materials develope a voltage under pure compression where no change in shape takes place. Lithium sulphate, barium titanate, and tourmaline all produce a voltage on pure compression. The sensitivities of two of these materials are tabulated in Table 3. The values given are the voltages produced between opposite faces of a 1 cm cube when subjected to a uniform pressure of 1 dyne/cm² on all six surfaces. Lithium sulphate has the obvious advantage of relatively great sensitivity although some of this must be sacrificed to obtain a reasonable value for the capacitance of the microphone. This is ac-

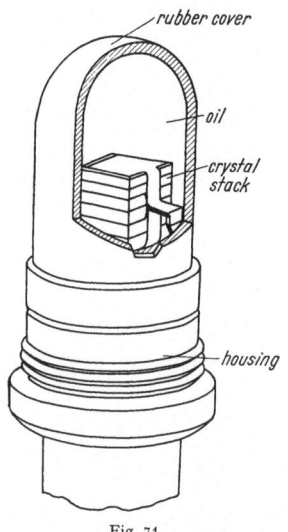

Fig. 71.
A lithium sulphate microphone.
(Courtesy of Brush Electronics Co.)

Table 3.

Piezoelectric Material	Sensitivity in volt/cm / dyne/cm²
Lithium Sulphate	148×10^{-6}
Barium Titanate "A" . .	2.2×10^{-6}

complished by a parallel connection of a number of thin plates stacked together as shown in Fig. 71. Lithium sulphate is highly soluble in water, and the crystal stack must be encased in an oil-filled envelope. Barium titanate and tourmaline require no such elaborate protection from moisture and may be used in water if coated with an insulating film of rubber or lacquer. Barium titanate, because of its high dielectric constant (1200 to 1700) provides a suitable capacitance without the complication of stacking slabs and, lends itself to the construction of microphones of very high impedance for use in the measurement of high acoustic pressures. Since the noise voltage varies as $C^{-\frac{1}{2}}$, the larger capacitance of the barium titanate microphone results in a reduction of the circuit noise of thermal origin. A more complete discussion of this type of circuit noise is given in Sect. 18 on condenser microphones.

20. Probe microphones. While it is possible to make very small pressure microphones such as those just described in the preceding section, their sensitivity is inferior to either moving coil or condenser microphones of conventional dimensions. The probe microphone attempts a solution of the small microphone problem by leading the sound picked up at the tip of the probe tube to a conventional microphone as shown in Fig. 72. Fig. 72b shows the equivalent circuit for a combination of probe tube (respresented as a section of transmission line) and the terminating microphone (represented as an acoustic impedance in parallel with the capacitance of the fronting volume V_f).

If a condenser microphone is used, its impedance may be represented as a capacitor over the useful range of the probe assembly. A low-loss transmission line terminated in a capacitative reactance may be expected to have a high standing wave ratio which will result in large variations in the input impedance of the transmission line and correspondingly large fluctuations in the ratio of the pressure at the microphone to the pressure at the open end. The accoustic resistance R is useful to insure that the minimal input acoustic impedance at the probe tip is not less than R. The fluctuations in the pressure sensitivity due to multiple reflections may be made to vanish by making R equal to Z_0 the characteristic accoustic impedance of the transmission line regardless of the terminating impedance at the microphone end of the probe tube.

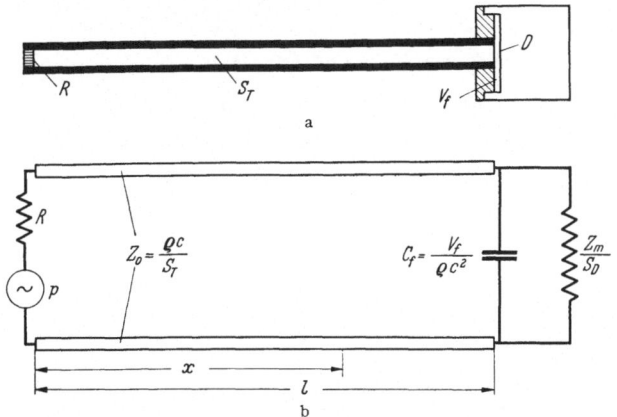

Fig. 72 a and b. Probe microphone (a) and equivalent circuit (b).

To analyze the response characteristics of a probe microphone in the most general case, we will start by considering wave propagation in the probe tube. The pressure at any distance x from the probe tip is given by

$$p(x) = P_i\, e^{j\omega t}\, (e^{-jkx - bx} - e^{-2\psi} e^{jkx + bx}) \tag{20.1}$$

where

$$e^{-2\psi} = -\frac{P_r}{P_i} \quad \text{at} \quad x = 0 \tag{20.2}$$

and P_r and P_i are the incident and reflected pressure amplitudes. It follows that

$$p(x) = 2 P_i\, e^{j\omega t} e^{-\psi}\, \mathrm{Sin}\,[\psi - bx - jkx]. \tag{20.3}$$

Since

$$U_i \approx \frac{P_i}{\varrho c} \quad \text{and} \quad U_r \approx \frac{P_r}{\varrho c} \quad \text{if} \quad b < k \tag{20.4}$$

where U_i and U_r are incident and reflected particle velocity amplitudes

$$u(x) = \frac{2 P_i}{\varrho c}\, e^{j\omega t} e^{-\psi}\, \mathrm{Cos}\,[\psi - bx - jkx], \tag{20.5}$$

the acoustic impedance may be written as

$$Z(x) = \frac{p(x)}{S_T\, u(x)} = \frac{\varrho c}{S_T}\, \mathrm{Tan}\,[\psi - bx - jkx] \tag{20.6}$$

where S_T is the cross-sectional area of the probe tube. Setting

$$\psi = \pi(\alpha_0 + j\beta_0) \tag{20.7}$$

and

$$\psi - bx - jkx = \pi(\alpha + j\beta) \tag{20.8}$$

we have

$$\alpha = \alpha_0 - \frac{b}{\pi}\, x, \quad \beta = \beta_0 - \frac{2x}{\lambda} \tag{20.9}$$

4*

and at $x = l$

$$\alpha_l = \alpha_0 - \frac{bl}{\pi}, \qquad \beta_l = \beta_0 - \frac{2l}{\lambda}. \tag{20.10}$$

Defining $p_0 = P_0 e^{j\omega t}$ as the pressure just outside of the acoustic resistance R, we may write for the pressure at $x = 0$ just inside R

$$p(0) = \frac{p_0 Z(0)}{R + Z(0)} = 2 P_i e^{j\omega t} e^{-v} \operatorname{Sin} \pi \left[\alpha_l + \frac{bl}{\pi} + j \left(\beta_l + \frac{2l}{\lambda} \right) \right] \tag{20.11}$$

and the pressure at the microphone end $(x = l)$ as

$$p(l) = 2 P_i e^{j\omega t} e^{-v} \operatorname{Sin} \pi \left[\alpha_l + j\beta_l \right] \tag{20.12}$$

and the last two equations yield the ratio of $p(l)$ to p_0 as

$$\frac{p(l)}{p_0} = \frac{Z(0)}{R + Z(0)} \cdot \frac{\operatorname{Sin} \pi \left[\alpha_l + j\beta_l \right]}{\operatorname{Sin} \pi \left[\alpha_l + \frac{bl}{\pi} + j \left(\beta_l + \frac{2l}{\lambda} \right) \right]} \tag{20.13}$$

and inserting

$$Z(0) = \frac{\varrho c}{S_T} \operatorname{Tan} \pi \left[\alpha_l + \frac{bl}{\pi} + j \left(\beta_l + \frac{2l}{\lambda} \right) \right] \tag{20.14}$$

we have, on simplification

$$\frac{p(l)}{p_0} = \frac{Z_0 \operatorname{Sin} \pi \left[\alpha_l + j\beta_l \right]}{R \operatorname{Cos} \pi \left[\alpha_l + \frac{bl}{\pi} + j \left(\beta_l + \frac{2l}{\lambda} \right) \right] + Z_0 \operatorname{Sin} \pi \left[\alpha_l + \frac{bl}{\pi} + j \left(\beta_l + \frac{2l}{\lambda} \right) \right]} \tag{20.15}$$

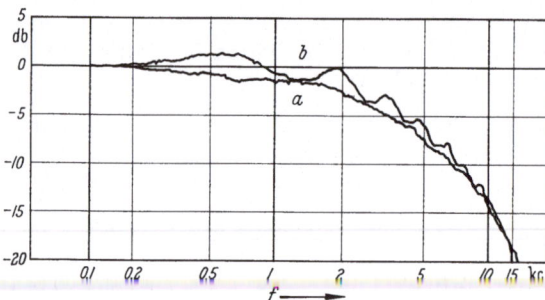

where $Z_0 = \frac{\varrho c}{S_T}$ the characteristic impedance of the probe tube. α_l and β_l are given by

$$Z_0 \operatorname{Tan} \pi \left[\alpha_l + j\beta_l \right] = Z_l \tag{20.16}$$

where Z_l is the terminating acoustic impedance at the microphone end including the acoustic impedance of the fronting volume in parallel with the acoustic impedance of the microphone.

Fig. 73. Frequency response curves for a resistively tipped probe microphone as computed on an electrical analogue. Curve (a) is for $R = Z_0$, and curve (b) is for $R = 0.7 Z_0$.

A significant simplification of Eq. (20.5) is possible if $R = Z_0$ yielding

$$\frac{p(l)}{p_0} = e^{-\pi} \left(\alpha_l + \frac{bl}{\pi} \right) e^{-j\pi} \left(\beta_l + \frac{2l}{\lambda} \right) \operatorname{Sin} \pi \left[\alpha_l + j\beta_l \right] \tag{20.17}$$

which is completely devoid of the effects of multiple reflections. α_l, b, β_l, and λ are all frequency dependent, but the pressure ratio as a function of frequency is a smooth curve. As an example, consider the case of total reflection at the microphone end of the probe for which Z_l may be the reactance of a pure capacitance C_l. This yields for α_l and β_l

$$\alpha_l = 0, \quad \beta_l = \frac{1}{\pi} \arctan \left(\frac{-1}{\omega C_l Z_0} \right) \tag{20.18}$$

where $\beta_l = \frac{1}{2}$ when $\omega = 0$ and $\beta_l = 1$ when $\omega = \infty$. The magnitude of the pressure ratio is given by

$$\left| \frac{p(l)}{p_0} \right| = e^{-bl} \sin(\pi\beta_l) \tag{20.19}$$

which is a monatonic decreasing function of frequency. Fig. 73 shows the probe pressure ratio for two values of R/Z_0 when the probe is terminated in a pure acoustic capacitance. Curve (a) is the case where $R=Z_0$ and (b) is for $R=0.7Z_0$. These curves were obtained from measurements on an elctrical analogue and not on an actual probe microphone.

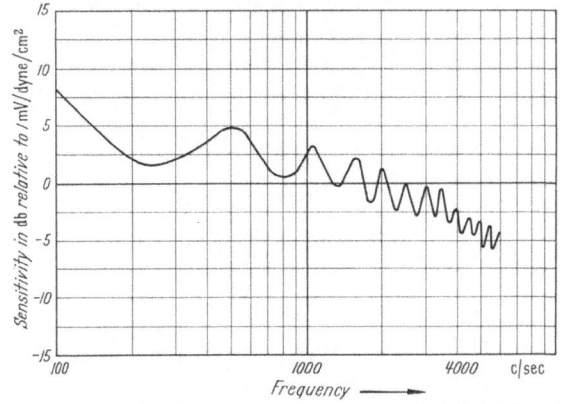

Fig. 74. The sensitivity vs. frequency for a resistively terminated probe microphone (after WEST).

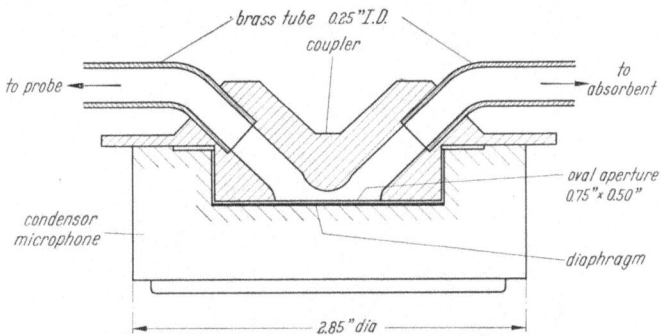

Fig. 75. Coupler detail for a resistively terminated probe microphone (after WEST

For the general case $R<Z_0$, Eq. (20.16) can be reduced to

$$\frac{p(l)}{p_0} = \frac{Z_0 \operatorname{Sin} \pi [\alpha_l + j\beta_l]}{Z_0^2 - R^2 \operatorname{Sin} \pi \left[\alpha_l + \frac{bl}{\pi} + \frac{a}{\pi} + j\left(\beta_l + \frac{2l}{\lambda}\right)\right]} \tag{20.20}$$

where
$$a = \text{Ar Tan}\, \frac{R}{Z_0}$$

and for the case where $R > Z_0$

$$\frac{p(l)}{p_0} = \frac{Z_0 \sin \pi [\alpha_l + j\beta_l]}{R^2 - Z_0^2 \cos \pi \left[\alpha_l + \dfrac{bl}{\pi} + \dfrac{a'}{\pi} + j\left(\beta_l + \dfrac{2l}{\lambda}\right)\right]} \tag{20.21}$$

where
$$a' = \text{Ar Cot}\, \frac{R}{Z_0}.$$

For the case of the open-tipped probe microphone $(R=0)$, Eq. (20.20) reduces to

$$\frac{p(l)}{p_0} = \frac{\sin \pi [\alpha_l + j\beta_l]}{\sin \pi \left[\alpha_l + \dfrac{bl}{\pi} + j\left(\beta_l + \dfrac{2l}{\lambda}\right)\right]}. \tag{20.22}$$

The magnitude of the denominator will fluctuate violently as the frequency is varied unless the real part of the argument of the hyperbolic sine is large compared to unity. If we choose $Z_l = Z_0$, the $\beta_l = 0$, and $\alpha_l = \infty$, and the pressure ratio for $R=0$ becomes

$$\frac{p(l)}{p_0} = e^{-bl - jkl} \tag{20.23}$$

showing the pase shift and amplitude decrease due to attenuation. This is the case of the resistively terminated probe microphone described first by West[1,2] who used a condenser microphone as a shunt element on a terminated transmission line. The shunt acoustic capacitance of the condenser microphone resulted in fluctuations of the pressure ratio of 1.5 db with a useful frequency range exceeding 6 kc as shown in Fig. 74. The bore of the probe tube used by West was 0.635 cm. The termination consisted of a yarnfilled tube about 1.5 m long. Fig. 75 shows the probe coupling assembly and the complete instrument is shown

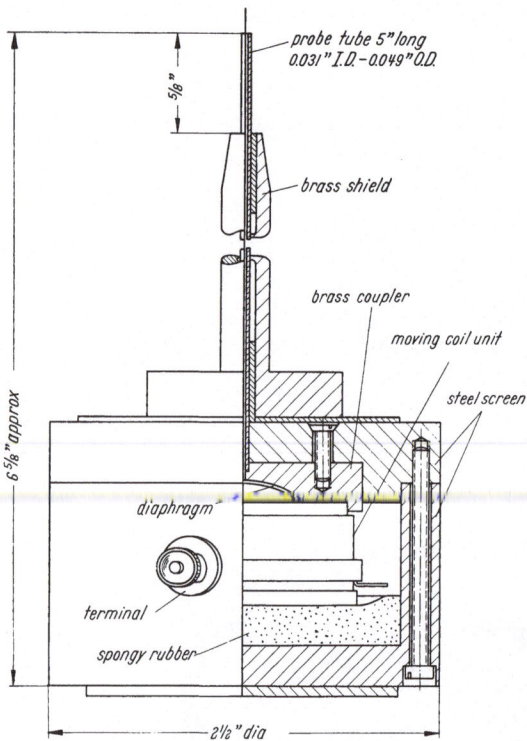

Fig. 77. The electrodynamic probe microphone (after Archbold).

in Fig. 76. Large bore probe microphones of this type are being studied for use at high acoustic pressures and high temperatures[3].

Archbold[4] has developed a terminated probe microphone using a moving coil microphone. The moving coil microphone, being resistively controlled, can

[1] W. West: Post Off. Electr. Engrs. J. **26**, 260 (1934).

[2] W. West: Acustica **3**, 132 (1954).

[3] K. W. Goff and D. M. A. Mercer: J. Acoust. Soc. Amer. **27**, 1133 (1955).

[4] R. B. Archbold: Post Off. Electr. Engrs. J. **45**, 145 (1933) [see also West: Acustica **3**, 132 (1954)].

be matched to a small probe tube by careful adjustment of the spacing between the rigid cap holding the probe tube in front of the domed diaphragm as shown in Fig. 77. The cap had the same radius of curvature as the dome of the diaphragm.

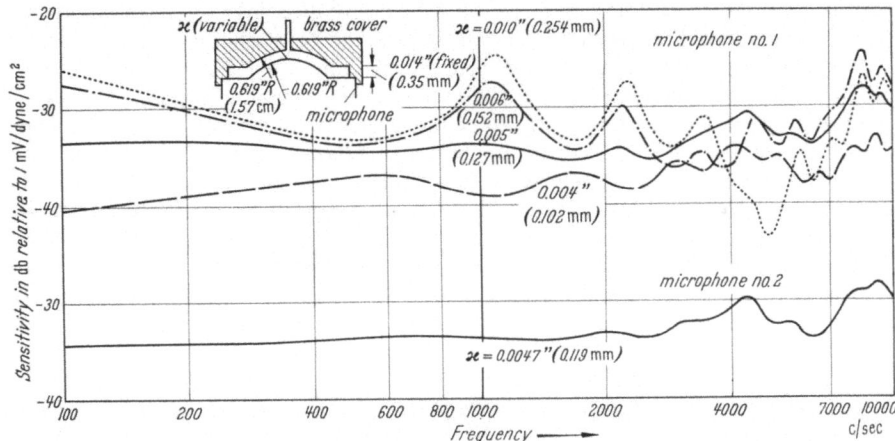

Fig. 78. The frequency dependence of the sensitivity of the electrodynamic probe microphone (after ARCHBOLD).

The inside diameter of the probe tube used was 0.79 mm. The diameter of the domed part of the diaphragm was about 2 cm. The spacing was optimized at 0.127 mm for which spacing the undulations in the frequency curve were minimized. Further decrease in the spacing increased the undulations and inter-

Fig. 79. The electrodynamic probe microphone assembly (after ARCHBOLD).

changed the positions of the troughs and peaks. It appears that the viscous damping between the diaphragm and the cap increased the acoustic impedance of the moving coil microphone until it matched the characteristic impedance of the probe tube. This combination had a sensitivity of about −95 db reference 1 volt per dyne at the output terminals of a transformer having a turns ratio of 1 to 50. This indicates a reduction in sensitivity due to the probe tube and additional damping of about 40 db compared to that of the moving coil microphone when exposed directly to the same pressure field. Fig. 78 shows the exceptionally uniform response obtained with this type of terminated probe microphone and Fig. 79 shows a photograph of components of the assembly.

The advantage of a resistively terminated probe microphone, apart from the uniformity of its response, lies in the fact that the acoustic impedance presented to the pressure field is essentially constant and may be relatively high for a small diameter probe. The probe tip acoustic impedance of a reactively terminate probe tube may dip to a relatively low value compared to that of a resistively terminated probe tube at one of its resonant response frequencies. A pressure microphone will not yield an accurate measure of the standing wave field in a tube unless the acoustic impedance of the microphone is large compared to that of the point in the standing wave where the measurement is made. The fluctuations in the impedance at the mouth of the open probe tube and undulations in pressure response are reduced by a large value of bl (the attenuation in nepers per tube length), but at a considerable loss in sensitivity at the high frequency end of the useful range of the probe microphone. It seems to the author the simplest means of attaining smooth response and reasonable tip impedances is to use a resistor $R = Z_0$ in the probe tip.

II. Velocity sensitive instruments.

21. The ribbon microphone. The ribbon microphone is sometimes referred to to as the velocity microphone since its output voltage may be proportional to the particle velocity in the incident sound wave. Olson[1-3] has treated the ribbon microphone fully. Only a simplified treatment will be presented here.

Consider the element of area of a free ribbon in a plane wave field propagating in the Z direction as shown in Fig. 80. The force on the ribbon element is given by

$$f = -S\, \frac{\partial p}{\partial z}\, d \cos \vartheta \qquad (21.1)$$

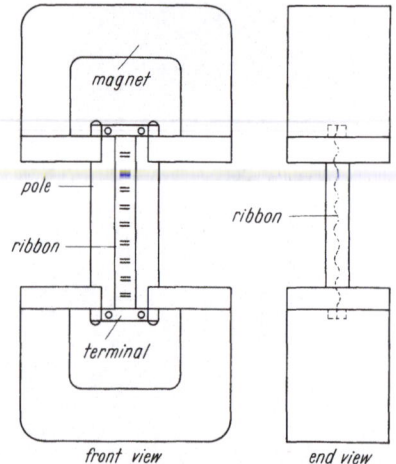

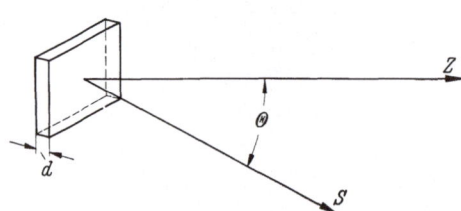

Fig. 80. Force diagram for a ribbon element.

Fig. 81. The essential components of a ribbon microphone (after Olson).

where ϑ is the angle between the wave normal and the normal to the element of area S and d is the thickness of the ribbon. The velocity of the ribbon element u_r

[1] Harry F. Olson: Proc. Inst. Radio Engrs. **21**, 655 (1932).
[2] Harry F. Olson: Elements of Acoustical Engineering, 2nd ed., p. 239. New York: D. Van Nostrand Co., Inc. 1947.
[3] Harry F. Olson: J. Acoust. Soc. Amer. **3**, 56 (1931).

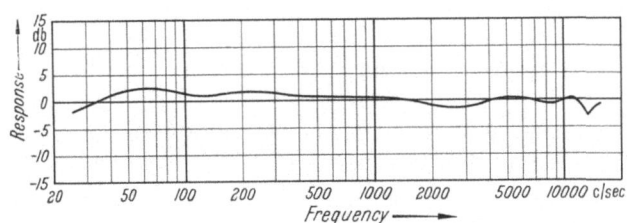

Fig. 82. The frequency response of high quality ribbon microphone (after OLSON).

is given by

$$u_r = -\frac{S d \cos \vartheta}{Z} \frac{\partial p}{\partial z} = \frac{j\omega S d p \cos \vartheta}{c Z_m} \qquad (21.2)$$

and the open-circuited voltage generated is

$$e = B u_r \text{ volts per meter} \qquad (21.3)$$

or

$$e = \frac{j\omega B S \varrho d \cos \vartheta}{Z_m} \cdot u \qquad (21.4)$$

where u is the particle velocity of the sound wave and B the flux density in webers per m². It is clear that the voltage e will be proportional to u in a frequency independent manner if Z_m is proportional to ω. Thus, the ribbon must be mass controlled. This is accomplished by placing its resonant frequency below the lowest frequency in its usable frequency range. Setting $Z_m = j\omega S d \varrho_r$ where ϱ_r is the density of the material in the ribbon, we have for e

$$e = B u \frac{\varrho}{\varrho_r} \cos \vartheta. \qquad (21.5)$$

The cosine type directivity holds as long as both dimensions of the ribbon are small compared to the wavelength. This results in zero sensitivity for waves having a direction of propagation in the plane of the ribbon. This directional characteristic makes the ribbon microphone useful in measurements where a single reflection must be eliminated. Fig. 81 shows a ribbon microphone complete with magnetic structure. The effect of the irregular baffle presented by the pole pieces and magnets has only a small effect on the directivity and sensitivity at the upper end of the useful frequency range of the microphone. Fig. 82 shows the frequency response of a high quality ribbon microphone.

Fig. 83. The ribbon microphone assembled with silk screen (after OLSON).

The constants of the microphone shown in Fig. 81 are as follows. The thickness of the ribbon d is 25 micron. Its resistance is 0.25 ohm. The magnetic flux density is 0.44 webers per m². The impedance level is raised to 250 ohm by a transformer in the base of the microphone. The sensitivity at the 250 ohm level is 200 microvolt per dyne/cm². The 250 ohm impedance makes possible the use of long cables (200 m) without appreciable attenuation of the signal. The ribbon and magnetic structure must be carefully screened from wind by a silk screen as shown in Fig. 83.

Combination of the ribbon microphone with a pressure sensitive microphone results in the interesting directivity pattern of the cardioid microphone[1].

22. The hot wire microphone. The hot wire microphone depends on the change in convective cooling on a fine wire placed in a sound field and heated by a steady electric current. The cooling effect is detected through the change in electrical resistance as in the case of a resistance thermometer. Tucker and Paris[2] first used this effect to measure the particle velocity of sound waves up to frequencies of 500 c.p.s. By combining a wire grid composed of Wollaston wire of 0.0006 cm diameter with a tuned resonator and followed by an amplifier, they obtained sensitivities exceeding that of the ear. The resistance of the grid was about 350 ohm at a temperature of about 600° C when carrying a safe working current of 25 milliamperes. Their study of the characteristics of the hot wire microphone was exceptionally well carried out. However, the response of a coplanar grid which laces back and forth across an opening is complicated by the interaction of the convective currents from one wire segment on that on a neighboring wire segment, especially if the plane of the grid is vertical. We will consider only the case of a single straight section of wire stretched horizontally between supports. (Sewing needles have proven quite satisfactory as shown in Fig. 84a). If such a wire is heated to a dull red color, convection currents will be set up around the wire as indicated in Fig. 84b. As pointed out by West[3], an alternating flow in the vertical direction will cause (in addition to a steady decrease in the temperature of the wire) a variation in the temperature of the wire at the same frequency as that of the alternating flow as long as the alternating flow is small compared to the vertical convective flow. This condition results in a modulation of the convective flow; and as long as the direction of the convection is not reversed, the second harmonic variation of the temperature will be small. In contrast to this, if the alternating flow is in a horizontal direction, that is perpendicular to the convective flow, there will be observed in addition to the steady decrease in temperature a variation at twice the frequency of the alternating flow and essentially none at the fundamental frequency. The double frequency is required by the symmetry of the convective flow with respect to the horizontal direction.

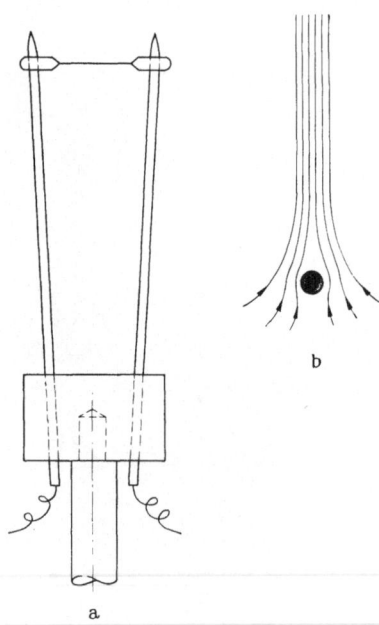

a   b

Fig. 84a and b.
The mounted hot wire microphone (a) and the convective flow near a hot wire (b).

Thus, if the hot wire microphone is to be linear, the particle motion measured must be parallel to the convective flow due to heat given off by the wire. Forced

[1] See footnote 2, p. 56.
[2] W. S. Tucker and E. T. Paris: Roy. Soc. Phil. Trans. A **22**, 389 (1921).
[3] W. West: Acoustical Engineering, p. 194. London: Pitman & Sons, Ltd. 1932.

convection may be used as suggested by WENTE[1]. At a given frequency, the alternating change in resistance at the fundamental frequency is proportional to the product of the steady flow velocity, the particle velocity, and the cosine of the angle between them. The hot wire is used as an anemometer and for studying turbulent fields. With proper equalization, turbulent spectra having frequencies as high as 10^5 c.p.s. can be measured[2].

RICHARDSON[3] has used the hot wire microphone in ultrasonic interferometers up to about 10^5 c.p.s. He used the steady drop in resistance as measured on a resistance bridge and thermal convection with wire temperatures as low as $50°$ C in progressive wave fields and $25°$ in standing wave fields. RICHARDSON remarks that the hot wire microphone must be calibrated in known displacement amplitudes at the frequency at which measurements are to be made. This is particularly true in enclosures where the walls may change the character of the thermally induced convection and especially true in regions very near to the boundaries of enclosures. Where only relative values of particle velocity are needed as in a standing wave measurement of phase velocity, the relative linearity of the hot wire may be determined by varying the voltage applied to the sound source setting up the field to be measured with the hot wire in a fixed position relative to the source. In measuring the standing wave field, it is important that the hot wire is at all times sufficiently separated from the vertical surfaces which might alter the thermal convection and thus change the sensitivity of the hot wire to alternating flow.

23. The Rayleigh disc. Lord RAYLEIGH[4] first observed that a disc immersed in a moving medium experienced a torque which tended to orient the plane of the disc normal to the direction of flow even in the case of alternating flow. A number of early instruments were devised using combinations of the Rayleigh discs and resonators such as the double resonator system of BOYS[5]. KÖNIG[6] first developed an expression for the torque on the discs due to an incompressible alternating flow. The torque L as derived by KÖNIG is given by

$$L = \tfrac{4}{2}\varrho\, U^2 a^3 \sin 2\vartheta \tag{23.1}$$

where ϱ is the density of the flowing fluid, U is the root mean square velocity of the laternating flow, ϑ is the angle between the normal to the plane of the disc and the direction of flow, and a is the radius of the disc. This expression does not take into account the compressibility of the fluid, diffraction of sound by the disc, nor the degree to which the disc moves in translation in the alternating flow. WOOD[7] observed that in using a RAYLEIGH disc in water, the correction to KÖNIG's formula for the motion of translation of the disc could amount to as much as a 2500-fold reduction. KING[8] derived a more accurate expression for the torque including the effects of translational motion and diffraction of sound by the disc. The expression derived by KING is

$$L = \frac{4}{3}\,\varrho\, a^3\, U^2 \sin\,(2\vartheta)\,\frac{m_1[1 + \tfrac{2}{5}(k\,a)^2\cos^2\vartheta]}{m_1 + m_0[1 + \tfrac{1}{5}(k\,a)^2]} \tag{23.2}$$

[1] E. C. WENTE: J. Acoust. Soc. Amer. **7**, 9 (1935).
[2] L. S. G. KOVASZNAY: J. Aernaut. Sci. **17**, 565 (1950).
[3] E. G. RICHARDSON: Ultrasonic Physics, p. 40. London: Elsevier Publ. Co. 1952.
[4] Lord RAYLEIGH: Proc. Roy. Soc. Lond. **32**, 110 (1881).
[5] C. V. BOYS: Nature, Lond. **42**, 604 (1890).
[6] W. KÖNIG: Wied. Ann. **43**, 43—60 (1891).
[7] A. B. WOOD: Proc. Phys. Soc. Lond. **47**, 779 (1935).
[8] L. V. KING: Proc. Roy. Soc. Lond., Ser. A **153**, 17 (1935).

where m_1 is the mass of the disc less the mass of the displaced fluid, and m_0 is the hydrodynamic mass of the disc given by

$$m_0 = \tfrac{8}{3}\varrho a^3 \qquad\qquad (23.3)$$

and k is $2\pi/\lambda$. For small values of ka, the correction term reduces to $m_1/(m_1 + m_0)$ as determined empirically by WOOD[1]. The terms involving $(ka)^2$ are seldom important in practical measurements since the disc is frequently used in a tube for pressure calibration of microphones and must be small compared to the diameter of the tube which in turn must be small compared to the wavelength to restrict the propagation to a plane wave field. This has been pointed out by SCOTT[2] who has made an excellent experimental study of the corrections to be applied to the König formula for the torque. He determined that the torque varied as $a^3 U^2$ within $\pm 1\%$ (an exception being noted only at 4250 c.p.s. where the error was 2% which would result in only a 0.1 db error in sound intensity measurement). He also studied the effect of the proximity of the tube wall surrounding the disc. The results may be summarized by noting that up to a ratio of disc radius to tube radius of 0.4 the increase in torque was essentially linear in the diameter ratio and may be expressed as

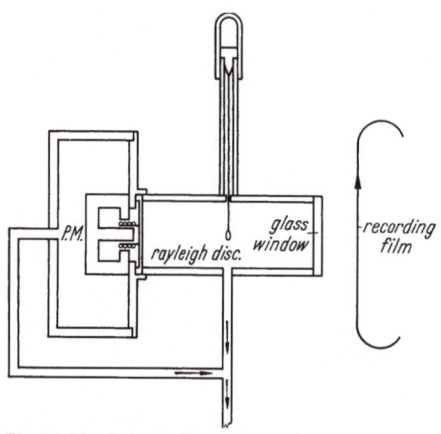

Fig. 85. The Rayleigh Disc mounted in apparatus used for measurement of the velocity of sound as a function of gas density (after DELSASSO).

$$\frac{L}{L_0} = 1 + 0.066\,\frac{a}{b}, \qquad \frac{a}{b} \leq 0.4 \qquad (23.4)$$

where a is the disc radius and b the tube radius. The dependence of the torque on uniform disc thickness observed by SCOTT may be approximated by the linearized expression

$$\frac{L}{L_0} = 1 + 0.66\,\frac{t}{a}, \qquad \frac{t}{a} \leq 0.13 \qquad (23.5)$$

where t is the disc thickness and a the radius of the disc. SCOTT made a careful evaluation of the numerical constant in the expression for the torque using smoke particles to determine the particle displacement amplitude at the velocity antinode in the standing wave field in which the disc was suspended. The results of this phase of his work indicated that the coefficient $(\tfrac{4}{3}\varrho)$ to be low by about 4%, but the frequency dependence of the observed correction seems, to the author, to be an anomalous one.

The Rayleigh disc is suspended by a fine fiber, and the sensitivity maximized by orienting it at 45° to the direction of the particle velocity to be measured. The torsional constant of the fiber may be determined by observing the tortional vibrations of a body of known moment of inertia attached to the fiber. If the disc, itself, is used, the measurement should be performed in an evacuated tube to avoid the effects due to the inertia of the air surrounding the disc. The measurements are simplified if the upper end of the suspension is rotated until the disc returns to the 45° position occupied in the absence of a sound field. The torque

[1] See footnote 7, p. 59.
[2] R. A. SCOTT: Proc. Roy. Soc. Lond., Ser. A **183**, 296 (1944).

is then the product of the angle of rotation of the fiber support and the torsional constant of the fiber, and the particle velocity may be computed from König's formula plus corrections by King and Scott, if required.

Fig. 85 shows the Rayleigh disc and mirror combination used by Delsasso[1] for the measurement of the velocity of sound at reduced pressures in gases.

To summarize, the Rayleigh disc seems a very dependable standard for the measurement of particle velocity capable of accuracies of the order of a few tenths of a decibel. For free field work, it must be protected from drafts and may require the use of King's diffraction correction since larger discs may be required to provide sufficient sensitivity. Correction may be necessary for the transmission loss through the silk screens used to reduce drafts. Barnes and West[2] have observed that the Rayleigh disc is not reliable at frequencies close to its natural resonance frequencies although this effect is noticeable only over a small frequency range. The Rayleigh disc is still used as a standard in Britain[3].

III. Temperature sensitive detectors.

24. Thermocouple microphones. The use of thermocouples as microphones was proposed by Johnson[4], who carried the analysis necessary to compute the frequency response of a coplanar thermocouple exposed to the alternating temperature field of a grazing sound wave, constructed thermocouples by a sputtering process, and tested them in a known sound field. The agreement between the theoretical response and the experimental calibration is excellent.

Johnson gives for V the ratio of the temperature amplitude at the center of the coplanar thermocouple to that in the air-borne sound wave.

$$V = \frac{1}{\sqrt{(1 + f^{\frac{1}{2}} d a)^2 + f d^2 a^2}} \tag{24.1}$$

where a is the half thickness of the coplanar thermocouple, f is the frequency of the sound wave, and d is a constant given by

$$d = \frac{\varrho_1 c_1 \sqrt{\pi}}{\sqrt{K_2 C_2 \varrho_2}} \tag{24.2}$$

where c_1 is the heat capacity per gram at constant pressure for air, ϱ_1 is the density of air, ϱ_2 the average of the metals forming the thermocouple, K_2 the average thermal conductivity of the metals, and C_2 the average heat capacity per gram of the couple metals.

It is clear from Eq. (24.1) that the smaller the thickness of the metal layer, $2a$, the larger the ratio V. The effect of the cellulose acetate film on which the metals of the couple were deposited has been neglected in the analysis leading to the temperature ratio V. In the actual computation of V, an allowance for this film was made by taking one-half the total thickness as a in computing the response. The metal films used were bismuth (1.2×10^{-5} cm thick) and antimony (0.6×10^{-5} cm thick). The construction of the thermocouple is shown in Fig. 86. The acetate cellulose film (10^{-5} cm thick) was supported by a mica frame (0.1 mm thick), the unsupported film covering a hole in the mica 0.5 cm in diameter. After the thermocouple has been formed, gold contacts are sputtered over the ends to provide a large heat capacity so that these additional junctions will

---

[1] Unpublished communication to the author.
[2] Barnes and West: J. Inst. Electr. Engrs. **65**, 871 (1927).
[3] R. S. Dadson and E. G. Butcher: Acustica **4**, 103 (1954).
[4] E. A. Johnson: Phys. Rev. **45**, 645 (1934).

remain at constant temperature in the sound field and in addition to provide surfaces to which wires may be soldered. Fig. 87 shows the relative frequency response in decibels for $a = 10^{-5}$ cm. The sensitivity falls very slowly with frequency at less then 3 db per octave at first. To obtain this frequency reponse characteristic, the narrow dimension of the thermocouple strip which is oriented in the direction of the wave

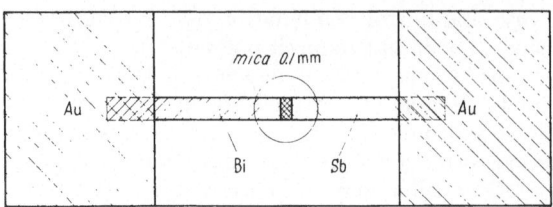

Fig. 86. Thermoelectric microphone (after Johnson).

normal must be small compared to the acoustic wavelength to avoid temperature cancellation. If the strip were one wavelength wide, the average instantaneous temperature of the strip would not vary.

As a wide band microphone, the thermocouple is limited to the measurement of relatively high sound pressures of the order of 10^4 dyne/cm² since sensitivities of the order of -180 db reference 1 volt/dyne/cm² are to be expected at 1000 c.p.s.

For use as a microphone in a narrow band measuring system, the thermocouple could be used to measure pressures of the order of 10 dyne/cm².

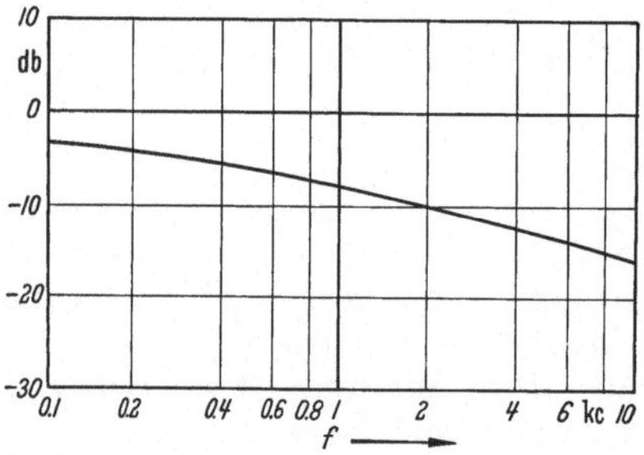

Fig. 87. Frequency response of thermoelectric microphone (after Johnson).

The most attractive feature about the thermocouple is its small size and insensitivity to velocity fields or solid-borne vibration. It is to be expected that it will find more use in the study of intense sounds in the future. Modern evaporation techniques may simplify the construction.

IV. Calibration of microphones.

25. The pistonphone. The pistonphone provides one of the simplest methods for the low-frequency calibration of pressure sensitive microphones. It consists of an oscillating piston of known or measurable volume displacement coupled to a known volume in which the microphone to be calibrated is exposed to the alternating acoustic pressure. Fig. 88 shows a simple crank driven pistophone. The volume displacement of the piston is determined

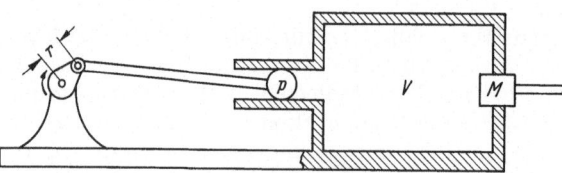

Fig. 88. A crank driven pistonphone.

by the radius of the crank pin r and the area of the piston S_p. The pressure p generated in the volume V is given by

$$p = \frac{r\,S_p\varrho\,c^2}{V(C+D)^{\frac{1}{2}}}\cos\omega t \tag{25.1}$$

where the quantity $(C+D)^{\frac{1}{2}}$ calculated by Daniels[1] is given in Fig. 89 as a

---

[1] Fred B. Daniels: J. Acoust. Soc. Amer. **19**, 569 (1947).

function of $\alpha V/S$ where S is the surface area of the chamber of volume V and α is given by

$$\alpha = \sqrt{\frac{\omega \varrho C_p}{2K}} \qquad (25.2)$$

where C_p is the specific heat and K the heat conductivity of the gas. Curve A is for a spherical volume, B for a long cylinder, and C for a narrow (essentially one-dimensional) rectangular box. Most practical cavities require a correction lying between that for the sphere and the cylinder. This correction takes into account the fact that there is a transition from isothermal compressibility at very low frequencies where the thermal wavelength is large compared to the dimensions of the cavity to adiabatic compressibility at high frequencies.

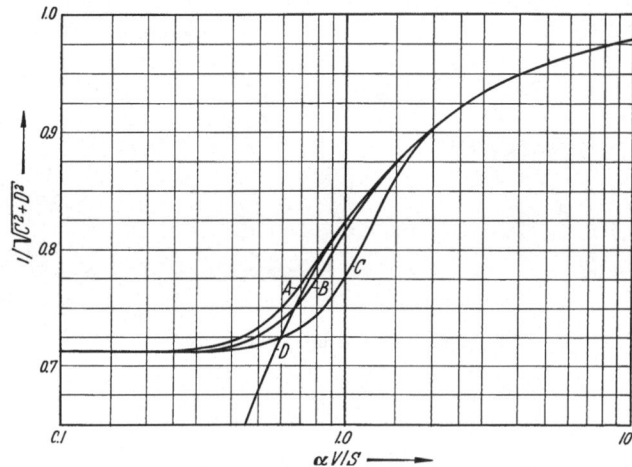

Fig. 89. Correction factor for the isothermal layer at the boundaries of a cavity (after DANIELS).

The crank driven pistonphone must have a stroke which is large compared to the clearance in the bearings and a long connecting rod between the crank and piston in order to produce sinusoidal motion at the piston. This type of pistonphone is limited to low frequency range below 60 c.p.s. if excessive vibration is to be avoided. It is difficult to isolate the microphone under calibration from the mechanical vibrations resulting from the mechanical drive.

An electrodynamic pistonphone has been described by GLOVER and BAUM-ZWEIGER[1] which eliminates most of the difficulties encountered with the mechanical drive. Fig. 90 shows the essential features of the electrodynamic pistonphone designed by the author. The amplitude of the piston is observed by means of a microscope and scale and stroboscopic illumination. The microscope is focussed on a ridge on the piston rod. The piston diameter is about 0.32 cm, and the volume of the cavity is variable depending on the height and diameter of the cylinder used. The top of the cylinder (not shown in the Figure) is a flat plate with a suitable opening for insertion of the microphone to be calibrated. With small condenser microphones and a correspondingly small cylindrical cavity, this pistonphone gives reliable calibration up to 500 c.p.s. The metal plate containing the piston is supported in rubber mounts to isolate it from the vibrations produced by the electromagnetic driver. The neon lamp, microscope, and scale are attached to the metal plate, and hence the correct amplitude is

---

[1] R. GLOVER and B. BAUMZWEIGER: J. Acoust. Soc. Amer. **10**, 200 (1939).

observed even in the presence of vibration of the base which is attached to the electrodynamic driver unit.

Stroboscopic illumination of adequate intensity is provided by viewing the ridge on the piston rod and the reference scale silhouetted against one of the

Fig. 90. Electrodynamic pistonphone.

glowing electrodes of a small neon lamp. The neon lamp is excited by an oscillator at a frequency slightly different from that of the pistonphone so that the excursions may be observed in slow motion.

26. The electrostatic actuator. The use of an auxilliary grid placed close to the diaphragm of a condenser microphone and suitably insulated permitting the diaphragm to be driven by a modulated d.c. electric field was described by Ballantine[1] and termed an electrostatic actuator. The geometrical arrangement is shown in Fig. 91. Ballantine[1] has calculated the force between a planar diaphragm and a slotted grid by a Schwartz transformation. Attempts to satisfy the requirements of this analysis have led to grid structures with excessively high acoustic impedances. It is important to recognize that the acoustic impedance of the actuator grid is in series with the microphone diaphragm as indicated in Fig. 92 The pressure generator representing an incident sound wave is labeled P_s and appears at a different point in the circuit as shown in Fig. 92. The acoustic impedance of the slots in the grid will become

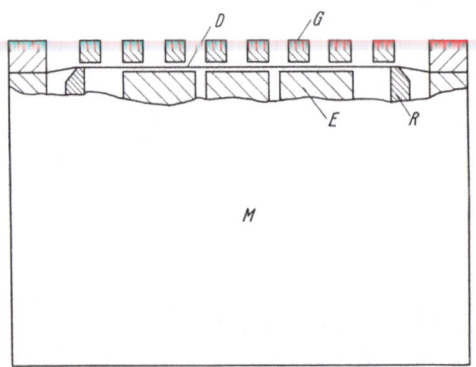

Fig. 91. The geometrical arrangement of electrostatic actuator grid G relative to diaphragm of a condenser microphone D. The stretching ring k and the back electrode E are also shown. The insulating support for the grid is not shown.

---

[1] Stuart Ballantine: J. Acoust. Soc. Amer. **3**, 319 (1932).

very large at a frequency near that for which the grid is $\lambda/4$ in thickness. As an example, consider a slotted grid 1.2 cm thick. This structure becomes a quarter wave in thickness at a frequency near 7000 c.p.s. and will introduce serious errors in the calibration at frequencies well below this. This type of error is increased if the microphone is used with the actuator grid in place due to the difference in the positions of the two pressure generators as indicated in Fig. 92. Thus, in the actuator calibration of a microphone at frequencies above 20 kc, thin open grids are necessary. The author has used a lucite grid identical in form to the metal protective grid on the Western Electric 640 AA microphone. A thin film of silver was evaporated onto the lower surface of the lucite grid over a circle equal in diameter to the stretching ring of the Western Electric 640 AA microphone (1.7 cm). Fig. 93 shows the plastic grid with the evaporated silver electrode on the

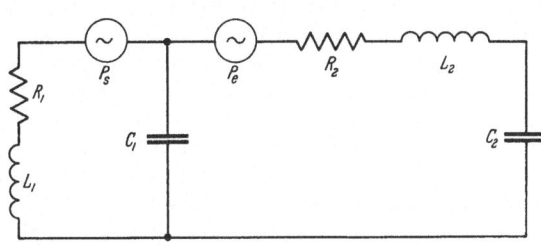

Fig. 92. The equivalent circuit showing the effect of the acoustic loading of the actuator grid on the condenser microphone P_e is the pressure generator associated with the actuator and P_s that associated with the acoustic pressure field.

surface which is placed next to the diaphragm when the grid is screwed in place on the microphone. While such a grid configuration does not lend itself to a computation of the absolute force exerted on the diaphragm by means of a Schwarz transformation, it has a negligible acoustic loading on the microphone up to about 11 kc. Even in the case where the configuration lends itself to exact computation, it is necessary to know the absolute spacing between the grid and diaphragm to an accuracy twice that expected in the calibration since the force depends inversely on the square of the spacing. With a spacing of the order 0.010 inches, this may require rather precise measurement (a 6% error in spacing introduces a calibration error of 1 db). The grid must be parallel to the diaphragm if a uniform electric field (equivalent to a uniform pressure) is to be obtained.

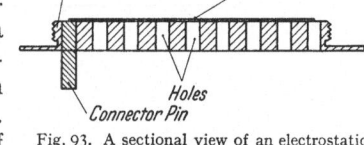

Fig. 93. A sectional view of an electrostatic actuator machined from lucite in a form identical to the metal protective grid for the Western Electric 640 AA condenser microphone. The evaporated silver electrode is in contact with the connector pin.

The author has devised a method of using the electrostatic actuator which, while it does not remove the requirement of parallelism between grid and diaphragm, requires only the measurements of electrical quantities. It is based on the fact that the force in the x direction on the electrodes of a capacitor is given by

$$F = \frac{1}{2} \frac{\partial C}{\partial x} V^2 \tag{26.1}$$

where C is the capacity in farads and V the potential difference in volts and F is given in newtons. The instantaneous pressure in newtons per m^2 is given by

$$p = \frac{1}{2S} \frac{\partial C}{\partial x} V^2_{\text{r.m.s.}} \, 2 \sin^2 \omega t \tag{26.2}$$

for

$$V = V_{\text{r.m.s.}} \sqrt{2} \sin \omega t \tag{26.3}$$

and S is the area of the diaphragm and the r.m.s. pressure of the double frequency alternating component is given by

$$P_{\text{r.m.s.}} = \frac{1}{2\sqrt{2} S} \frac{\partial C}{\partial x} V^2_{\text{r.m.s.}} \tag{26.4}$$

when no d.c. polarizing voltage is used. The derivative $\partial C/\partial x$ is evaluated by measurement of the grid-diaphragm capacitance as the grid-diaphragm spacing is changed by known amounts. Absolute pressure calibrations accurate to about 1 db have been made by this method using the plastic grid structure shown in Fig. 93. The known displacement was obtained by screwing the grid in and out of the microphone shell by known fractions of a turn.

The electrostatic actuator method is capable of highly accurate calibration only when great care is exercised. However, relative pressure sensitivities may be readily compared through its use. If the absolute pressure sensitivity is known for one frequency, the actuator calibration will supply the relative sensitivities at other frequencies with adequate precision if the grid is parallel to the diaphragm and has an acoustic impedance small compared to that of the microphone diaphragm. The electrostatic actuator with its high electrical impedance in the audio frequency may be used to test the square wave response of a condenser microphone thus determining its transient characteristics. No d.c. polarization is necessary with square wave excitation on the actuator grid since squaring the instantaneous voltage in a square wave still yields a square force wave at the diaphragm. The author sees no reason why the electrostatic actuator cannot be applied to the testing of other types of diaphragm actuated pressure microphones in addition to condenser microphones except for the difficulties in making non-planar grids.

27. Reciprocity relations. The application of the reciprocity theorem to electro-mechanical transducers was first stated by Schottky[1]. The validity of the reciprocity theorem as stated by Schottky has received considerable attention since MacLean[2] pointed out its importance in the calibration of acoustic trans-ducers. Primakoff and Foldy[3,4] have made a meticulous study of the symmetry conditions required for the validity of the theorem.

The reciprocity theorem may be expressed most simply as given by McMillan[5] in the form of three equations

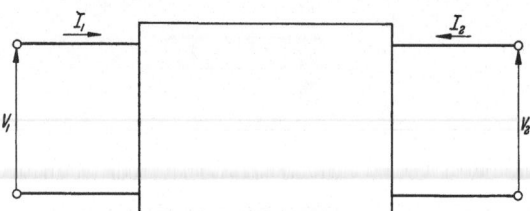

Fig. 94. A generalized four-terminal network showing the directions of currents and voltages assumed in Eqs. (27.1).

$$V_1 = aI_1 + bI_2,$$
$$V_2 = cI_1 + dI_2, \qquad (27.1)$$
$$b = c$$

where the quantities in the equations are related as shown in Fig. 94.

For four-terminal networks made up of electrical impedances only or me-chanical impedances only, $b=c$ and such networks always obey reciprocity. When the four-terminal network includes an electromechanical transducer, satisfaction of the reciprocity theorem will depend on the electromechanical analogue used. Electrostatic and piezoelectric transducers will satisfy reci-procity if voltage and force or pressure are analogous and if current and velocity are analogous. Electrodynamic and magnetostrictive transducers are anti-reciprocal[5] ($b=-c$) unless the mobility analogue is used where current and force, or pressure are analogous and where voltage is analogous to velocity.

[1] F. Schottky: Z. Physik **36**, 689 (1926).
[2] W. R. MacLean: J. Acoust. Soc. Amer. **12**, 140 (1940).
[3] L. Foldy and H. Primakoff: J. Acoust. Soc. Amer. **17**, 109 (1945).
[4] H. Primakoff and L. Foldy: J. Acoust. Soc. Amer. **19**, 50 (1947).
[5] E. M. McMillan: J. Acoust. Soc. Amer. **18**, 344 (1946).

Linearity and reversibility of the transducer are implied in Eq. (27.1) as a necessary condition for satisfaction of the reciprocity theorem.

A useful and interesting relationship may be derived from a four-terminal network composed of two reciprocal transducers of similar electroacoustic coupling (both electrostatic, for example) as shown in Fig. 95. Such a combination of reciprocal elements obeys reciprocity as a whole. The sensitivities of the transducers as microphones are defined as the ratio of the open-circuited voltage at the terminals of the transducers to the pressure at a reference point fixed with respect to the transducer. The reference points may be chosen

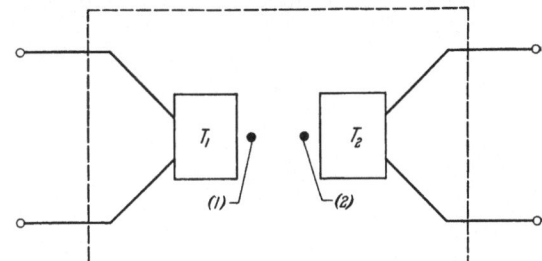

Fig. 95. A four-terminal network composed of two similar electroacoustic transducers T_1 and T_2 with their corresponding fixed reference points (1) and (2) located in the coupling medium.

near the centers of the transducer diaphragms as shown in Fig. 95, or at any other point in the coupling medium as far as the following derivation is concerned. The sensitivities are given by

$$M_1 = \frac{V_{0C1}}{P(1)}, \quad M_2 = \frac{V_{0C2}}{P(2)} \tag{27.2}$$

where $P(1)$ and $P(2)$ refer to the pressures at reference points (1) and (2).

The source constants S_1 and S_2 are defined in terms of the pressure produced at reference point (2) by a current flowing in transducer T_1 and vice versa. They are given by

$$S_1 = \frac{P(2)}{I_1}, \quad S_2 = \frac{P(1)}{I_2} \tag{27.3}$$

where $P(1)$ and $P(2)$ have the same meaning as before. Making use of Eq. (27.1), (27.2), and (27.3) we may write

$$\left. \begin{array}{l} V_{0C1} = bI_2 = M_1 P(1) = M_1 S_2 I_2, \\ V_{0C2} = cI_1 = M_2 P(2) = M_2 S_1 I_1 \end{array} \right\} \tag{27.4}$$

and since $b=c$, we have

$$M_1 S_2 = M_2 S_1 \tag{27.5}$$

or

$$\frac{M_1}{S_1} = \frac{M_2}{S_2}. \tag{27.6}$$

Thus, for each transducer, the ratio of microphone sensitivity to source constant is independent of the strength of electroacoustic coupling and depends only on the acoustic environment of each transducer. The acoustic environment in which a transducer finds itself will, in the most general case, depend on its own acoustic impedance, that of the coupling medium, the impedance of the other transducer, and the exact geometrical relationship between the two transducers as well as the location of the boundaries of the coupling medium. The uniqueness of this ratio in a prescribed environment makes it possible to compute it for simple environments by an application of the reciprocity theorem to a transducers with only simple restrictions on the characteristics of the transducer in addition to the requirement that it be reciprocal. This calculation will be illustrated in the next

5*

two sections where it will be applied to the pressure and free-field calibration of reciprocal transducers.

It has been indicated by Foldy and Primakoff[1] that for non-identical transducers of similar type coupling, satisfaction of Eq. (27.6) establishes the fact that both transducers involved satisfy the reciprocity condition. Eqs. (27.4) may be rewritten as

$$\left.\begin{array}{l} M_1 S_2 = \dfrac{V_{0C1}}{I_2}, \\[2mm] M_2 S_1 = \dfrac{V_{0C2}}{I_1} \end{array}\right\} \tag{27.7}$$

and using Eq. (27.5), we have

$$I_1 V_{0C1} = I_2 V_{0C2} \tag{27.8}$$

as a test as to the reciprocal nature of the transducers. This test has been employed by Rudnick and Stein[2] to establish the fact that the Western Electric 640 AA condenser microphone satisfies the reciprocity condition.

28. Pressure calibration by the reciprocity method. To obtain a pressure calibration by the reciprocity method, two microphones, a sound source, and a rigid-walled coupling cavity are required. One of the microphones T_1 must be a reciprocal transducer. The other microphone T_2 and the source T_3 need not be reciprocal devices. The dimensions of the coupler must be small compared to the acoustic wavelength in order that the pressure be uniform throughout its volume. Fig. 96 shows one form of coupler used in the calibration of condenser microphones. A detailed description of coupler design for the calibration of Western Electric 640 AA condenser microphones has been given by Beranek[3]. Beranek indicates that a cylindrical cavity having a diameter to length ratio of 3.5 is about optimal for a given volume.

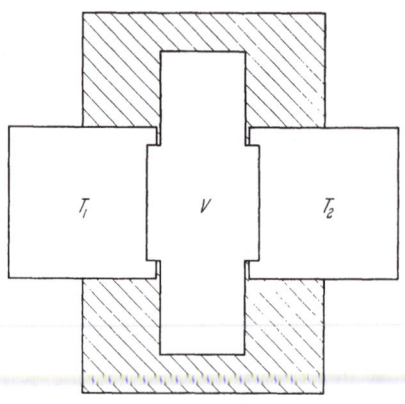

Fig. 96. One form of acoustic coupler used in the pressure calibration of condenser microphones by the reciprocity method.

Before indicating the steps involved in a coupler reciprocity calibration, it is necessary to calculate the value of M_1/S_1 for the reversible transducer T_1 coupled to microphone T_2 by the coupler. This is done by applying the reciprocity conditions written as

$$\left.\begin{array}{l} V = aI + bU, \\ P = bI + dU \end{array}\right\} \tag{28.1}$$

to T_1 using the definitions of M and S given by

$$M = \frac{V_{0C}}{P}, \qquad S = \frac{P}{I}. \tag{28.2}$$

From Eq. (28.1), the blocked pressure $(U=0)$ is given by

$$P_{0C} = bI \tag{28.3}$$

[1] See footnote 4, p. 66.
[2] I. Rudnick and M. N. Stein: J. Acoust. Soc. Amer. 20, 818 (1948).
[3] L. L. Beranek: Acoustic Measurements, p. 137. New York: John Wiley & Sons 1949.

and the pressure in the cavity by

$$P = \frac{P_{0C} Z_{C2}}{Z_{C2} + Z_1} = bI \frac{Z_{C2}}{Z_{C2} + Z_1} \tag{28.4}$$

where Z_1 is the acoustic impedance of T_1 and Z_{C2} represents the impedance of the cavity Z_C in parallel with Z_2 the impedance of T_2. Eq. (28.4) yields for S_1

$$S_1 = \frac{b Z_{C2}}{Z_{C2} + Z_1}. \tag{28.5}$$

M_1 is obtained from (28.1) by noting that

$$V_{0C} = bU, \tag{28.6}$$

and U, the volume velocity of T_1 is given by

$$U = \frac{P}{Z_1} \tag{28.7}$$

yielding

$$M_1 = \frac{b}{Z_1}. \tag{28.8}$$

Eqs. (28.5) and (28.8) give for the ratio

$$\frac{M_1}{S_1} = \frac{Z_{C2} + Z_1}{Z_{C2} Z_1} = \frac{1}{Z_{C12}} \tag{28.9}$$

where Z_{C12} is the impedance of the cavity Z_C in parallel with the transducer impedances Z_1 and Z_2. The ratio M_1/S_1 combined with the product $M_1 S_1$ permits the calculation of M_1 the required sensitivity.

The product $M_1 S_1$ is obtained from three measurements taken on pairs of transducers coupled by the coupler.

α) T_1 *coupled to* T_3. Source T_3 of acoustic impedance Z_3 is characterized by a blocked pressure

$$P_{0C} = IA. \tag{28.10}$$

The pressure in the coupler produced by a current I_1 in T_3 is given by

$$P_1 = AI_1 \frac{Z_{C1}}{Z_{C1} + Z_3} \tag{28.11}$$

where Z_{C1} represents the parallel combination of the acoustic impedance Z_C of the coupler and Z_1 the acoustic impedance of T_1. The open-circuit voltage V_1 measured at the terminals of T_1 is given by

$$V_1 = M_1 P_1 = \frac{M_1 A I_1 Z_{C1}}{Z_{C1} + Z_3} \tag{28.12}$$

from which

$$\frac{M_1 A}{Z_3} = \left(\frac{V_1}{I_1}\right) \frac{Z_{C1} + Z_3}{Z_3 Z_{C1}} = \left(\frac{V_1}{I_1}\right) \frac{1}{Z_{C13}} \tag{28.13}$$

where Z_{C13} represents Z_C, Z_1, and Z_3 in parallel.

β) T_2 *coupled to* T_3. When T_3 is driven by a current I_2, the open-circuit voltage measured at the terminals of T_2 is given by

$$V_2 = M_2 P_2 = \frac{M_2 I_2 A Z_{C2}}{Z_{C2} + Z_3} \tag{28.14}$$

from which

$$\frac{M_2 A}{Z_3} = \left(\frac{V_2}{I_2}\right) \frac{Z_{C2} + Z_3}{Z_3 Z_{C2}} = \left(\frac{V_2}{I_2}\right) \frac{1}{Z_{C23}}. \tag{28.15}$$

Combining Eqs. (28.13) and (28.15), we obtain

$$\frac{M_1}{M_2} = \left(\frac{V_1}{I_1}\right)\left(\frac{I_2}{V_2}\right)\frac{Z_{C23}}{Z_{C13}} \tag{28.16}$$

as the ratio of the sensitivities of the two transducers T_1 and T_2 as microphones.

γ) T_2 coupled to T_1. When T_1 is driven by a current I_3, the pressure in the cavity is given by

$$P_3 = S_1 I_3 \tag{28.17}$$

and the open-circuit voltage at the terminals of T_2 is given by

$$V_3 = M_2 S_1 I_3 = \left(\frac{M_2}{M_1}\right) M_1 S_1 I_3 . \tag{28.18}$$

Combining Eq. (28.18) with (28.16) yields

$$M_1 S_1 = \left(\frac{V_3}{I_3}\right)\left(\frac{M_1}{M_2}\right) = \left(\frac{V_3}{I_3}\right)\left(\frac{V_1}{I_1}\right)\left(\frac{I_2}{V_2}\right)\frac{Z_{C23}}{Z_{C13}} \tag{28.19}$$

as a final result of the three sets of measurements.

The open-circuit voltage per unit pressure for the reciprocal transducer T_1 is given by

$$M_1 = \sqrt{\left(\frac{M_1}{S_1}\right) M_1 S_1} = \sqrt{\left(\frac{V_1}{I_1}\right)\left(\frac{I_2}{V_2}\right)\left(\frac{V_3}{I_3}\right)\frac{Z_{C23}}{Z_{C13} Z_{C12}}} \tag{28.20}$$

where M_1 is in open-circuit volts per newton per square meter and all acoustic impedances are in M.K.S. units.

If transducers T_1 and T_2 are identical, Eq. (28.20) simplifies to

$$M_1 = \sqrt{\left(\frac{V_1}{I_1}\right)\left(\frac{I_2}{V_2}\right)\left(\frac{V_3}{I_3}\right)\frac{1}{Z_{C12}}} . \tag{28.21}$$

Eqs. (28.20) or (28.21) permit the determination of both the magnitude and phase angle[1] of M_1 provided the phase angles of the ratios of voltage to current are measured as well as the magnitudes of the ratios. In addition, it is necessary to be able to compute the phase angle and magnitude of Z_{C12}. This computation is greatly simplified if $Z_C \ll Z_1 \approx Z_2$ for which condition Z_{C12} may be replaced by Z_0 the impedance of the cavity alone. If the above inequality is not satisfied, a detailed knowledge of the acoustic impedances of the transducers involved will be required even though only the magnitude of M_1 is evaluated. A coupler of sufficient volume to satisfy the above requirement may put a severe limitation on the highest frequency for which the dimensions of the coupler remain small compared to the wavelength. Hydrogen-filled couplers have been used to increase the frequency range of pressure calibration[1, 2]. NIELSEN[3] has described a very convenient method of measuring the currents and voltages required in a coupler reciprocity calibration. DANIELS[4] has described a precise method of calculating the impedance of the cavity at low frequencies where the transition from adiabatic to isothermal compressibility takes place.

29. Free field reciprocity calibration of microphones. The free field reciprocity calibration of condenser microphones in air has been carried out over a frequency range extending up to 100 kc by RUDNICK[5] and by KUHL[6].

[1] FRANCIS M. WIENER: J. Acoust. Soc. Amer. **20**, 707 (1948).
[2] A. L. DIMATTIA and F. M. WIENER: J. Acoust. Soc. Amer. **18**, 341 (1946).
[3] A. K. NIELSEN: Acustica **2**, 112 (1952).
[4] F. B. DANIELS: J. Acoust. Soc. Amer. **19**, 569 (1947).
[5] I. RUDNICK and M. N. STEIN: J. Acoust. Soc. Amer. **20**, 818 (1948).
[6] W. KUHL, G. R. SCHODDER and F. K. SCHRÖDER: Acustica **4**, 519 (1954).

The essential conditions for this type of calibration are the following.

1. The calibration must be made in a free field room where reflections from the room boundaries are negligible compared to the direct radiated field of the source transducer at the position of the receiving transducer.

2. There must be no effect on the radiation impedance of either transducer due to the presence of the other.

3. The receiving transducer must be separated from the source by a distance large enough compared to the transverse dimensions of the source so as to be in a spherical wave field. In addition, the radius of curvature of the wave sufaces at the receiver must be large enough so that phase differences due to curvature of the wavefront are negligible over the surface of the receiving transducer.

The first condition depends only on the separation while the second involves the wavelength. These restrictions may be represented by the two inequalities given by FOLDY[1]

$$L \ll d, \qquad \frac{L^2}{\lambda} \ll d \tag{29.1}$$

where d is the separation of two transducers, L is the largest transverse dimension of either transducer, and λ is the wavelength of sound.

The separation d is measured between the acoustic centers of the two transducers. While the location of the acoustic center relative to the surface of the transducer is arbitrary[2], a position close to the center of curvature of the radiated spherical wave field on the axis of symmetry represents a good choice. RUDNICK[3] has found a point 2 mm in front of the center of the radiating surface of the transducer to be satisfactory for the Western Electric 640 AA Condenser microphone.

In addition to the acoustic center, it is necessary to specify an axis for each transducer. The normal to the radiating surface at its center is usually chosen for this purpose. For many transducers this is also the axis of symmetry. The transducer axis is used to specify the angular orientation of the transducer relative to the line between acoustic centers. Free field reciprocity calibrations are usually made for normally incident sound.

α) *Determination of reciprocity parameter.* Since, as shown in Sect. 27, the ratio of microphone sensitivity M to source constant S is independent of the detailed construction of a reciprocal transducer and depends only on its acoustic environment, we may calculate it for an idealized spherical transducer. This transducer will be characterized by a radius $R \ll \lambda$ and an acoustic impedance Z. The pressure generated by the transducer will be computed at a distance d from the acoustic center of the transducer and be designated as $P(d)$. The source constant is given by

$$S = \frac{P(d)}{I} \tag{29.2}$$

where I is the current supplied to the transducer. The microphone sensitivity is given by

$$M = \frac{V_0 C}{P} \tag{29.3}$$

where P is the pressure at the surface of the reciprocal transducer. P is uniform over the surface of the transducer since $R \ll \lambda$. Writing the reciprocity condition

---

[1] L. FOLDY and H. PRIMAKOFF: J. Acoust. Soc. Amer. **17**, 109 (1945).

[2] See footnote 2, p. 66.

[3] See footnote 5, p. 66.

for the idealized transducer

$$V = a\dot{I} + b\dot{U},$$
$$P = bI + dU. \qquad (29.4)$$

Where P is the pressure at the surface of the transducer and U the volume velocity of the surface, we have for the open-circuit voltage at the electrical terminals

$$V_{0C} = bU = \frac{bP}{Z + Z_R} \qquad (29.5)$$

where P is the pressure driving the surface of the transducer, Z the acoustic impedance of the transducer, and Z_R the acoustic radiation impedance of the transducer. From Eq. (29.5), M is given by

$$M = \frac{V_{0C}}{P} = \frac{b}{Z + Z_R}. \qquad (29.6)$$

From Eqs. (29.4), we have for the blocked pressure generated by the transducer as a source

$$P_{0C} = bI \qquad (29.7)$$

from which the pressure in the medium at the surface of the transducer due to the current I is given by

$$P = \frac{P_{0C} Z_R}{Z + Z_R} = \frac{bI Z_R}{Z + Z_R}. \qquad (29.8)$$

Correcting for spherical divergence, phase shift, and attenuation we have for $P(d)$

$$P(d) = \frac{bI Z_R}{Z + Z_R} \left(\frac{R}{d}\right) e^{-\alpha d} e^{-jkd} \qquad (29.9)$$

and S is given by

$$S = \frac{P(d)}{I} = \frac{bZ_R R}{(Z + Z_R) d} e^{-\alpha d} e^{-jkd}. \qquad (29.10)$$

Combining Eqs. (29.6) and (29.10), we have

$$\frac{M}{S} = \frac{d}{R Z_R} e^{\alpha d} e^{jkd}, \qquad (29.11)$$

For $R \ll \lambda$ the acoustic radiation impedance is given by

$$Z_R = \frac{j \varrho c}{2 R \lambda}. \qquad (29.12)$$

Inserting Eq. (29.12) in (29.11) and dropping all phase factors, we have for the magnitude of the reciprocity parameter

$$\frac{M}{S} = \frac{2 d \lambda}{\varrho c} e^{\alpha d}. \qquad (29.13)$$

Eq. (29.13) holds for any reciprocal transducer regardless of its geometrical form as long as d is chosen large enough so that the radiated pressure field is spherical in form. The minimal satisfactory value of d may be determined experimentally by measuring $P(d)$ with a second transducer as d is increased until

$$P(d) \propto \frac{e^{-\alpha d}}{d} \qquad (29.14)$$

establishing the spherical character of the field and permitting the evaluation of the amplitude attenuation factor α. The determination of this minimal value of

d for each source and receiver is an essential step in a free field reciprocity calibration.

β) *Calibration procedure.* Transducer T_1 will be assumed to obey reciprocity. The comparison microphone T_2 and the source T_3 need not be reciprocal. All three may be identical for convenience. The two sets of measurements required for the calibration are the following.

1. **Determination of M_1/M_2.** The source T_3 is set up in the free field room and a point on its axis selected for the comparison of the microphone sensitivities of T_1 and T_2. T_1 is placed with its acoustic center at the comparison position and oriented to receive normally incident sound. With a known current in T_3, the open-circuit voltage V_{13} is measured at the terminals of T_1. Transducer T_1 is now replaced by T_2 with its acoustic center at the comparison point and oriented to receive normally incident sound. With the same current driving T_3, the open-circuit voltage V_{23} is measured at the terminals of T_2. The ratio of their sensitivities as microphones for normally incident sound is given by

$$\frac{M_1}{M_2} = \frac{V_{13}}{V_{23}}. \tag{29.15}$$

2. **Determination of $M_2 S_1$.** Transducers T_1 and T_2 are placed with their acoustic centers at a distance d satisfying the conditions of Eq. (29.1). They are oriented with their receiving surfaces perpendicular to the line through their acoustic centers to satisfy the normal incidence condition. T_1 is driven by a current I_1, and an open-circuit voltage V_{21} is measured at the terminals of T_2. V_{21} is related to I_1 by the following equation

$$V_{21} = M_2 P_1(d) = M_2 S_1 I_1 \tag{29.16}$$

from which we obtain the product

$$M_2 S_1 = \frac{V_{21}}{I_1} \tag{29.17}$$

in terms of the measured quantities V_{21} and I_1.

The magnitude of the sensitivity of the reciprocal transducer T_1 as a microphone is obtained by combining Eqs. (29.13), (29.15), and (29.17) as follows.

$$M_1 = \sqrt{\left(\frac{M_1}{M_2}\right) M_2 S_1 \left(\frac{M_1}{S_1}\right)} = \sqrt{\left(\frac{V_{13}}{V_{23}}\right)\left(\frac{V_{21}}{I_1}\right) \frac{2 d \lambda}{\varrho c}} \; e^{\frac{\alpha d}{2}} \tag{29.18}$$

where M_1 is given in volts per newton/m² and all quantities in (29.18) are expressed in M.K.S. units. The proper value of d in (29.18) is the separation in meters used in the measurement determining $M_2 S_1$. It is essential that the voltage V_{13} and the current I_1 be measured at the actual terminals of T_1. This is particulary important when condenser microphones having low electrical capacity are calibrated. The introduction of the shunt capacity of a short length of cable between the current measuring resistor and the microphone when used as a source can introduce a serious error in I_1 if the cable is not used during the measurement of V_{13}. Accuracies of the order of 0.3 db are obtainable in free field reciprocity calibrations.

Generation, Detection and Measurement of Ultrasound.

By

A. BARONE.

With 99 Figures.

In this article a general exposition of the production of ultrasound is given and some methods of measurement, based on the physical phenomena which accompany the propagation of ultrasonic waves of small amplitude are discussed. The physical effects produced by high energies and the use of ultrasound in the investigation of the molecular and structural properties of matter are dealt with in other chapters of this Encyclopedia.

I. Generation of ultrasound.

1. Basic concepts. Elastic waves whose frequency exceeds the range within which they cause sound sensations in the human ear, are called ultrasound waves or ultrasonics.

The frequency range of ultrasound extends in practice from about 16000 Hz to about 10^9 Hz. The upper limit of ultrasound does not depend on any intrinsic physical property of ultrasonics of such high frequencies, but on the technical possibility to produce such waves. However the upper frequency limit in current use is of the order of 100 MHz.

Ultrasound is employed as a means of investigation in physical research. Its modes of propagation can yield information on the properties of the medium in oscillation and on its structure.

Ultrasonic waves, as all elastic waves, can vary in type according to the nature and the dimension of the body in which they propagate. The velocity of propagation depends on the elastic properties and density of the medium, and it varies according to the type of wave.

From a general point of view, the waves can be subdivided in longitudinal (L) and transverse or shear waves (S). In the case of longitudinal waves, particles oscillate in the direction of the propagation. They can propagate in a medium possessing bulk elasticity, hence in matter in any state of aggregation.

In transverse waves the particles oscillate in a direction normal to that of the wave propagation. In this case the medium must possess elasticity of form. S waves can therefore propagate only in solids; these waves are sometime polarized in a plane or circularly. However, propagation of shear waves in highly viscous liquids is also possible.

In solids, under certain conditions, waves which at the same time possess both longitudinal and transverse character can be observed.

A further distinction between various wave types must therefore be made, of which Fig. 1 may help to get a general idea. Fig. 1a shows *purely longitudinal waves*, propagating in an unbounded homogeneous medium in any state of aggregation. In gases and liquids their velocity of propagation is

$$c = \left(\frac{1}{\varrho\, B_{ad}}\right)^{\frac{1}{2}}$$ (1.1)

with ϱ the density and B_{ad} the adiabatic compressibility. In solids we have

$$c = \left(\frac{\delta + 2\mu}{\varrho}\right)^{\frac{1}{2}} = \frac{E}{\varrho}\, \frac{1 - \sigma}{(1 + \sigma)\,(1 - 2\sigma)}\,,$$ (1.2)

where δ and μ are the two Lamé constants, σ is the POISSON's ratio and E YOUNG's modulus. *Purely transverse waves* (Fig. 1 b) may propagate in an unbounded

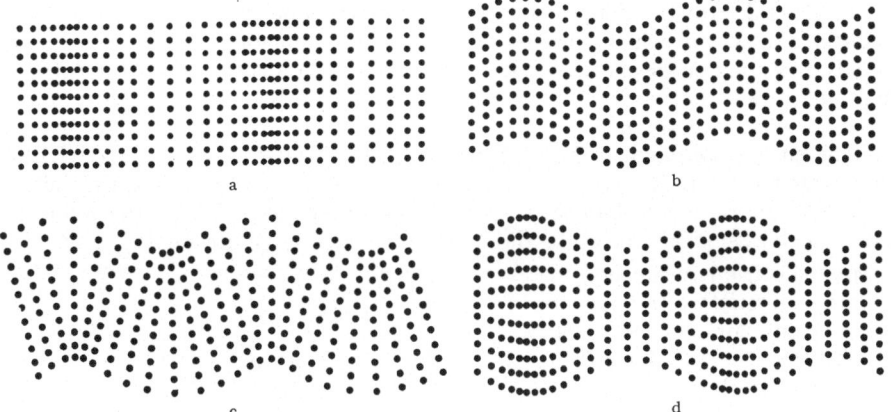

a   b

c   d

Fig. 1 a—d. Configuration of displacements in the various types of elastic waves.

solid medium. Their velocity of propagation is

$$c = \left(\frac{\mu}{\varrho}\right)^{\frac{1}{2}} = \left[\frac{E}{\varrho}\, \frac{1}{2\,(1 + \sigma)}\right]^{\frac{1}{2}}.$$ (1.3)

Flexural waves of frequency f (Fig. 1 c), may propagate in a solid medium having the shape of an infinitely long rod of radius R which is small in comparison with the wavelength Λ, or having the shape of a plate of thickness $d \ll \Lambda$. Their velocity of propagation is:

$$c = \frac{\pi R}{\Lambda}\left(\frac{E}{\varrho}\right)^{\frac{1}{2}} = (\pi R f)^{\frac{1}{2}}\left(\frac{E}{\varrho}\right)^{\frac{1}{4}}$$ (1.4a)

or, respectively,

$$c = \frac{\pi d}{\Lambda\sqrt{3}}\left[\frac{E}{(1 + \sigma^2)\,\varrho}\right]^{\frac{1}{2}} = (\pi d f)^{\frac{1}{2}}\left[\frac{E}{3\,(1 - \sigma^2)\,\varrho}\right]^{\frac{1}{4}}.$$ (1.4b)

Extensional waves (Fig. 1 d) may propagate in a solid medium shaped like a rod of a diameter small as compared with the wavelength. Their velocity of propagation is:

$$c = \left[\frac{\mu\,(3\delta + 2\mu)}{(\delta + \mu)\,\varrho}\right]^{\frac{1}{2}} = \left(\frac{E}{\varrho}\right)^{\frac{1}{2}}.$$ (1.5)

Torsional waves may propagate in a solid bar or tube with the velocity of propagation

$$c = \left(\frac{\mu}{\varrho}\right)^{\frac{1}{2}} = \left[\frac{E}{\varrho}\, \frac{1}{2\,(1 + \sigma)}\right]^{\frac{1}{2}}.$$ (1.6)

Surface waves, the so-called Rayleigh waves, may propagate along the surface of an unbounded medium with the velocity of propagation

$$c = \frac{0.87 + 1.12\,\sigma}{1 + \sigma} \left(\frac{\mu}{\sigma}\right)^{\frac{1}{2}}. \tag{1.7}$$

Of course, in order to generate these various types of wave the source of sound must, in each case, transfer the suitable oscillations to the medium; thus, e.g., in order to generate longitudinal and extensional waves, the radiating surface of the source of sound must oscillate in a direction perpendicular and, to generate shear waves, parallel to its own extension.

Any mechanical system capable of vibrating within the ultrasonic frequency range practically constitutes a source of ultrasound.

In general, a distinction must be made in devices producing ultrasonic waves between the system which supplies energy and an element receiving this energy and transferring it to another system (medium of propagation) in the form of high frequency mechanical vibration. This element belongs to a category of devices called *transducers*.

The energy supplied to the transducer can be electric or mechanical. Ultrasonic transducers may thus be divided in two chief groups according to whether they use electric or mechanical power. To the first group belong the following transducers, listed according to the extent in which they are used:

piezoelectric transducers,
magnetostrictive transducers,
electrodynamic transducers,
electrostatic transducers,
spark transducers.

The second group includes such transducers in which pressure or speed of the fluid receiving the energy, is high-frequency modulated either:

by means of fixed mechanical elements, or
by means of moving mechanical elements.

a) Transducers using electric power supply.

2. Piezoelectric transducers. These transducers are based on the piezoelectric effect.

Some crystalline materials are polarized when subjected to a stress. By inverting the direction of the stress the polarization changes sign. On the other hand, if such a material is subjected to the action of an electric field by applying a potential difference between two distinct parts of it, the material is deformed and the strain changes sign when the direction of the electric field is inverted (Curie 1880).

An element of piezoelectric material can, therefore, be employed to generate elastic waves. Indeed, if the potential difference applied is alternating, the element vibrates, and its mechanical vibrations are transferred to the surrounding medium.

The most important piezoelectric materials to be applied in acoustics are: quartz (SiO_2) and tourmaline (trigonal crystalline system), lithium sulphate ($Li_2SO_4 + H_2O$) and ammonium di-hydrogen phosphate ($NH_4H_2PO_4$) (tetragonal), Rochelle salt ($NaKC_4H_4O_6 + 4H_2O$) (rhombic) and polarized barium titanate ($BaTiO_3$) (tetragonal). Quartz and barium titanate are chiefly used because of the stability of their mechanical properties, even when the vibrations are of considerable amplitude. The quartz transducer consists of a plate cut from the natural crystal. The barium titanate transducer is a ceramic in which piezoelectric properties are artificially induced.

α) *Quartz.* Quartz crystallizes in the trigonal system forming a hexagonal prism with two rhombohedric ends (Fig. 2). The three axes connecting opposite

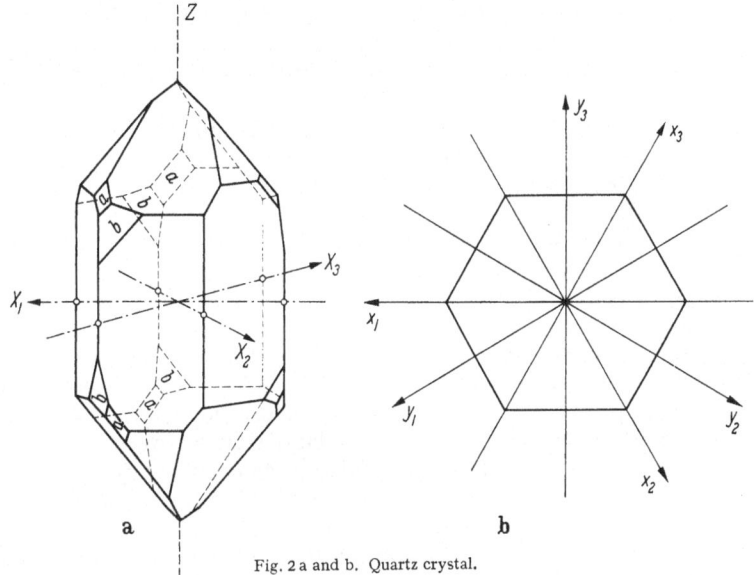

Fig. 2 a and b. Quartz crystal.

apexes of a cross section of the prism are called piezoelectric axes (x_1, x_2, x_3). The axes perpendicular to the opposite sides of such a section are the mechanical axes (y_1, y_2, y_3). The axis of the prism passing through the centre of the section is an axis of optical symmetry called the optical axis of the crystal (z).

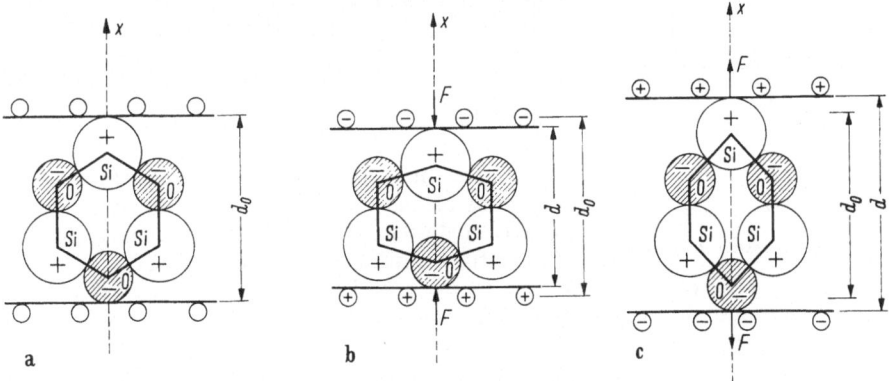

Fig. 3 a—c. Schematic representation of the behaviour of the quartz crystal under stress.

For the generation of longitudinal waves the transducer plate is cut from the crystal at a right angle to one of the piezoelectric axes (*X*-cut) and has both opposite faces covered with a thin conducting layer to enable a voltage to be applied.

A simple diagram of an undeformed crystal lattice is shown in Fig. 3 a. Owing to their position with respect to the two opposite faces of the element, the positive ionic charges of silicon and the negative ones of oxygen neutralize the electric effect on these faces and no charge of opposite sign is observed on the outer surface of the two electrodes.

When a stress is applied to the surfaces in the direction x (Fig. 3b), then, owing to the strain $\left(\frac{d - d_0}{d_0}\right)$ thus caused, the silicon ions are shifted towards the upper electrode and the negative oxygen ions towards the lower electrode. The electric equilibrium on the surfaces of the electrodes is disturbed and the presence of the charge indicated leads to a potential difference V between the electrodes. If the direction of the stress is inverted (Fig. 3c), the sign of the strain is also inverted and, for reasons similar to those just mentioned, the charges on the electrodes change sign and the potential difference becomes $- V$. This is called the longitudinal piezoelectric effect.

The charge distribution in Fig. 3b also follows if tension is applied at right angles to x (i.e. in direction y), and that in Fig. 3c by applying a compression. This is the transverse piezoelectric effect. Conversely, if the element is subjected to an electric field in the direction x by applying a potential difference between the electrodes, ionic charges are caused to move in the direction of the field, giving rise to a strain of the element according to thickness (reciprocal piezoelectric effect).

These strains of course produce mechanical and electrical effects also in the other directions of the element. A general description of the phenomenon is given by the equations which establish the relationships between the piezoelectric, elastic and dielectric constants of quartz. These equations are fully discussed in the fundamental writings of Cady[1] and Mason[2].

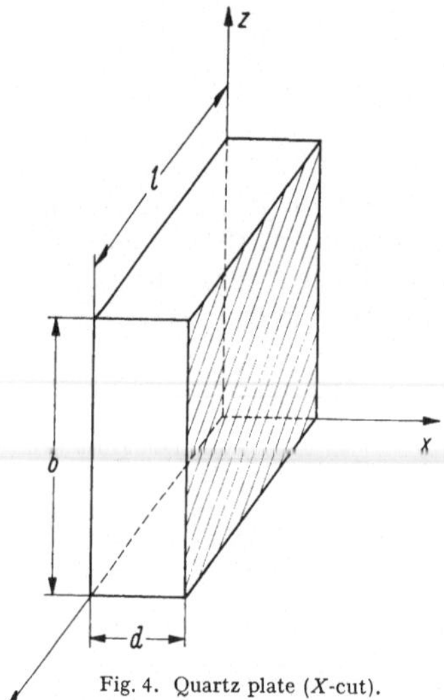

Following Voigt's notation[3] we can, in any case, represent by a few tensorial equations[4] in a phenomenological way, the behaviour of a plate cut in parallelopipedon shape with respect to the axes of the crystal, as indicated in Fig. 4.

If S_h is the tensor of the elastic strain, having as components the three longitudinal strains x_x, y_y, z_z and the three tangential strains y_z, z_x, x_y we get:

$$S_h = \sum_i d_{ih} E_i \qquad (2.1)$$

where the d_{ik} represent the piezoelectric strain constants (with i running from 1 to 3, h and k from 1 to 6), and E_i the three components of the electric field vector. This means that each component of the electric field contributes to the production of longitudinal and tangential strains along the axes x, y, z.

Fig. 4. Quartz plate (X-cut).

In the case of the X-cut quartz plate of Fig. 4 many coefficients are zero and others are simplified by symmetry; thus, the system of the six equations represented by (2.1) is reduced to the following simple relations:

$$x_x = d_{11} E_x ,$$
$$y_y = - d_{11} E_x ,$$
$$y_z = d_{14} E_x ,$$
$$z_x = - d_{14} E_y ,$$
$$x_y = - 2 d_{11} E_y$$

[1] W. G. Cady [4].
[2] W. P. Mason [5].
[3] W. Voigt: Abh. Ges. Wiss. Göttingen 36, 1 (1890).
[4] T. F. Hueter and R. H. Bolt [3].

where the constants d_{11} and d_{14} have the numerical values

$$d_{11} = 2{,}3 \times 10^{-12} \text{ Coul/Nt}$$

and

$$d_{14} = 5.66 \times 10^{-12} \text{ Coul/Nt}.$$

The piezoelectric strains indicated by Eq. (2.1) are, in their turn, connected with stresses from another tensorial equation representing HOOKE'S law in its most general form,

$$T_k = \sum_h c_{hk} S_h \qquad (2.2)$$

where T_k is the tensor of the elastic stresses and has the three normal stresses X_x, Y_y, Z_z and the three tangential stresses Y_z, Z_x, X_y as components.

c_{hk} are the 36 elastic stiffness constants.

By introducing a new equation between the piezoelectric stress constants e_{ik} and the piezoelectric strain constants d_{ih},

$$e_{ik} = \sum_h c_{hk} d_{ih} \qquad (2.3)$$

the relations between the components of stress T_k and the components E_i of the electric field can be established. For the quartz plate of Fig. 4 the number of the independent constants e_{ik} is considerably reduced and the system of equations following from Eqs. (2.2) and (2.3) by elimination of the coefficients c_{hk}, becomes simply:

$$X_x = e_{11} E_x,$$
$$Y_y = -e_{11} E_x,$$
$$Y_z = e_{14} E_x,$$
$$Z_x = -e_{14} E_y,$$
$$X_y = -e_{11} E_y.$$

The transducer is obviously excited to one of its mechanical normal modes. Because the plate is vibrating in the direction of the thickness, the natural frequency of elastic oscillation is that which determines maximum elongation of the faces in the two opposite directions, a situation described by a standing elastic wave with displacement antinodes on both faces.

In the case of the first normal mode of vibration there is only a single nodal plane, and the thickness d is equal to half a wavelength. In that case the oscillation frequency becomes

$$f = \frac{c}{\varLambda} = \frac{1}{2d} \left(\frac{c_{11}}{\varrho} \right)^{\frac{1}{2}},$$

where c_{11} is the elastic modulus in direction x and ϱ the density of the quartz.

In practice, the measured value of f differs slightly from its calculated value owing to the limited size of the plate. As a matter of fact, under standard conditions the other oscillations in directions other than x, interact with the thickness oscillations and modify the natural frequency of the plate.

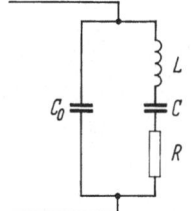

Fig. 5. Equivalent electrical circuit of quartz transducer.

β) *Equivalent circuit.* From the electrical viewpoint, the transducer behaves as a circuit possessing resistance, capacitance and inductance[1] such as shown in Fig. 5. Let C_0 be the capacitance (under static conditions) of the condenser having quartz as dielectric and the surfaces of the crystal as its plates.

---

[1] K. S. VAN DYKE: Phys. Rev. **5**, 895 (1925).

The other circuit elements (L, C, R) are the electric equivalents of mass, compliance, and friction of the mechanical system. Let us for the moment pay no attention to radiation and replace the plate by the simple mechanical model of two masses connected at the nodal plane by means of an elastic bond with friction. For either mass the differential equation of the motion impressed by a force F, is:

$$M \frac{du}{dt} + R_m u + G \int u \, dt = F \qquad (2.4)$$

where u is the velocity of mass M, R_m the coefficient of the force of friction and G the constant of the elastic restoring force.

The equivalent electric circuit can have its constituting elements arranged in series or in parallel. In the series circuit, the intensity of current corresponds to the velocity. The differential equation which in such a case connects current i and voltage V is:

$$L \frac{di}{dt} + R i + \frac{1}{C} \int i \, dt = V. \qquad (2.5)$$

From a comparison of Eqs. (2.4) and (2.5) a correspondence is seen between L and M, R and R_m, $1/C$ and G.

In order to establish a quantitative relation, let us remember that the force applied to the plate of surface S is the piezoelectric force due to the potential difference V, or:

$$F = K V = \frac{e_{11} S}{d} V. \qquad (2.6)$$

Moreover the total charge induced on the electrodes by V is $q = q_0 + q_1$, where $q_0 = C_0 V$, and $q_1 = K x$ represents the contribution due to the piezoelectric effect. We thus can write:

$$\left. \begin{aligned} i_1 &= \frac{dq_1}{dt} = K \frac{dx}{dt} = K u; \\ \frac{di_1}{dt} &= K \frac{du}{dt}. \end{aligned} \right\} \qquad (2.7)$$

Introducing Eqs. (2.6) and (2.7) in (2.5), we get:

$$L K^2 \frac{du}{dt} + R K^2 u + \frac{K^2}{C} \int u \, dt = F.$$

From a comparison of this equation with (2.4) we obtain the quantitative correspondences:

$$L = \frac{M}{K^2}, \qquad R = \frac{R_m}{K^2}, \qquad C = \frac{K^2}{G}.$$

Under dynamic conditions, the mechanical quantities can be calculated and, remembering that the forces acting on each element of mass are distributed along the thickness according to a sinusoidal law, we obtain:

$$M = \frac{1}{2} \varrho S d, \qquad G = \frac{\pi^2}{2} \frac{c_{11} S}{d}, \qquad R_m = \frac{\pi^2}{2} \frac{\varXi S}{d}.$$

Here M is half the mass of the plate, and $\varXi$ represents an internal friction coefficient of the quartz.

All the constants of the equivalent electrical circuit can thus be determined.

The capacitance of the clamped plate is $C_0 = \varepsilon \frac{S}{d}$ where ε is the dielectric constant of the quartz. The other electric circuit constants are:

$$L = \frac{\varrho \, d S}{2 K^2}, \qquad C = \frac{2 K^2 d}{\pi^2 c_{11} S}, \qquad R = \frac{\pi^2 \varXi S}{2 K^2 d}.$$

At resonance, capacitance and inductance cancel each other, and the equivalent circuit is reduced to the capacitance C_0 in parallel with resistance R. The latter will be of the order of a few ohms and the energy lost in it amounts to very little for an unloaded quartz plate. But, if the transducer is placed in a medium to which it transfers a considerable part of its vibration energy, then the equivalent resistance is determined almost exclusively by the external load and may be much larger.

The power radiated into the medium from the vibrating surfaces of the plate is:

$$W = u_{\text{eff}}^2 Z_r, \tag{2.8}$$

where u_{eff} is the effective vibration velocity, and $Z_r = 2\varrho c S$ the total mechanical radiation impedance, S being the surface of each face of the plate.

Since the velocity u of the surfaces depends on the potential difference V applied to the plate:

$$u = \frac{KV}{\varrho c S},$$

Eq. (2.8) becomes:

$$W = \frac{K^2 V_{\text{eff}}^2}{2\varrho c S} = \frac{2e_{11}^2 S}{d^2 \varrho c} V_{\text{eff}}^2 = 2W_0,$$

where W_0 is the power emitted from each face.

In practice it is often convenient to use the ultrasonic beam emitted by one side only of the plate; therefore, to reduce energy radiation from the opposite side to a minimum, the other face is placed in contact with a medium of zero (vacuum) or very little specific acoustic resistance (air).

Under ideal conditions, i.e. if one of the faces is placed in vacuum, the wave propagating inside the plate is completely reflected at the non-radiating surface and returns to the radiating one in phase with the vibration of this surface. The amplitude and, hence, the velocity of vibration of the interface in contact with the transducer medium are thus doubled. In unilateral radiation, therefore, the total radiated power W_1 being proportional to the square of the amplitude is quadrupled:

$$W_1 = 4W_0.$$

If, as assumed above, dissipation inside the quartz is negligible, the total power, at resonance, is completely supplied to a resistive load

$$R_e'' = \frac{d^2 \varrho c}{2 e_{11}^2 S}$$

in the case of bilateral radiation, or

$$R_e' = \frac{d^2 \varrho c}{4 e_{11}^2 S}$$

in the case of unilateral radiation. These equations can also be written in terms of the frequency f and the ratio between the specific acoustic resistances ϱc of the medium and $\varrho' c'$ of the quartz:

$$R_e'' = \frac{A}{f^2 S} \frac{\varrho c}{\varrho' c'}; \qquad R_e' = \frac{1}{2} \frac{A}{f^2 S} \frac{\varrho c}{\varrho' c'}$$

where

$$A = \frac{c' c_{11}}{8 e_{11}^2}$$

is a characteristic constant of the quartz.

If the plate is equally loaded on both sides, the plane of symmetry, normal to the axis, is a nodal plane in which the amplitude of displacement is zero; if, however, the load is dissymmetric, as in the case of unilateral radiation, the amplitude of displacement in the plane of symmetry will still be a minimum, but never zero.

At mechanical resonance, the transducer behaves as an electric load consisting of the resistance R'_e or R''_e, as the case may be, and the electrostatic capacitance C_0 of the quartz plate arranged in parallel.

Fig. 6 gives the values (as depending on the fundamental resonance frequency) of the resistance of an X-cut quartz plate of unit surface, radiating unilaterally

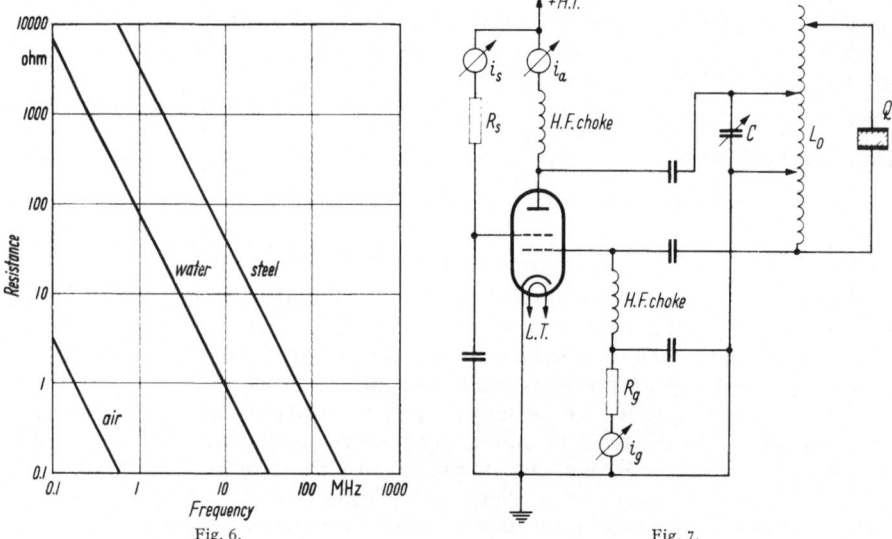

Fig. 6.

Fig. 7.

Fig. 6. Resistance of a unit surface (1 m²) quartz transducer at fundamental frequency for a gas, a liquid and a solid (unilateral radiation).

Fig. 7. Typical ultrasonic generator.

in a gas, a liquid, and a solid. These data are very important in designing ultrasound generators because they enable the output impedance to be suitably matched to the quartz.

The reactance of C_0 can be cancelled by means of a coil of inductance

$$L_0 = \frac{1}{4\pi^2 f_0^2 C_0}$$

where f_0 is the resonance frequency of the crystal.

Fig. 7 shows a simple ultrasonic generator for experimental purposes. Movable contacts on the coil L_0 help the operator find the best working conditions.

X-cut quartz plates are suitable for generation of longitudinal or extensional waves. In order to generate shear waves in solids and viscous liquids other modes of vibration are used and, generally, the plate is cut normally to the y axis.

Torsional vibrations can be excited in a quartz cylinder if its axis is parallel to the x-axis of the crystal and the voltage is applied to electrodes arranged as in Fig. 8. The fundamental frequency of torsional oscillations is given by

$$f = \frac{1}{2l}\left(\frac{\mu}{\varrho}\right)^{\frac{1}{2}}$$

where l is the length of the cylinder, μ the shear modulus, and ϱ the density of the quartz.

A quantity which generally has to be held in consideration in the analysis of a piezoelectric transducer, is the so-called electromechanical coupling factor k. It is defined as the square root of the ratio of the energy stored in mechanical form, for a given type of strain, to the total electrical energy supplied to the crystal. In each particular application, valuable hints may be obtained from it, wheter the piezoelectric crystal may be used profitably as an ultrasonic source.

This factor, for a given mode of vibration, is related to the dielectric constant ε, to the piezoelectric strain constant d_{ik} and to the elastic stiffness constant c_{hk}. Especially, in the case of the quartz plate, it becomes:

$$k = d_{11} (4\pi c_{11}/\varepsilon)^{\frac{1}{2}} = 0.1 .$$

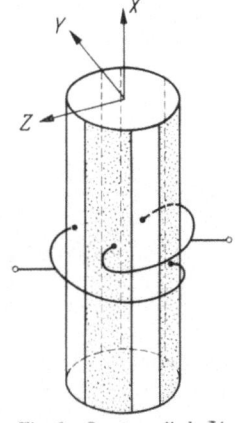

$\gamma)$ *Barium titanate.* The so-called piezoelectricity of barium titanate can be traced to the more general phenomenon of electrostriction—a property exhibited by all dielectric materials—and consists in a strain proportional to the square of the applied electric field.

This effect is a very small one in ordinary dielectric materials, but reaches appreciable values in barium titanate ceramics. These ceramics are obtained by grinding barium titanate crystals together with other salts possessing electrostrictive properties and heating the mixture at a temperature of about 1300 to 1400° C. This yields a material in which an electric field of 10^6 volts/m produces a strain of approximately 3×10^{-4}.

Fig. 8. Quartz cylinder in torsional vibration (after BERGMANN).

Elements of whatever shape—plates, spherical bowls, cylinders, etc.—can now be manufactured. For use as transducers, ceramics should be suitably polarized. If an alternating component is added to the polarization field, a periodical strain is obtained which, within certain limits, increases linearly with the applied voltage. The direct and alternating components of the electric field are directed according to the thickness of the transducer. If, on the other hand, a force is applied to the polarized element in the direction of the field, then a variation of the electric field proportional to the stress is observed inside the material.

These properties give rise to effects similar to those observed in piezoelectric materials, and for this reason barium titanate transducers are usually regarded as piezoelectric transducers.

One of the fundamental characteristics of these ceramics consists in the fact that permanent polarization ensues if a d.c. field of about 2000 volts/cm is applied through the thickness of the element at a temperature above the transition point (about 120° C) and is maintained during the gradual cooling. In this way there is no longer any need to add a d.c. field to the alternating voltage.

Since the residual polarization depends on the tendency of the domains of the dipoles to remain aligned in the direction of the inducing field, it is necessary to insure that the alternating field and the thermal effect will not appreciably modify this tendency. The addition of small quantities of lead titanate serves the purpose by producing an elevation of the transition point.

This fact, at any rate, imposes limitations on the values of the alternating voltage and upon temperature. On the other hand, the piezoelectric properties of barium titanate are so superior to those of quartz (with the same voltage applied the effect is 100 times greater), that keeping transducers at a temperature below

70° C, intensities of plane waves of about 100 W/cm² can be obtained, which are very high if compared with those obtainable with other kinds of piezoelectric transducers. The surfaces of the transducer between which the alternating voltage is applied are generally covered with a thin layer of silver applied by baking. The modes of vibration of the transducers depend on shape, direction of prepolarization field and configuration of the electrodes.

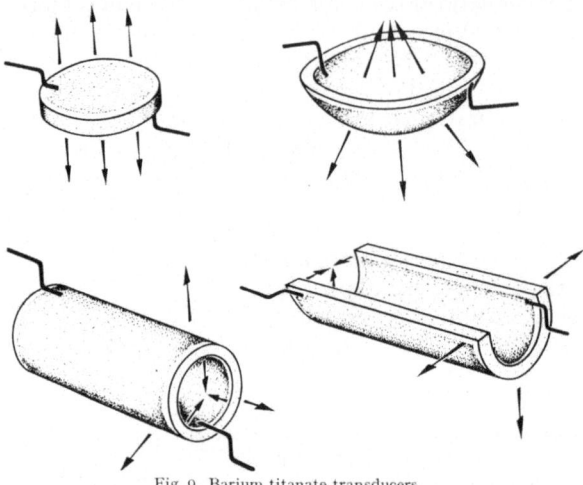

Fig. 9. Barium titanate transducers.

Some types of barium titanate transducer are shown in Fig. 9. The arrows indicate the direction of the ultrasonic radiation. Frequencies obtainable with barium titanate transducers are lower than those of quartz, in the first place because of losses due to dielectric hysteresis and owing to the difficulty of obtaining very small thicknesses. Plate transducers are generally used for frequencies from a few hundred kHz to several MHz.

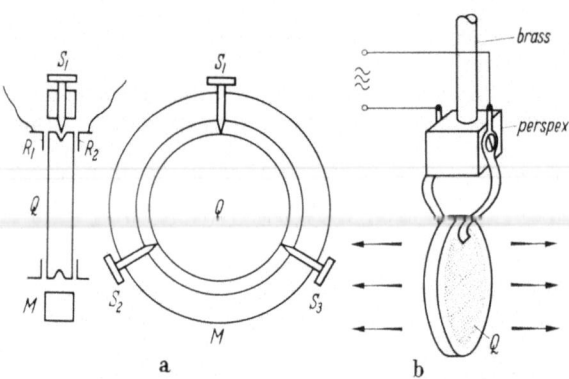

Fig. 10. Quartz holders for bilateral radiation in an insulating liquid.

Electric impedance is mainly determined by the capacitance of the transducers which is always very great owing to the high dielectric constant of barium titanate (about 270 times that of quartz).

At resonance, assuming equal characteristic impedance of the medium, the contribution of the motional impedance is greater than in the case of quartz because mechanical losses are much larger.

In brief, barium titanate transducers have a lower impedance than quartz transducers so that in order to obtain the same radiated acoustic power, they require much smaller driving voltages.

δ) *Crystal holders.* The efficiency of a transducer depends largely on the manner how it is kept in contact with the medium to which it must communicate acoustic power.

For bilateral radiation in a fluid the plate can be supported in one of the ways indicated in Fig. 10. In Fig. 10a, the quartz Q is supported by the screws $S_1 S_2 S_3$ at only three points of its nodal plane[1]. The connection of the electrodes with the

[1] R. Bechmann: Telefunkenztg. **14**, H. 36, 17 (1934).

two conductors of the high frequency generator is assured by means of either direct contact or the capacitance existing between the electrodes and two metal rings R_1 and R_2 near the faces of the plate. In Fig. 10b a very simple solution

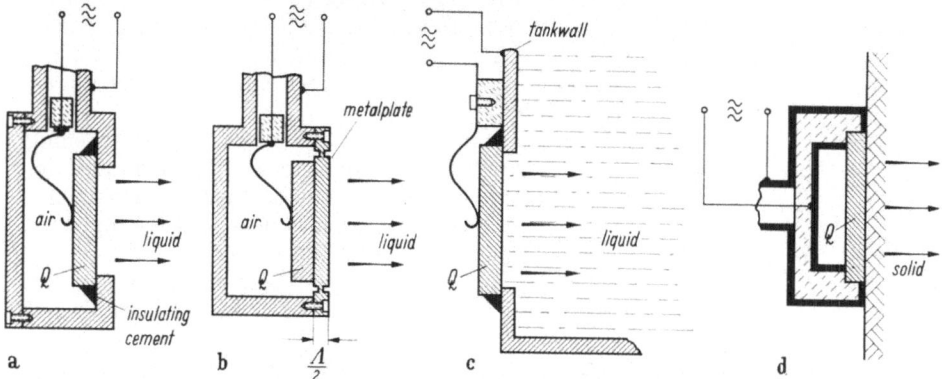

Fig. 11 a—d. Quartz holders for unilateral radiation. a in a liquid; b in a liquid through a resonant plate; c in a liquid through a wall of the container; d in a solid.

is indicated which can be used also with thin plates (higher frequencies). The plate is supported by a kind of pincers formed by two insulated metal springs. The same springs apply the high frequency voltage to the electrodes plated on the faces of the crystal.

For these types of support the liquid into which the transducer radiates must be insulating to prevent short circuiting.

In the case of unilateral radiation the transducer radiates power into a liquid or a solid. One face is therefore in contact with the propagation medium and the other is in contact with air. Some practical solutions are indicated in Fig. 11. In order to radiate considerable power into a liquid, holders such as shown in Fig. 12 are used.

3. Magnetostrictive transducers. If ferromagnetic material is subjected to the action of a magnetic field, strain occurs in the material. Conversely, if the

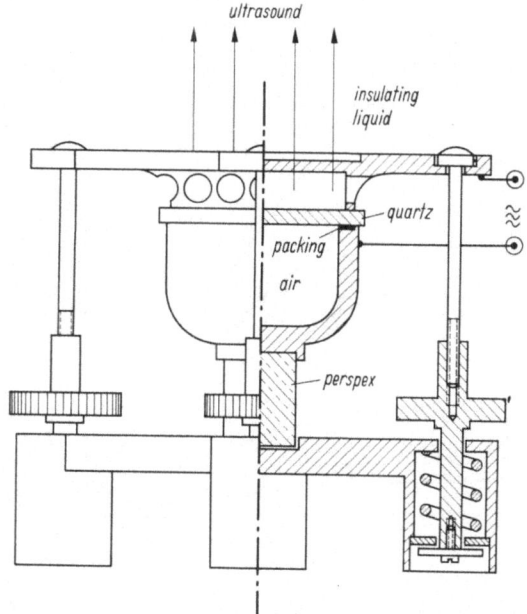

Fig. 12. Quartz holder for unilateral radiation of high ultrasonic power.

same material is deformed as a result of external forces, a variation of its magnetic properties is observed. These properties, viewed as a whole, are called magnetostriction. According to the methods used in the experiments carried out to investigate the various aspects of these phenomena, a number of effects were observed and these were named after the workers who first studied them; thus, e.g., the variation in length observed in a bar when the magnetic field coaxial to it, varies, is known by the name of Joule effect. The converse effect, showing the influence of strain on the magnetic permeability of the material is called the Villari effect.

The principal metals exhibiting the magnetostrictive effect are nickel, iron, cobalt, and their alloys in various combinations. Such alloys often contain small quantities of other metals, e.g. copper, chromium, vanadium, etc.

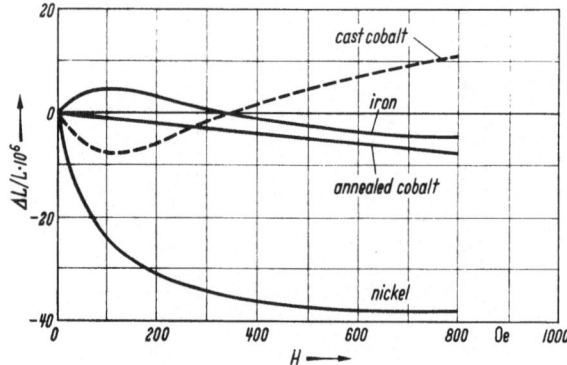

Fig. 13. Magnetostrictive effect in nickel (after Crawford).

In a bar, the lengthening due to magnetostriction depends on the type of meta and is not linear with the intensity of the magnetic field.

As seen in Fig. 13, in the case of nickel, a shortening of the bar occurs at whichever intensity of the field. An iron bar, on the other hand, lengthens in a weak field and shortens in a strong field.

The magnetostrictive effect also depends on temperature and becomes zero at the Curie point where the magnetic properties of the material disappear.

Fig. 14 shows the strain in a nickel bar when temperature is increased. The curves are drawn for several relative intensities of the magnetic field.

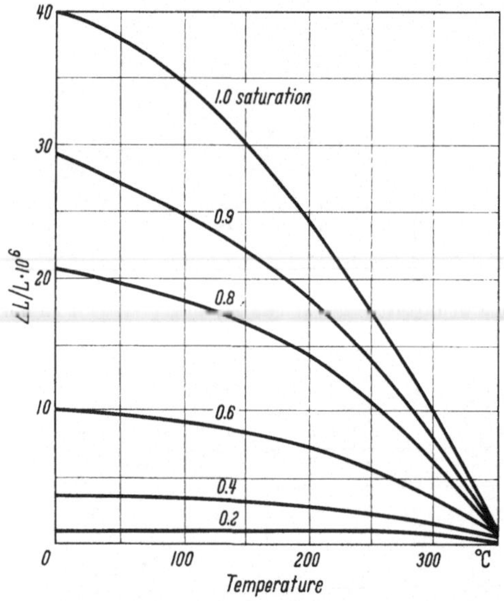

A simple equation correlates relative lengthening $\Delta L/L$ and magnetic induction B in the material. It holds well for small ΔB variations, starting from a given value B_0:

$$\frac{\Delta L}{L} = K\, B_0\, \Delta B$$

The coefficient K is not constant but depends on B_0. For example, in the case of nickel, $K = -2 \cdot 10^{-3}$ if $B_0 = 0.5$ Weber/m^2.

In constructing a magnetostrictive transducer the value of B_0 will be chosen in such a way that, for equal variations of ΔB, the variation of the mechanical strain becomes maximum.

Fig. 14. Dependence of the magnetostrictive effect on temperature in nickel (after Crawford).

In its simplest form, a transducer consists of a rod of magnetostrictive material inserted in a coil fed by an electric current composed of a d.c. component of such intensity as to produce in the metal the suitable magnetic induction B_0 and of an a.c. component able to produce the vibration of the rod. The variable component of the current is sinusoidal and its frequency must coincide with one of the resonance frequencies of the rod so that oscillation reaches a considerable amplitude.

Polarization of the rod, due to the d.c. component, is also essential in order to prevent the rod from oscillating at a frequency twice that of the excitation.

Given a substance, the magnetostrictive effect always occurs in the same direction, even if the polarity of the external magnetic field inverts the sign. During each period of exciting oscillation, therefore, the rod contracts and dilates twice in the same manner.

Pure nickel, which—as we have seen—has the best magnetostrictive properties, is often employed; however, its Curie point at 360° C and the low resistance to fatigue render nickel little suitable for the production of great acoustic power.

Pure iron exhibits an even weaker effect than nickel and, moreover, at a certain value of the magnetic field, magnetostriction changes sign. But the magnetostrictive properties of iron can be greatly improved in alloys with other metals. For instance, addition of small quantities of nickel is sufficient to produce positive magnetostriction for all values of the magnetic field that may be required. Alloys of iron and cobalt exhibit a pronounced magnetostrictive effect. One such alloy (Permendur) is widely used today and consists of equal parts of cobalt and iron plus about 2% of vanadium to improve the mechanical qualities.

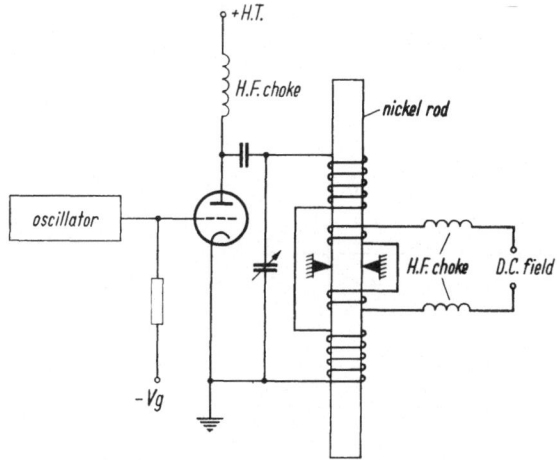

Fig. 15. Basic circuit of a magnetostriction generator.

The essentials of a magnetostrictive generator are illustrated in Fig. 15. The magnetostrictive element (e.g. nickel) in the shape of a rod or tube, is held at the centre, in a nodal line, by a rigid support in order to allow it to oscillate freely in its fundamental mode of extensional vibration. The element is placed at the inside of the coil L of the resonant circuit of a vacuum tube fed at the mechanical resonance frequency of the rod.

Polarization of the magnetostrictive element can be obtained in various ways: (1) by using the d.c. component of the anodic current of the vacuum tube; (2) by means of a separate power supply sending through a high-frequency choke the required current into the coil L; (3) by means of a separate coil coaxial with L (see Fig. 15); (4) by placing a permanent magnet or an electromagnet near the rod so that the magnetic flux issuing from the polar expansion penetrates the rod lengthwise.

In many generators of simple design the output tube which transmits the oscillations to the transducer is self-excited by means of a suitable feddback and acts simply as an oscillator whose frequency is controlled by the vibration of the magnetostrictive element.

In setting up the generator, attention should be paid to the fact that magnetic induction modifies, however slightly, the value of Young's modulus E in the magnetostrictive element. This leads to a variation of the speed of sound in the metal and, hence, of the frequency of its mechanical vibration.

The maximum frequency obtainable is limited by electrical losses due to hysteresis and eddy currents. These losses are connected with magnetic permeability and resistivity. A useful device by which to diminish the eddy currents consists in making cuts along the rod by which their electric circuit is interrupted. Better still, a lamination core may be used, in the same manner as in transformers.

$\alpha)$ *Equivalent circuit.* As in the case of piezoelectric transducers, the working of magnetostrictors can be investigated by indicating an equivalent circuit which, when replacing the transducer, behaves in the same way as whith respect to the electric generator. In practice, the equivalent circuit, at working frequency, will have the same impedance, as the magnetostrictor.

A very suitable circuit is that proposed by VIGOUREUX[1] which is shown in Fig. 16. The resistance r represents the load equivalent to electrical losses occurring in the transducer. This resistance could be disregarded in the equivalent circuit of the piezoelectric transducer since in quartz dielectric losses are negligible.

L_0 is the value of the inductance of the coil wound around the magnetostrictive

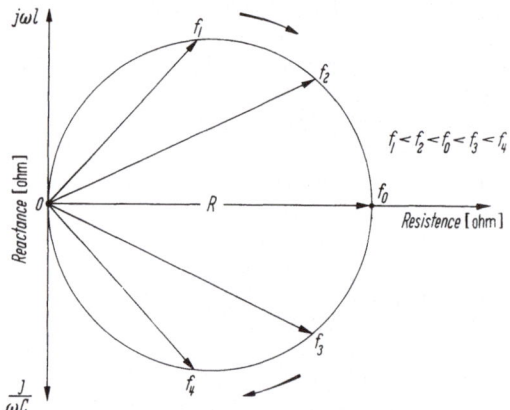

Fig. 17. Motional impedance circle.

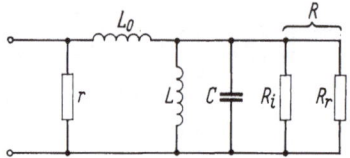

Fig. 16. Equivalent electrical circuit of magnetostrictor (after VIGOUREUX).

element, in absence of vibration. L, C, R represent the motional properties of the transducer. The resistance R consists of two resistances in parallel R_i and R_r the first of which takes account of the mechanical losses in the vibrating element due to internal friction, whereas the second represents the losses due to radiation of sound in the medium. When the transducer is fed off resonance, the core does not vibrate and no electric reaction will modify the current in the exciting coil, so that the total impendance will simply be that of an inductance L_0 with losses.

Normal impedance measurements will then furnish, in series arrangement, the inductance and resistance values of the coil. By converting, in the usual way, the data obtained into the equivalent parallel values, we find L_0 and r.

If we now repeat the measurements at frequencies in the neighbourhood of mechanical resonance, the core starts to vibrate and to modify the impedance previously measured. We thus arrive at two new values for the series reactive and resistive components of the total impedance. Expressing this impedance in the parallel form we simply subtract L_0 and r from the respective components, thus obtaining the reactance and resistance introduced by the motional factors. Plotting a diagram of the motional impedance with the resistance as abscissa and the reactance as ordinate, a point is obtained at each frequency which represents the end of the motional impedance vector applied at the coordinate origin. The line connecting all the points found is a circle passing through the origin (Fig. 17); its diameter is the resistance R. The end point of R corresponds

[1] P. VIGOUREUX: Proc. Phys. Soc. Lond. **59**, 19 (1947).

to the resonance frequency f_0 at which the two reactances ωL and $\dfrac{1}{\omega C}$ cancel themselves out.

In the equivalent circuit of Fig. 16, the two branches L and C taken together, exhibit an infinite impedance, and the circuit is reduced as shown in Fig. 18, where the only motional element is represented by R. In this case two points of the motional circle are sufficient to determine the values of L and C. In other words, it is sufficient to measure the motional impedance at only two frequencies. The resistance R_r depends exclusively upon the radiation resistance $\varrho c S$ where S is the radiating surface of the magnetostrictor.

In the operation in vacuo, R_r is eliminated and the whole energy supplied to motional elements is dissipated in R_i. The value of R_r can therefore be derived by means of comparative measurements in vacuo and in the medium in which the transducer is to operate.

In the case of most gases these measurements lack precision, but in liquids—in which magnetostrictors are usually employed—they are useful. If

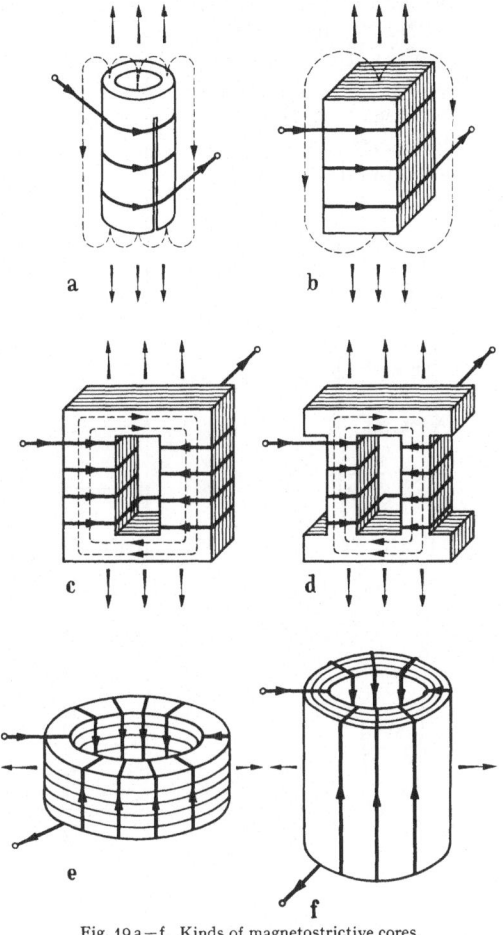

Fig. 19a—f. Kinds of magnetostrictive cores.

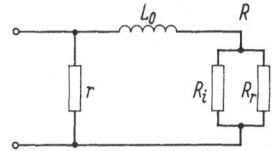

Fig. 18. Equivalent electrical circuit of magnetostrictor at resonance.

ultrasound of considerable energy is produced, electrical losses can be neglected in comparison with radiation, and the ultrasonic power radiated is approximately:

$$W = R_i^2 \, R_r \, i^2/(R_i + R_r)^2$$

as is obvious from the circuit in Fig. 18; i is the current in the transducer circuit at resonance.

The laminations used to avoid parasite currents are covered with suitable insulating material and riveted or soldered together to form a pack. The thickness of the laminations affects the amount of losses at the various frequencies. Thicknesses of about 0.3 to 0.4 mm are used up to 30 kc/s, and of about 0.1 to 0.15 mm at higher frequencies.

The usual forms of core are shown in Fig. 19, also the exciting coils as well as the lines of the magnetic induction field B and the direction of the sound radiation emitted. Fig. 19a represents a non-laminated tube with longitudinal cuts; the ends of the tube are the radiating surfaces which are usually closed by a circular plate. Fig. 19b shows a laminated bar. Since in both types considerable magnetic flux leakage occurs, the so-called "window" types, shown in Figs. 19c and d, are preferable, in which the magnetic circuit is entirely closed within the core. The fundamental mechanical resonance frequency of the above mentioned cores is about that which corresponds to vibration with half a wavelength, the

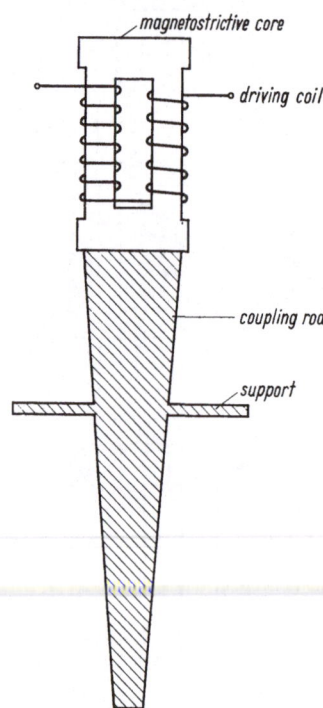

Fig. 20. Magnetostrictive transducer with coupling rod.

frequency kHz	profile	l mm	h mm
125	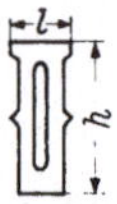	58	152
15		46	98
20	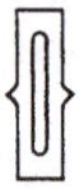	30	112
22	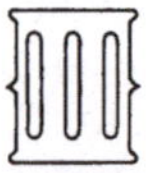	85	98
23	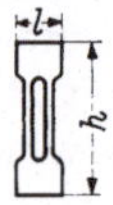	23	69
30		82	68
30		140	50
38	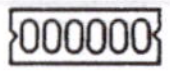	112	40,5
80	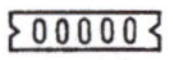	85	25
175	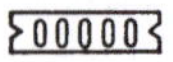	40	11

Fig. 21. Magnetostrictive cores and frequencies produced (after Atlas Werke).

cores being fixed at the central point of their length. The precise resonance frequency value depends on the particular shape of the core and the load on it.

In the window-type, the core can also vibrate at a right angle to the direction indicated since that is the direction in which the inductive flux runs through the ends of the core. Moreover, if the length of the lateral mountings is very different from the dimension of the ends, the resonance frequencies of the two vibrational modes differ rather from one another and the normal operation will not be disturbed.

Figs. 19e and f show transducers of toroidal shape having a square or rectangular cross section. The laminations are clearly to be distinguished. Cores oscillate radially at a fundamental frequency $f = \dfrac{c}{\pi D}$ ($D =$ average diameter), and the ultrasonic power is radiated in form of almost cylindrical waves.

In almost all uses of transducers of the types shown in Figs. 19a to d radiation of ultrasound from one end only of the vibrator is desired. Hence, as in the case of piezoelectric plates, the other end is placed in a medium with a specific acoustic resistance much below that of the medium of propagation. The ideal medium would be the vacuum, but if the transducer is to emit unilaterally in a liquid or solid, the non-radiating end may conveniently be left in contact with air.

In the modern transducers, especially in those working in liquids, the radiating face communicates the vibration to the medium through a coupling bar which has the double purpose of avoiding contact of the liquid with the laminated

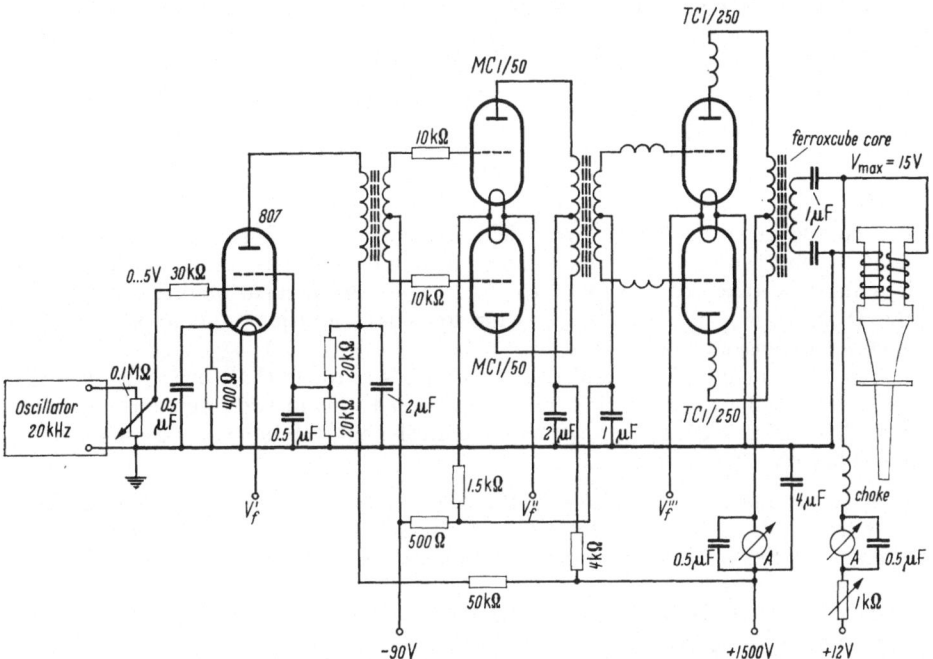

Fig. 22. Typical magnetostriction generator.

surface and of transforming the impedance and movement so as to match them to the medium. The bar is soldered on to the end of the vibrator. Its dimensions are such that the resonance frequency of the vibrator is not appreciably affected. The whole does not vibrate at half wavelength, but has several vibration nodes. In the place of one of these the support of the transducer is fixed. A typical example[1] is shown in Fig. 20.

The coupling bar can be cylindrical, conical or exponential. In this latter case the ratio of the vibration amplitudes along the bar is equal to the inverse ratio of the diameters.

Frequencies obtainable with ordinary magnetostrictive generators range from a few kHz to about 200 kHz. Fig. 21 illustrates some types of magnetostrictive cores suitable for use at the frequency range indicated[2] in the margin.

In some transducer polarization of the core can be accomplished by means of the residual magnetism. For example, the Permendur transducer can work in this way, a direct current being circulated initially in the exciting coil so as

---

[1] These transducers are constructed by Mullard Ltd. London.
[2] After Atlas Werke, Bremen.

to bring the flux density in the core to saturation value. A magnetizing force of about 10^4 ampere-turns per meter is sufficient for the purpose; after this the core remains permanently polarized.

Since the impedance of a magnetostrictive transducer is generally lower than that of the electric generator, a transformer is required for matching impedances.

An example of a typical magnetostriction ultrasound generator is shown in Fig. 22.

β) Other magnetostrictive materials. With the object in mind of finding new ferromagnetic materials for use as coil cores for radiofrequencies, there have been studied ceramic materials consisting of cubic ferrites[1], which were obtained by means of processes similar to those used in the production of insulating ceramics. The best known material of this kind is Ferroxcube. Its principal characteristics are a high initial permeability, low coercitive force and a high resistivity, all qualities which command its use at high frequencies. It is produced in two different forms: "Ferroxcube A" containing zinc-manganese ferrite and "Ferroxcube B" containing zinc-nickel ferrite. The form B has pronounced magneto-strictive properties. Magnetostriction is negative, as in pure nickel, and the equation correlating strain to magnetic field, is identical with Eq. (3.1). The greatest relative lengthening obtainable under optimum conditions of magnetic values is within the range of $-4 \cdot 10^{-6}$ to $-22 \cdot 10^{-6}$. Mechanical qualities are good; the Q factor is approximatly 5000 for torsional vibrations and approximately 4000 for extensional vibrations.

An advantage of Ferroxcube as a magnetostrictor is that, owing to its high resistivity, it does not require core lamination to avoid eddy currents. On the other hand, Ferroxcube is a very fragile material and can be used only to produce low intensity ultrasound.

4. Electromagnetic transducers. Piezoelectricity and magnetostriction are not the only effects used in the generation of ultrasound. The well-known interaction of magnetic field and electric current is used in the design of other types of ultrasonic transducers.

Saint Clair's electromagnetic transducer[2] is a typical example and has proved very suitable to generate ultrasound in gases. A longitudinal section of this transducer is shown in Fig. 23. A solid metal cylinder of considerable diameter is put in longitudinal oscillation along its axis. It is supported in the middle of its length by a flat ring cut in one piece with the cylinder itself. In the lower part of the cylinder a short thin-walled tube (driving tube) has been lathe-hollowed. This is lodged in the ring-shaped air gap of an electromagnet, very similar to that used in electrodynamic loudspeakers. Around the end of the central core of the electromagnet a fixed coil is wound in which an alternating current circulates, having the mechanical resonance frequency of the vibrating cylinder.

The current induced in the driving tube interacts with the radial magnetic field and comunicates the resulting force to the cylinder.

Since this kind of transducer is used to generate ultrasound in gases, the radiation impedance is low and the Q factor is very high. In the case of a duraluminum vibrator perfectly fixed at its nodal section, Q may reach a value of about 15000. For this reason small variations of the exciting frequency may bring the system beyond resonance so that the amplitude of oscillations becomes negligible.

In practice, the thermal energy dissipated in the vibrator changes its elastic constants yielding a variation of the resonance frequency. The frequency of the

---

[1] C.M. van der Burgt: J. Acoust. Soc. Amer. **28**, 1020 (1956).
[2] H.W. Saint Clair: Rev. Sci. Instrum. **12**, 250 (1941).

driving current is then controlled by the oscillation of the vibration system. This can be obtained by means of a pick-up condenser[1] consisting of an insulated electrode placed on the central core of the electromagnet opposite to the oscillating surface of the vibrator. The circuit employed is shown in Fig. 24.

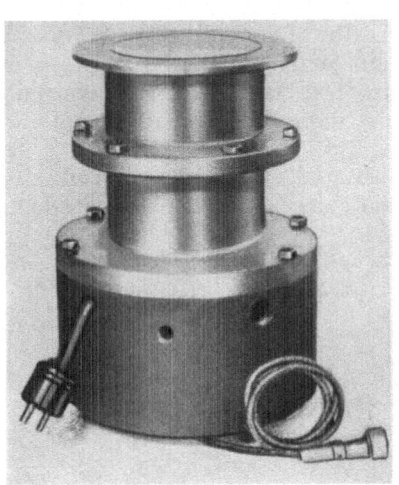

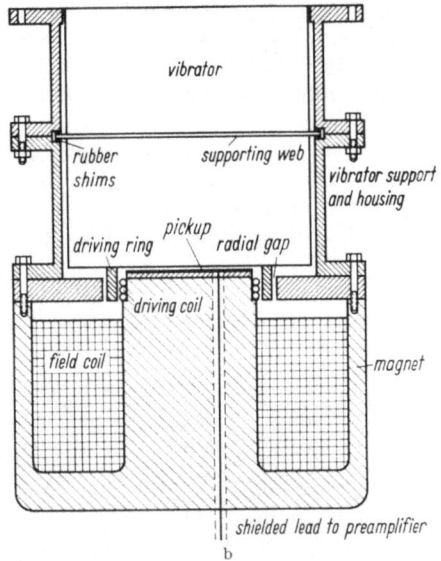

a

b

Fig. 23 a and b. Saint Clair transducer.

The variation of capacitance is transformed into a variation of voltage, as in an ordinary condenser microphone. This oscillating voltage, maintained at a constant value, controls by its phase the frequency of the oscillator which feeds the driving coil. If for any reason the oscillation of the vibrator tends to change frequency, the phase of the pick-up varies and, consequently the frequency of the oscillator changes until the phase relation between the driving current and the oscillation of the vibrator returns to a value at which the oscillations can be maintained. Usually the diameter of the vibrator is much greater than the wavelength of the radiated ultrasound, so that the beam emitted consists of almost plane waves. Consequently, the system is very suitable for use in the formation of standing waves by means of a reflector.

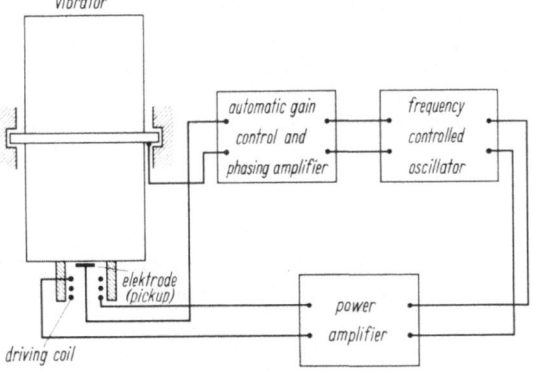

Fig. 24. Self excited Saint Clair generator.

The resonance frequency for this type of vibrator, whose diameter is approximately equal to its length, can be calculated by RAYLEIGH's formula

$$f = \frac{1 - \sigma^2 \pi^2 R^2}{4 l^2} \frac{1}{2l} \left(\frac{E}{\varrho}\right)^{\frac{1}{2}} \tag{4.1}$$

where σ is POISSON's ratio, R is the radius, l the length of the vibrator.

[1] E.V. POTTER: Rev. Sci. Instrum. **14**, 207 (1943).

Transducers so far constructed have a frequency range of between 15 and 20 kHz, and an efficiency of about 30 percent.

Into the class of electromagnetic transducers for ultrasound also the following types have to be included:

—the moving coil systems[1] which are very similar to electrodynamic loudspeakers, with the difference that the paper membrane is replaced by a metal circular membrane or plate firmly fixed along the circumference;

—the moving iron systems[2] which are similar in principle to old type telephon receivers. Frequencies obtainable with good efficiency are not very high. The use of these systems has until now been confined to special tasks of an industrial character.

A method of causing bars of insulating material to vibrate also at ultrasonic frequencies[3] is outlined in Fig. 25. A glass bar is metallized to render conductive

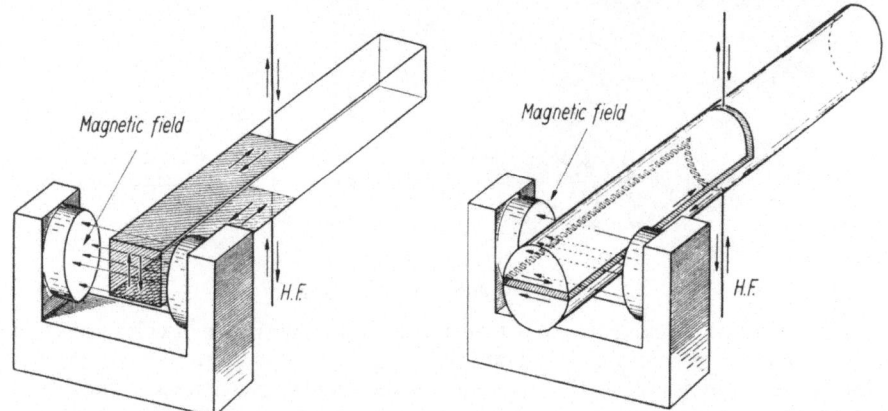

Fig. 25. Glass bar in extensional vibration (after BARONE and GIACOMINI).   Fig. 26. Glass cylinder in torsional vibration (after BARONE and GIACOMINI).

those parts shaded in the figure. The alternating driving current interacts with a constant magnetic field in which there is placed one end of the bar.

The vibromotive force of electrodynamic origin, proportional to the magnetic induction and to the intensity of the current has the same frequency as the latter. Choosing a suitable frequency, it is possible to excite, with considerable intensity, extensional vibrations corresponding to the various normal modes of the bar.

By this simple method any insulating solid can be put in extensional vibration at frequencies of the order of 10 kHz. The ultrasound radiation emitted from the ends of the bar is of low intensity but can easily be observed both in liquids and gases.

In order to produce torsional vibrations the bar may be arranged as shown in Fig. 26.

To apply this electrodynamic method to metallic rather than insulating materials it is sufficient to connect the conductors of the driving current at two diametrically opposite points at the end of the bar immersed in the magnetic field so that the current flows perpendicular to the field.

5. Spark generators[4]. In these transducers an electric discharge generates in the medium a practically spherical shock-wave.

[1] These transducers are constructed by de Havilland Propellers Ltd.

[2] E. ACKERMANN: Rev. Sci. Instrum. **22**, 649 (1951).

[3] A. BARONE and A. GIACOMINI: Acustica **4**, 182 (1954).

[4] F. FRÜNGEL: Optik **3**, 124 (1948).

Strictly speaking, they should not be classified as ultrasonic generators, but it is convenient to mention them here since the wave produced contains a very large number of components of different frequencies, which are distributed in a continuous spectrum extending beyond 100 kHz.

Fig. 27 shows such a device for operation in fluids. A condenser C is charged, through the resistance R, by a rectifier vacuum tube at a voltage of several thousand volts. At a given potential a discharge takes place between the two electrodes S_2 immersed in the liquid. The spark gap S_1 in series with S_2 serves to avoid electric losses due to the conductivity of the liquid during the charge of the condenser. The energy dissipated in the discharge consists largely of thermal energy, the remainder is mechanical and light energy. It is difficult to calculate the ratios between the various energies, but, even if a very low efficiency of conversion into acoustic energy is assumed, e.g. 1 percent, it is found

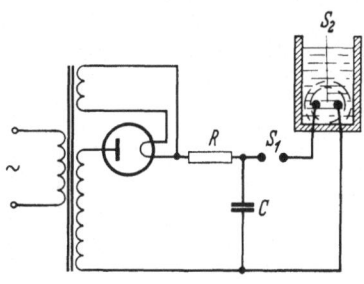

Fig. 27. Spark generator.

that the power of the shock wave can reach several hundred kW using condensers of approximately 10 μF and voltages of between 7 and 10 kV. This system has found promising applications in industry[1] and in underwater acoustics[2].

b) Transducers supplied by mechanical energy.

Under this heading we deal with ultrasound generators not requiring the use of an electric oscillator but a compressor, a pump or a motor, as the case may be. Most of them are fluid jet generators. The principal types are described in the following sections.

6. GALTON's whistle[3]. The principle is illustrated in Fig. 28. Air from a compressor is introduced at A and issues at high speed from the ring-shaped duct B, which is formed by arranging inside the exit nozzle and co-axial to it a cylindrical body C of slightly smaller diameter. The air jet encounters a resonance cavity V whose orifice D starts to

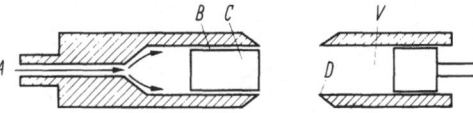

Fig. 28. Essentials of GALTON's whistle.

vibrate owing to the shock of the fluid jet producing an edge tone. The air contained in the cavity oscillates at $\frac{1}{4}$ of the wavelength, and the corresponding sound propagates in the surrounding air. The other components of the edge tone are of course also present, but the main part of the energy is transferred to the resonance component. Independently of the jet velocity, the best conditions for obtaining the greatest effect depend on the distance between nozzle and cavity, and for this reason the possible variation of this distance is provided in the models constructed; furthermore, the depth of the resonance cavity can be changed to obtain ultrasound of various frequencies.

In Fig. 29 a commercial type of GALTON's whistle is shown.

The frequency of the sound wave emitted also depends on the gas pressure[4], and it is

$$f = c \sqrt{\frac{1 + t/273}{4l + K}}, \qquad (6.1)$$

[1] E. ZÉDET: Bull. A.F.T.P. No. 115 (1956).
[2] H. H. DRUST and H. DRUBBA: Z. angew. Phys. **5**, 251 (1953).
[3] F. GALTON: Nature, Lond. **27**, 491 (1883).
[4] L. BERGMANN [1].

where t is the temperature in °C, c the speed of sound in mm/sec, l the length of the cavity in mm, and K a constant which depends on the pressure and is approximately 6 under good efficiency conditions. The practically obtainable frequency range lies between a few kHz and an upper limit of about 40 to 50 kHz.

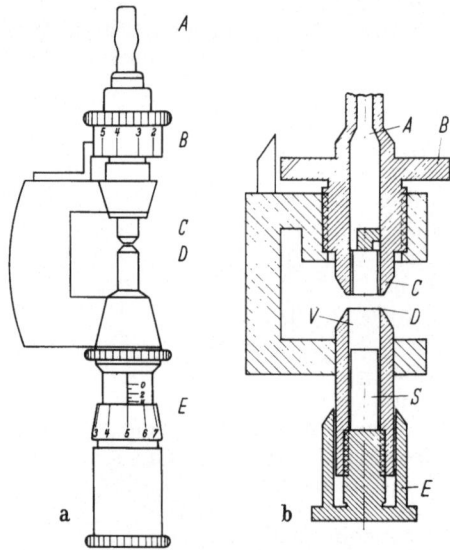

Fig. 29 a and b. GALTON's whistle.

7. HARTMANN's whistle[1]. The principle is illustrated in Fig 30. The configuration of the device resembles GALTON's whistle but it works in a different way. The compressed gas is introduced through the duct A and leaves the apparatus through the narrowing of conical profile, B. At the exit, the pressure of the gas decreases and its speed increases, following BERNOULLI's principle.

It can be shown that, if in A the excess of pressure exceeds a threshold value of approximately $9.2 \cdot 10^4$ Nt/m² ($\sim$0.9 atm), the speed v of the issuing jet reaches supersonic values ($v \geq c$). The behaviour of the pressure has been plotted in Fig. 31. It rapidly falls off along the way of the jet to below the external pressure P_0 until it reaches a minimum at the point x_1 where the velocities of jet and sound become equal. At this point the surrounding gas precipitates and gives rise to a shock-wave. At the following points the pressure rises again to a maximum, and then the cycle restarts as before.

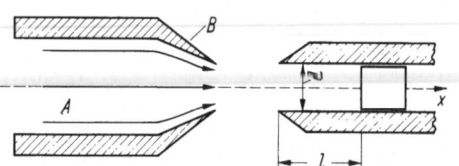

Fig. 30. Essentials of HARTMANN's whistle.

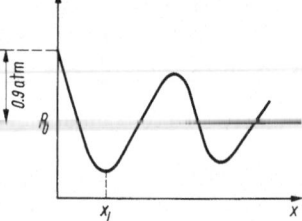

Fig. 31. Pressure variation in the course of the jet.

The regions where a shock wave is formed are unstable and if a cavity is placed there, the gas contained oscillates with a frequency

$$f = \frac{c}{4(l + 0.3\,d)},$$

where l and d are the length and the diameter of the resonant cavity. The maximum frequency obtainable is higher than that of GALTON's whistle. The efficiency of this type of whistle is about 5 percent, but considerable acoustic power can easily be reached. If pressure P_e of the fluid current issuing exceeds the threshold value, the output power[2] can be expressed by the formula

$$W = 3\,d^2\,(P_e - 0.9)^{\frac{1}{2}} \text{ Watts},$$

where d is the diameter of jet and cavity in mm, and P_e is expressed in atmospheres.

[1] T. HARTMANN: J. Sci. Instrum. **16**, 146 (1939).
[2] F.T. HUETER and R.H. BOLT [3].

For instance, at a frequency of 10 kHz, a pressure $P_e = 3.4$ atm and a diameter $d = 4.5$ mm an acoustic power of about 100 Watts can be obtained.

8. Vortex whistle[1]. The principle is outlined in Fig. 32. A fluid current penetrates tangentially into a cylindrical cavity where it causes a vortex. At the centre of the lateral wall of the cavity a short tube is inserted which opens outwards. The fluid continues to rotate while passing the tube, but, since the diameter of the vortex must here be smaller, the angular velocity increases according to the law of conservation of angular momentum. The sound emitted is attributed to the instability of the vortex at the tube exit. The tangential velocity of the fluid in the tube is very near to that of sound. The frequency of the sound wave generated can fairly well be computed from the equation:

$$f = a \frac{c}{\pi d} \sqrt{\frac{P_1 - P_2}{\varrho_2}}, \qquad (8.1)$$

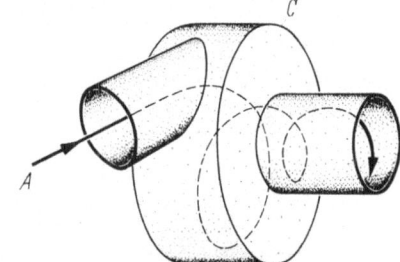

where d is the diameter of the tube, and P_1 and P_2 are the input and output pressure of the device. The numerical coefficient a has a value <1 and occurs in consequence of the decrease in angular velocity of the fluid as a result of its friction against the walls of the tube. Various devices of different

Fig. 32. Essentials of vortex whistle.

form, but working on the same principle, have been designed. Vortex whistles are equally suitable for gases and liquids. So far, models operating at a frequency of about 15 kHz have been constructed.

9. Jet-edge systems[2]. When a fluid current passes through an orifice or hits an edge, vortices may form. Their origin is due to forces of a viscous nature in the layer of the fluid jet in immediate contact with the solid wall.

These vortices cause, in the surrounding medium, local pressure variations which give rise to a sound wave. The frequency f of the sound is given by the number of vortices which pass a given point per unit time. If v is the velocity at which the vortices move along the jet and s is the distance between two successive vortices, we have:

$$f = \frac{v}{s}. \qquad (9.1)$$

The stability of the sound regime can be explained as follows. The section of a hole of diameter d, made in a wall of thickness $s \approx d$, is indicated in Fig. 33. A fluid current, from the left, crosses the hole; the layer of the fluid in contact with edge A starts a vortex motion which at the point B forms a closed vortex. If it is assumed that this vortex has been completed while rolling along the path s, its diameter will be s/π. At the point B the vortex produces in the free atmosphere a pressure pulse propagating in all directions. Arriving at A, it provokes in the layer in contact with the wall the starting of another vortex, etc. Under these conditions, the distance separating two successive vortices is approximately equal to s, so that the frequency of the sound generated by successive pulses must mainly lie around $f = v/s$, as indicated in Eq. (9.1).

The working of this generator, as has been seen, requires a feedback on the fluid current to control the rhythm of production of the vortices. The stability

[1] B. VONNEGUT: J. Acoust. Soc. Amer. **26**, 18 (1954).
[2] G. B. THURSTON and C. E. MARTIN: J. Acoust. Soc. Amer. **25**, 26 (1953).

of the generator can therefore be improved by enhancing this feedback by means of a sharp edge suitably arranged near the outlet aperture.

These systems are used to generate ultrasound both, in gases and liquids. One of the best-known apparatus based on this principle is that proposed by W. JANOVSKY and R. POHLMANN[1] for use in liquids. As indicated in Fig. 34, the liquid current issuing from a rectangular slit hits the edge of a thick blade capable of oscillating in a flexural mode. Its resonance frequency corresponds to a wavelength, since the blade is fixed at two nodal points. In the special case of Fig. 34 the fundamental resonance frequency becomes

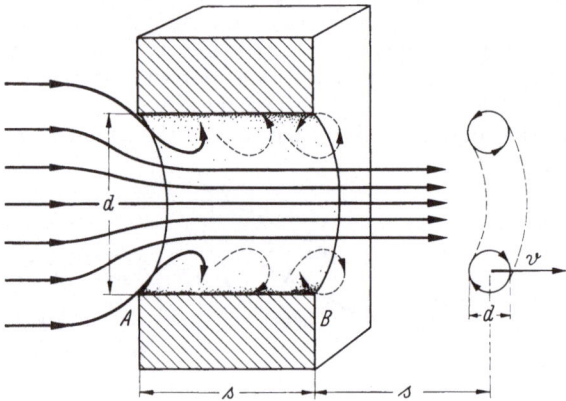

Fig. 33. Essentials of the "jet-edge" system.

$$f = 2.8 \frac{d}{l^2} c_l$$

where d and l are the thickness and length of the blade and c_l is the speed of sound in the blade. The blade can as well oscillate with one nodal plane only, situated at the centre where it has been firmly fixed.

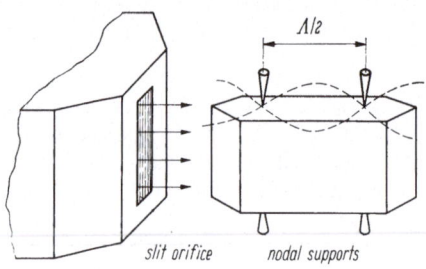

Fig. 34. Janovsky-Pohlmann whistle.

The frequency of these devices is of the order of 10 kHz and the intensity of the ultrasound generated in the liquids amounts to a few watt/cm² near the source.

10. Sirens. The design of a siren is based on a well-known principle in acoustics. A gaseous jet, e.g. of air, is rhythmically interrupted by a rotating device. The interruptions cause periodical pressure variations in the medium which propagate in the form sound waves. The classic instruments based on this principle have been variously modified recently so as to make them capable of generating ultrasound. The advantage of the siren compared with all other mechanical transducers for ultrasound radiation in gases, consists in the possibility of reaching acoustic power of several houndred watts at frequencies of the order of 10 kHz.

A typical example[2] of an ultrasonic siren is shown in Fig. 35. Through one or several ducts A, compressed air penetrates into a ring-shaped chamber from which it passes into the free atmosphere through a series of holes P, made along a circumference of the stator S. Between the deflecting plate and the stator a small air gap is left, in which there is inserted the tooth rim of the rotor R driven at high speed by an electric motor. The width of the teeth is slightly greater than the diameter of the holes, to ensure maximum variation of pressure. By means of a screw C it is possible to ensure the proper distance between rotor and stator.

[1] W. JANOVSKY and R. POHLMANN: Z. angew. Phys. **1**, 222 (1948).
[2] C.H. ALLEN and I. RUDNICK: J. Acoust. Soc. Amer. **19**, 857 (1947). — Cf. also p. 26, this volume.

Ordinary models with 100 holes and a 12000 r.p.m. motor, have a frequency of 20 kHz. If the gas pressure is increased, the acoustic power generated, increases too, however, the motor will have to work harder to interrupt the gas jet and a higher electric power will therefore be required.

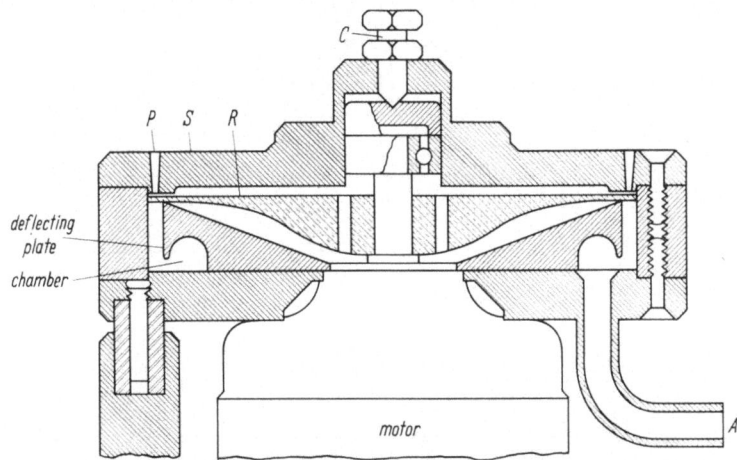

Fig. 35. Ultrasonic siren (after ALLEN and RUDNICK).

In some cases, especially when working at low ultrasonic frequencies, the siren can be furnished with an exponential horn which improves matching the impedance to the medium and increases its directivity.

11. Static sirens. Another type of apparatus must be mentioned whose principle has characteristics in common with almost all the fluid current generators so far discussed, including the siren, but which has no mobile part. Such a siren, designed by LEVAVASSEUR[1], works in air. Its principle of operation is shown in Fig. 36.

Compressed air, A, penetrates through a narrow ring-shaped duct B into a chamber consisting of two toroidal cavities 1 and 2 and passes into the atmosphere through C. Near the upper edge of 1 the jet is partly deflected towards the interior of the cavity, where it starts a vortical movement, and then returns to B. Here it interupts the air jet A, and, leaving at C, produces a change of pressure

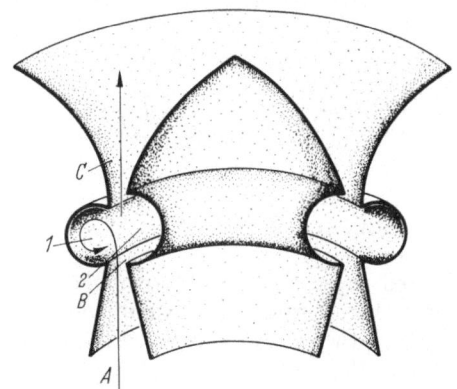

Fig. 36. Static siren (after LEVAVASSEUR).

in the outside atmosphere. The cycle repeats itself and the successive variations generate a sound whose frequency is controlled by the rhythm of the interruptions. Cavity 2 acts in push-pull with cavity 1 and contributes to the efficiency of the mechanism of the interruptions. At aperture C an exponential horn is connected for matching to the outside medium.

The frequency does not depend on the diameter of the toroidal cavity but is correlated only to the dimensions of the cross-section of the cavity. An experi-

---

[1] G. LEVAVASSEUR: Note techn. No. 37 Centre de Rech. Sci. Ind. et Mar., Marseilles.

Ordinary models with 100 holes and a 12000 r.p.m. motor, have a frequency of 20 kHz. If the gas pressure is increased, the acoustic power generated, increases too, however, the motor will have to work harder to interrupt the gas jet and a higher electric power will therefore be required.

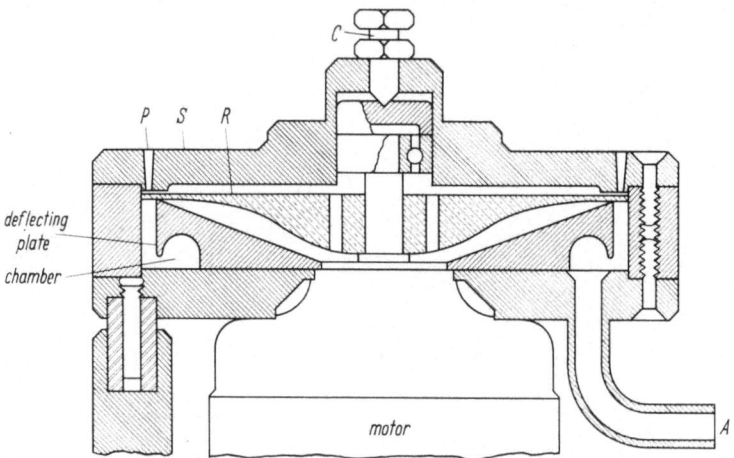

Fig. 35. Ultrasonic siren (after ALLEN and RUDNICK).

In some cases, especially when working at low ultrasonic frequencies, the siren can be furnished with an exponential horn which improves matching the impedance to the medium and increases its directivity.

11. Static sirens. Another type of apparatus must be mentioned whose principle has characteristics in common with almost all the fluid current generators so far discussed, including the siren, but which has no mobile part. Such a siren, designed by LEVAVASSEUR[1], works in air. Its principle of operation is shown in Fig. 36.

Compressed air, A, penetrates through a narrow ring-shaped duct B into a chamber consisting of two toroidal cavities 1 and 2 and passes into the atmosphere through C. Near the upper edge of 1 the jet is partly deflected towards the interior of the cavity, where it starts a vortical movement, and then returns to B. Here it interupts the air jet A, and, leaving at C, produces a change of pressure

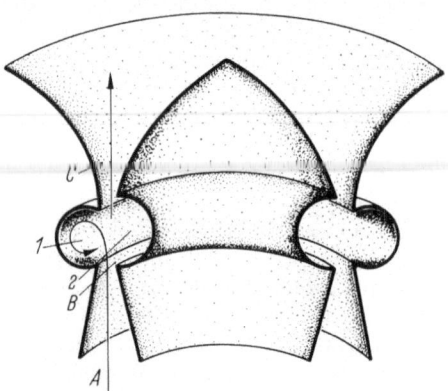

Fig. 36. Static siren (after LEVAVASSEUR).

in the outside atmosphere. The cycle repeats itself and the successive variations generate a sound whose frequency is controlled by the rhythm of the interruptions. Cavity 2 acts in push-pull with cavity 1 and contributes to the efficiency of the mechanism of the interruptions. At aperture C an exponential horn is connected for matching to the outside medium.

The frequency does not depend on the diameter of the toroidal cavity but is correlated only to the dimensions of the cross-section of the cavity. An experi-

---

[1] G. LEVAVASSEUR: Note techn. No. 37 Centre de Rech. Sci. Ind. et Mar., Marseilles.

parallel to the x axis. The other part of the bar is outside the cavity. By exciting the cavity at 1000 MHz frequency, an elastic wave is generated in the bar.

By means of the schlieren method it is possible to observe the ultrasonic beam which appears clearly defined in the direction of the axis. Using this method, it has been possible to measure the ultrasound absorption in quartz at this very high frequency.

As seen in Fig. 38, the direction of the sound beam depends on the orientation of the terminal face of the bar with respect to the x axis. A variation of 10° in this orientation produces a variation of 25° in the direction of the ultrasonic beam. Placing the free end of the bar

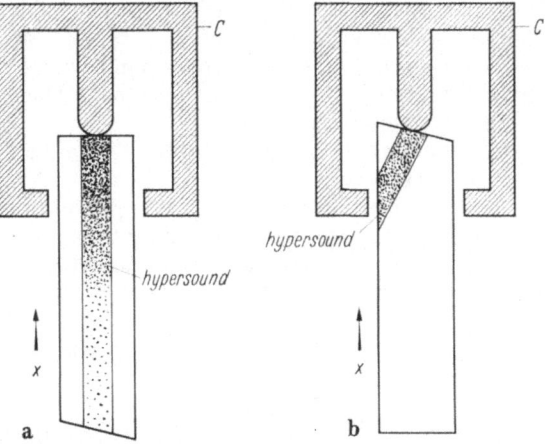

Fig. 38 a and b. Excitation of hypersound in a quartz bar (after BÖMMEL and DRANSFELD).

into another identical cavity, it is possible to detect, as electric signal, the wave propagated in the quartz. The excitation is pulsed in order to distinguish the

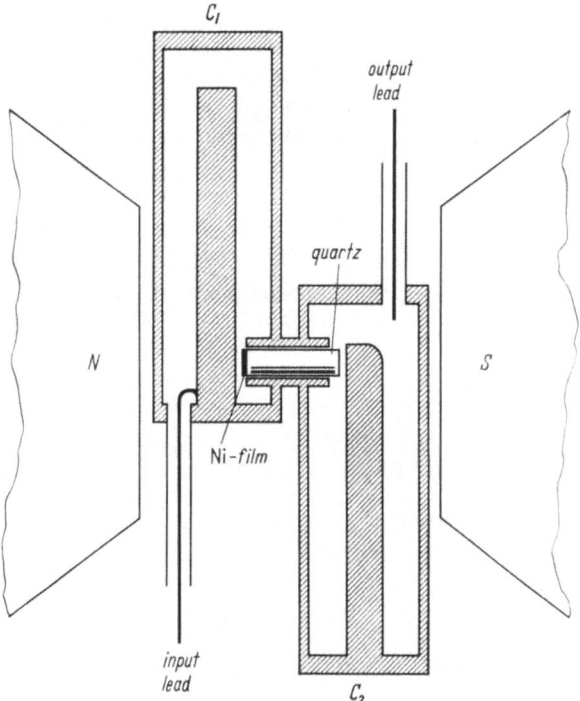

Fig. 39. Excitation of hypersound in a nickel film (after BÖMMEL and DRANSFELD).

signal transmitted electromagnetically from the acoustic signal which takes a certain time to travel along the bar.

The highest frequency achieved in these experiments has been 2500 MHz. It seems that such hypersound waves are excited at the surface of the crystal. Here the piezoelectric stress gives rise to a displacement which propagates in the form of a wave inside the crystal. BÖMMEL and DRANSFELD[1] have also observed the emission of transversal hpyersound waves from a thin nickel film deposited by evaporation in vacuo on the terminal surface of a small quartz bar and excited at ferromagnetic resonance.

The quartz bar was cut from the crystal in a direction suitable for the generation of shear waves.

As shown in Fig. 39, the bar was placed in a d.c. magnetic field between two cavities C_1 and C_2. The end bearing the nickel film, in cavity C_1, was exposed to a high frequency magnetic field parallel to the nickel surface. The other end was in the electric field of cavity C_2. The two cavities were tuned to the same frequency of 1000 MHz. Cavity C_1 was fed at resonance frequency from a pulsed generator, and cavity C_2 was coupled to a receiver.

The curve in Fig. 40 represents the power of the hypersonic pulses generated by the nickel vs. the intensity of the d.c. magnetic field.

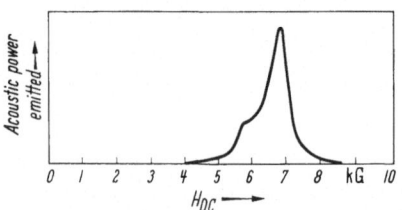

Fig. 40. Hypersonic emission depending on the intensity of the magnetic field.

The production mechanism of the hypersonic pulse from the nickel film would be as follows: A uniform precession of the magnetization in the film produces by magnetostriction an ellipsoid of stress which, in its turn, performs a precession motion around the d.c. magnetic field. If the direction of the field is normal to the film, at magnetic resonance, the surface will be subjected to circularly polarized transversal vibrations which are then communicated to the quartz bar.

Further particulars on the production of elastic waves of this kind can be gathered from the articles by KAGANOV and ZULCERNIK[2], and KITTEL[3].

d) The radiated field.

14. General remarks. The medium in contact with the source represents a load to which the source transfers vibration energy. This load is determined by the impedance offered by the medium to the radiating surface. If we consider the simplest type of source, viz. a monopole source (which like a pulsating sphere, radiates spherical waves into the medium), the specific radiation impedance is:

$$Z_r = \frac{p}{u} = \varrho\,c\left[\frac{(k\,a)^2}{1 + (k\,a)^2} + j\,\frac{1}{1 + k\,a}\right], \tag{14.1}$$

where $k = 2\pi/\Lambda$ is the wave number and a the radius of the source.

Z_r is the therefore a complex quantity whose real part

$$R = \frac{\varrho\,c\,k^2\,a^2}{1 + k^2 a^2} \tag{14.2a}$$

represents the resistive component and the coefficient of the imaginary part

$$X = \frac{\varrho\,c\,k\,a}{1 + k^2 a^2} \tag{14.2b}$$

[1] H. E. BÖMMEL and K. DRANSFELD: Phys. Rev. Letters **3**, 83 (1959).
[2] N. I. KAGANOV and Z. N. ZULCERNIK: Ž. eksp. teor. Fiz. **36**, 224 (1959).
[3] C. KITTEL: Phys. Rev. **110**, 836 (1958).

the reactive component. The dimensionless quantity $ka=2\pi a/\Lambda$ is a measure of the ratio between the linear dimensions of the source and the wavelength of the sound emitted.

If the wavelength is very large compared to the linear dimensions of the source ($ka \ll 1$), the acoustic resistance and reactance become, in a good approximation:

$$R=\varrho c\,(k\,a)^2,\qquad X=\varrho\,c\,k\,a. \tag{14.3a}$$

For a wavelength equal to the circumference of the source ($ka=1$) the two components have the same modulus

$$R=X=\varrho\,c/2. \tag{14.3b}$$

In the case of waves very short compared to the source ($ka \gg 1$) the moduli of the two components tend towards the values:

$$R=\varrho\,c,\qquad X=0. \tag{14.3c}$$

In that case, the specific radiation impedance is therefore reduced exclusively to the real component and the source operates on a purely resistive load equal to the characteristic impedance of the medium. This can be regarded as the case of a source radiating plane waves in an unlimited homogeneous medium.

This condition is well satisfied for most ultrasonic sources usually consisting of a plane face (of a diameter much larger than the wavelength) in longitudinal oscillation.

The ideal case, of course, is that of unbounded plane surface whose irradiated field is formed by perfectly plane waves. In such a field, the distribution of acoustic energy is uniform. In practice the limited extension of the source can originate remarkable effects of diffraction in the configuration of the wave fronts bearing also on the distribution of the energy in the ultrasonic field.

According to the HUYGENS' principle we can regard the radiating plane face of the source as consisting of an infinite number of monopole sources of infinitely small radii, each emitting an elementary spherical wave in phase with, and of equal amplitude as, all the other elementary sources. At every point of the halfspace in contact with the radiating face interference between the various elementary waves will occur, but the greater part of radiated energy will be contained in a cone with an aperture will depending, for each prefixed wavelength, on the diameter of the source. If the plane source is unbounded, the elementary waves interfere in such a way as to form a perfectly plane wave front.

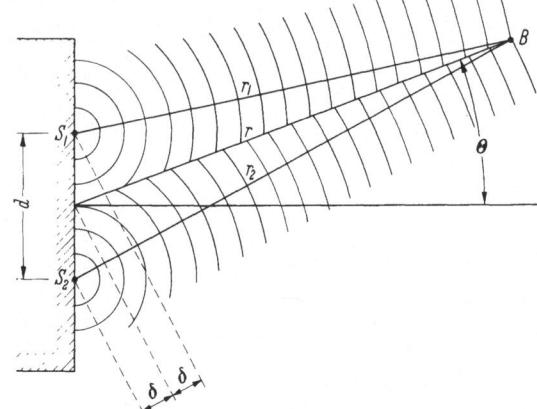

Fig. 41. Emission from two elementary sources.

15. Directivity. Given a radiating plate of limited surface, let us choose from among the infinite number of elementary sources which compose it, two, S_1, and S_2, separated from each other by a distance d and equidistant from the center of the plate (Fig. 41)[1]. We can easily calculate the acoustic pressure at any point B if we know the contribution from the two sources.

---

[1] T. F. HUETER and R. H. BOLT [3].

Since the acoustic pressure, produced at distance r from any single monopole source of radius a, is

$$p = \frac{1}{r} P_a \, a \, e^{j(\omega t - kr)},$$

where P_a is the variable pressure amplitude at the source, the acoustic pressure produced by S_1 and S_2 at the point B will be:

$$p = \frac{1}{r} P_a \, a \, e^{j(\omega t - kr)} \left(e^{jk\delta} + e^{-jk\delta} \right) = \frac{2}{r} P_a \, a \, e^{j(\omega t - kr)} \cos \left(\frac{k \, d}{2} \sin \Theta \right),$$

where δ and Θ are as indicated in Fig. 41. This equation is a good approximation if B is very far from the sources $(r \gg d)$. In this case we can assume:

$$r_1 \approx r_2 \approx r$$

and

$$\delta \approx r - r_1 \approx r_2 - r \approx \frac{d}{2} \sin \Theta.$$

This does not appreciably alter the value of the individual pressures, but the cosine takes into account the phase relation. The factor

$$\cos \left(\frac{k \, d}{2} \sin \Theta \right) = \varDelta$$

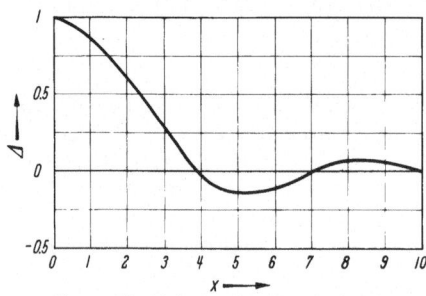

Fig. 42. Directivity function for a circular piston.

is the "directivity function" for this particular case. By the same procedure, if the number of sources is increased, other directivity functions[1] are found which enable us to draw a polar diagram of the angular distribution of the pressure or the energy, or the intensity of the beam before the ultrasonic source.

When the number of elementary sources to be taken account of becomes very large, as in a normal radiating plate, a solution will easily be obtained by calculating an integral.

The directivity functions for the most ordinary sources are as follows:

For the plane circular plate (circular piston) of radius a:

$$\varLambda = 2 J_1 \left(\frac{2\pi \, a}{\varLambda} \sin \Theta \right) \Big/ \frac{2\pi \, a}{\varLambda} \sin \Theta$$

where J_1 is the first order Bessel function.

For the plane rectangular plate with sides l_a and l_b (rectangular piston):

$$\varDelta = \frac{\sin \left(\dfrac{\pi \, l_a}{\varLambda} \sin \Theta \right) \sin \left(\dfrac{\pi \, l_b}{\varLambda} \sin \varphi \right)}{\dfrac{\pi^2}{\varLambda^2} \, l_a \, l_b \sin \Theta \sin \varphi}$$

where Θ and φ are the angular coordinates on the two planes normal to the plate, parallel to sides l_a and l_b.

For the plane circular ring of radius a:

$$\varDelta = J_0 \left(\frac{2\pi \, a}{\varLambda} \sin \Theta \right)$$

where J_0 is the zero order Bessel function.

As an example, Fig. 42 shows the directivity function for a circular piston. On the axis of the abscissae the parameter $x = k \, a \sin \Theta$ is indicated. The function becomes zero at the points $x = 3.83, 702, 10.15$, etc. In the directions which, with

[1] E.M. J. HERREY: J. Acoust. Soc. Amer. 25, 154 (1953).

the axis of the source, form the angles corresponding to these values, radiation vanishes. A typical polar diagram of the directivity of a circular source drawn by means of the directivity function, is given in Fig. 43. It presents a central lobe in which the greater part of the energy is distributed, and several lateral lobes separated from the central one by the zeros in the directivity function. The semiangle of aperture of the central lobe is that for which there is the first zero in the directivity function, i.e. for

$x = 3.83$ or $\sin \Theta = 1.22 \dfrac{\Lambda}{d}$ which is FRAUNHOFER'S well-known formula.

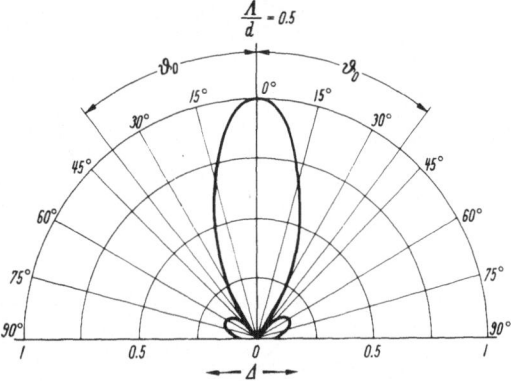

Fig. 43. Directivity diagram for circular piston.

16. Energy distribution. Let us now consider the distribution of the energy, or ultrasound intensity, along the central axis of the radiating plate. At distances from the source large compared to the diameter, as has been assumed above, all the elementary waves coming from the surface of the source arrive at every point of the axis in phase, with an approximation which is the better the smaller is the ratio d/Λ. Close to the source, however, this does not apply, but then the distances of elementary sources from the same point of the axis may be very different from

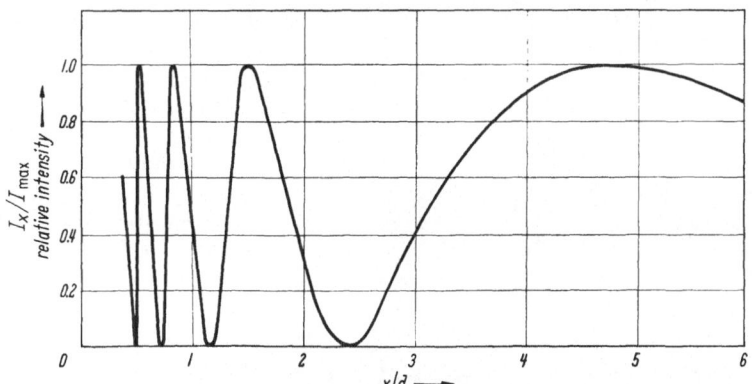

Fig. 44. Axial distribution of intensity for circular piston.

one another. Thus, at such a point the individual waves may arrive with such phase relations as to cancel out their global effect or to strengthen it.

Close to the source, therefore, the radiated energy shows an interference pattern with a succession of maxima and minima.

The distribution, along the x-axis, of the relative intensity for a circular plate of diameter d becomes[1]

$$I_x/I_{\max} = \sin^2 \left\{ \frac{k}{2} \left[\left(\frac{d^2}{4} + x^2 \right)^{\frac{1}{2}} - x \right] \right\}. \tag{16.1}$$

In Fig. 44 the distribution of the relative intensity on the central axis of the plate is shown as depending on the distance x/d if $d/\Lambda = 19$.

---

[1] H. BACKHAUS and F. TRENDELENBURG: Z. techn. Phys. **7**, 630 (1926).

Beyond the farthest maximum, at $x/d = 4.5$, all elementary waves arrive on the axis practically in phase and the intensity decreases with distance following the inverse square law because of the divergence of the beam as determined by FRAUNHOFER's formula.

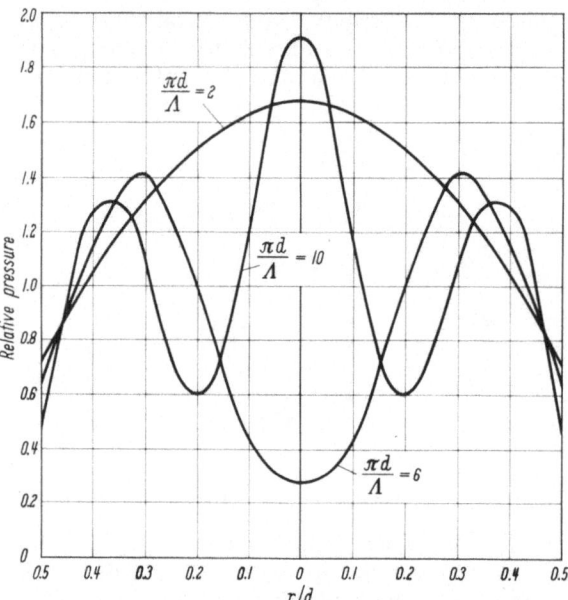

Fig. 45. Radial distribution of acoustic pressure in FRESNEL's region (after HUETER and BOLT).

This part of the field, far from the source, is called FRAUNHOFER's region. The part of the field nearest to the source is characterized by the presence of maxima and minima, and is called FRESNEL's region.

For the same reasons which determine the variation of axial intensity, the energy distribution in FRESNEL's region exhibits maxima and minima also along the diameter of the beam.

Fig. 45 shows, in arbitrary units, the values of the acoustic pressure along a diameter as depending on the relative distance r/d from the axis of the beam, calculated for three different values of the parameter $\pi d/\Lambda$. As the values of this parameter decrease, the FRAUNHOFER's region draws nearer to the source. When $d/\Lambda \leqq 1$ the lateral energy distribution shows no more than one single maximum.

a   b

Fig. 46a and b. Ultrasonic field near a quartz plate (after OSTERHAMMEL).
a $d/\Lambda = 8.31$ b $d/\Lambda = 10.38$.

The photographs Fig. 46[1], taken by the schlieren method (see Sect. 30 below), give two pictures of FRESNEL's region in the ultrasonic field radiated in xylol by rectangular quartzes of width 8.31Λ and 10.38Λ. In the complex interference pattern the maxima and minima of the axial and lateral distribution of the ultrasonic intensity are clearly to be seen.

17. Reflection and refraction. When a sound wave hits an obstacle of dimensions large in comparison with the wavelength, it is in part reflected by, and in part transmitted through the obstacle.

Let us consider a plane ultrasonic wave propagating in a first medium of specific acoustic resistance $\varrho_1 c_1$ which hits at an angle of incidence φ_i the plane

[1] H. OSTERHAMMEL: Akust. Z. **6**, 73 (1941).

separation surface of a second medium of specific acoustic resistance $\varrho_2 c_2$. In the following let us assume that there are only longitudinal waves. In reality, the incidence will give rise to changes in wave-type, as we shall see later (Sect. 19).

The directions of the reflected and transmitted wave are governed by the laws of optical reflection and refraction. Thus, we have:

$$\varphi_i = \varphi_r ; \qquad \frac{\sin \varphi_i}{\sin \varphi_t} = \frac{c_1}{c_2}$$

where φ_r is the angle of reflection and φ_t is the angle of refraction defining the direction of the transmitted wave.

The specific acoustic resistances determine the ratio of the reflected and transmitted energies.

If we postulate, at the separation surface, continuity of acoustic pressure p and of the normal component of particle velocity u, we can write at first:

$$p_i + p_r = p_t$$

where p_t is the pressure of the wave transmitted in the second medium. Since $p = \pm \varrho c u$, it follows:

$$\varrho_1 c_1 u_i - \varrho_1 c_1 u_r = \varrho_2 c_2 u_t, \tag{17.1}$$

the minus sign indicating the fact that the reflected wave has inverted direction. Further we have:

$$u_i \cos \varphi_i + u_r \cos \varphi_r = u_t \cos \varphi_t . \tag{17.2}$$

From (17.1) and (17.2) we get:

$$\frac{u_r}{u_i} = - \frac{\varrho_2 c_2 \cos \varphi_i - \varrho_1 c_1 \cos \varphi_t}{\varrho_2 c_2 \cos \varphi_r + \varrho_1 c_1 \cos \varphi_t} \tag{17.3}$$

as well as:

$$\frac{u_t}{u_i} = \frac{2 \varrho_1 c_1 \cos \varphi_i}{\varrho_2 c_2 \cos \varphi_r + \varrho_1 c_1 \cos \varphi_t} . \tag{17.4}$$

If the incidence is normal ($\varphi_i = 0$, $\varphi_t = 0$) we have:

$$r \equiv \frac{u_r}{u_i} = \frac{\varrho_1 c_1 - \varrho_2 c_2}{\varrho_1 c_1 + \varrho_2 c_2} , \qquad t \equiv \frac{u_t}{u_i} = \frac{2 \varrho c_1}{\varrho_1 c_1 + \varrho_2 c_2} . \tag{17.5}$$

If we consider the intensities of the incident, reflected and transmitted waves, the relation

$$I_i = I_r + I_t$$

applies when there are no losses in the medium and, since in a plane wave $I = \frac{1}{2} \varrho c u^2$, we can write:

$$\varrho_1 c_1 u_i^2 = \varrho_1 c_1 u_r^2 + \varrho_2 c_2 u_i^2$$

or, introducing the abbreviations of Eq. (17.5)

$$r^2 + \frac{\varrho_2 c_2}{\varrho_1 c_1} t^2 = 1 .$$

r^2 is simply the ratio R between the intensities of reflected and incident wave:

$$R = \frac{I_r}{I_i} = \left(\frac{\varrho_1 c_1 - \varrho_2 c_2}{\varrho_1 c_1 + \varrho_2 c_2} \right)^2 . \tag{17.6}$$

$\frac{\varrho_2 c_2}{\varrho_1 c_1} t^2$ is, therefore, the ratio D of the intensities of the transmitted and the incident wave:

$$D = \frac{I_t}{I_i} = \frac{\varrho_2 c_2}{\varrho_1 c_1} \left(\frac{2 \varrho_1 c_1}{\varrho_1 c_1 + \varrho_2 c_2} \right)^2 . \tag{17.7}$$

R and D are called respectively "reflection coefficient" and "transmission coefficient" for normal incidence.

R and D are obviously correlated by the equation:

$$D = 1 - R. \tag{17.8}$$

Fig. 47 shows an experiment carried out to demonstrate the above phenomena[1]. The ultrasonic beam, coming (in the figure) from above, propagates in two different liquids and is made visible using the schlieren method (see Sect. 30).

In Fig. 47a the ultrasonic beam passes from water $(c_1 = 1497$ m/sec; $\varrho_1 c_1 = 148.6 \cdot 10^4$ kg/m² sec) to carbon tetrachloride $(c_2 = 938$ m/sec; $\varrho_2 c_2 = 150 \cdot 10^4$ kg

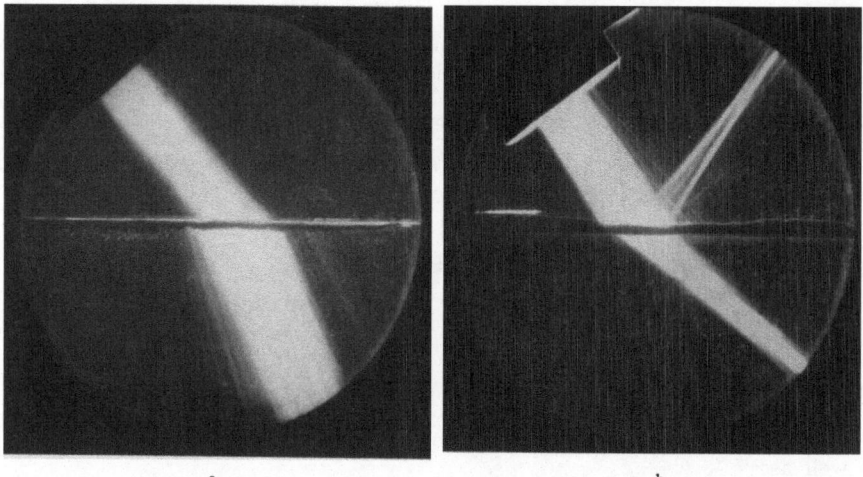

a   b

Fig. 47 a and b. Refraction and reflection of ultrasound on the interface of two liquids (after Giacomini). a beam traveling from water to carbon tetrachloride; b beam traveling from petroleum-ether to water.

per m² sec). Since $\varrho_1 c_1 \approx \varrho_2 c_2$, the whole energy passes into the second medium, but since $c_1 > c_2$ the refracted beam is deflected towards the normal to the separating surface.

In Fig. 47b the beam passes from petroleum $(c_1 \approx 1330$ m/sec; $\varrho_1 c_1 \approx 93 \cdot 10^4$ kg per m² sec) into the water (the constants of which, given above, are now marked with index 2).

Since $\varrho_1 c_1 \neq \varrho_2 c_2$ the beam is in part reflected and in part transmitted, but since $c_1 < c_2$, the refracted beam is deflected away from the normal.

18. Transmission through plates. If the second medium is bounded by two plane parallel surfaces, such important problems as that of the passage of ultrasound through a plane solid plate immersed in a liquid, can be resolved. The energy is several times reflected at the separating surfaces.

The fractions of the intensity, or energy, reflected and transmitted through the plate, according to the theory of Rayleigh[1], are:

$$\frac{I_r}{I_i} = \frac{\left(m - \dfrac{1}{m}\right)^2}{4 \cot^2 2\pi \dfrac{d}{\varLambda} + \left(m + \dfrac{1}{m}\right)^2}; \tag{18.1}$$

$$\frac{I_t}{I_i} = 1 - \frac{I_r}{I_i}, \tag{18.2}$$

---

[1] A. Giacomini: Nuovo Cim. **6**, 39 (1949).

where d is the thickness of the plate and m is the ratio between the specific acoustic resistance of the plate and of the liquid in which it is immerged:

$$m = \frac{\varrho_2\, c_2}{\varrho_1\, c_1}.$$

The transparency of the plate (I_t/I_i) is a periodical function of the ratio d/Λ of the plate thickness to the wavelength inside the plate. Fig. 48 shows the behaviour of this function for an aluminium plate and a plexiglas plate in water. The transparency is perfect (100%) if the plate thick-

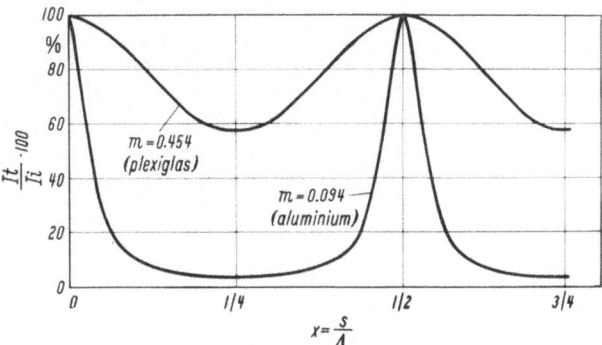

Fig. 48. Ultrasonic transparency of a plate (after GIACOMINI).

ness is an even multiple of $\Lambda/4$ (resonance) and a minimum when the thickness is an odd multiple of $\Lambda/4$ (antiresonance). Fig. 49 illustrates an experimental demonstration[1].

The minimum transparency value depends on m. The nearer the specific acoustic resistance of the liquid is to that of the plate, the less pronounced are the

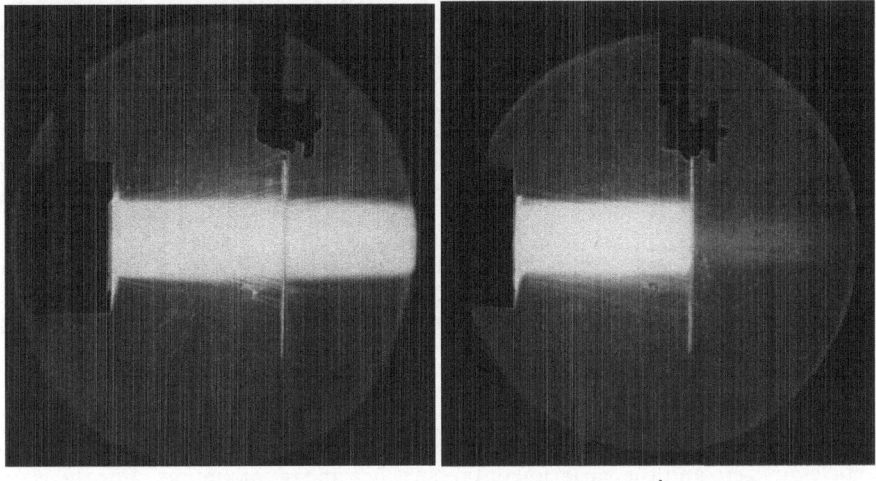

Fig. 49a and b. Ultrasonic beam through a plate (after GIACOMINI). a plate thickness $= \Lambda/2$; b plate thickness $= \Lambda/4$.

resonances. In the case of $m=1$, transparency is complete whatever the thickness that was penetrated.

19. Transformations of wave type. Let us consider a plane wave of a given type, e.g. a longitudinal wave obliquely incident on the plane interface of two different media.

If the normal and the tangential components of the stress produced on the interface by the wave are not zero in both media, then longitudinal (L) and shear (S) waves will be reflected in the first medium and transmitted to the second, with an energy repartition corresponding to the ratio of the specific acoustic resistances.

---

[1] See footnote 1, p. 108.

The directions of the various waves are determined by the following equations:

$$\varphi_{L_i} = \varphi_{L_r}, \qquad \frac{\sin \varphi_{L_i}}{c_{L_1}} = \frac{\sin \varphi_{S_r}}{c_{S_1}} = \frac{\sin \varphi_{L_t}}{c_{L_2}} = \frac{\sin \varphi_{S_t}}{c_{S_2}} \qquad (19.1)$$

where c_{L_1}, c_{S_1} are the velocity of longitudinal and transversal waves, in medium 1: c_{L_2}, c_{S_2} are the corresponding velocities in medium 2.

Fig. 50 illustrates the case where $c_{L_2} > c_{L_1}$ and takes account of the fact that in each medium $c_L > c_S$. The angles are clearly shown in the figure.

When medium 1 is a liquid and medium 2 a solid, there is no reflected shear wave S_1 since in the liquid the tangential component of the stress is generally zero.

When the angle of incidence φ_{L_i} has such a value $\left(\varphi_{L_i} = \text{arc}\sin \frac{c_{L_1}}{c_{L_2}}\right)$ that the refraction angle for the longitudinal wave is $\varphi_{L_t} = \pi/2$, only the shear wave S_2 propagates in the solid. For an angle of incidence $\varphi_{L_i} \geqq \text{arc}\sin \frac{c_{L_1}}{c_{S_2}}$, no wave can penetrate the solid and the whole energy is reflected into the first medium.

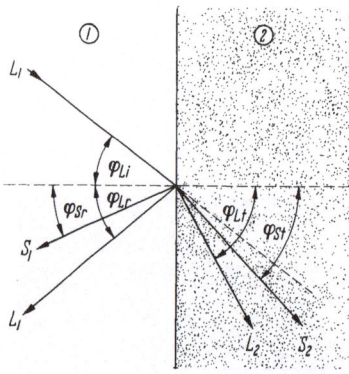

Fig. 50. Direction of reflected and transmitted L and S waves.

As in the optical case, total reflection can occur only if $c_{L_2} > c_{L_1}$. For normal incidence, the stress components at the interface are reduced to the normal component in the case of longitudinal waves, and to the tangential one in the case of shear waves. Thus there is no change of wave type.

These concepts, in combination with what has been said about the distribution of energy in different media, may be applied to studying the transmission of ultrasonic waves through plates, when oblique incidence occurs.

20. Standing waves. An ultrasonic field of special interest is formed when plane waves with incidence normal to the interface are reflected and interfere with the incident waves following. If the two media have very different characteristic impedances, $\varrho_1 c_1$ and $\varrho_2 c_2$, perfect reflection will occur $(R = 1)$. However, we distinguish three cases[1].

First case. The wave propagates in a gas and is reflected by a liquid or solid $(\varrho_1 c_1 \ll \varrho_2 c_2)$. On the interface $(x = 0)$ the displacement ξ and the velocity of the particles are zero and the variation of pressure becomes a maximum. For the wave resulting by interference in the first medium we can write the displacement equation:

$$\xi = a_1 \cos(\omega t - k x) - a_2 \cos(\omega t + k x)$$

where a_1 and a_2 are the vibration amplitudes of the incident wave and the reflected wave, respectively. If reflection is perfect and there is no energy dissipation in the medium, we have $a_1 \approx a_2 \approx a$, and hence approximately:

$$\xi = 2a \sin \omega t \sin k x.$$

This is the equation of a standing wave with amplitude $2a$ at the points $x = \Lambda/4$, $3\Lambda/4$, $5\Lambda/4$, ... and with zero amplitude at $x = 0$, $\Lambda/2$, Λ, ...

---

[1] A.B. Wood [12].

Second case. The wave propagates in a liquid or solid and is reflected by a gas $(\varrho_1 c_1 \gg \varrho_2 c_2)$. On the interface, displacement as well as velocity has a maximum. For the resulting wave we find:

$$\xi = a_1 \cos (\omega t - k x) + a_2 \cos (\omega t + k x)$$

and, using the same assumptions as before:

$$\xi = 2a \cos \omega t \cos k x.$$

In this standing wave, maximum amplitude $2a$ will occur at the interface distances $x = 0, \Lambda/2, \Lambda \ldots$ and zero amplitude at the distances $x = \Lambda/4, 3\Lambda/4, 5\Lambda/4, \ldots$.

Third case. If the characteristic acoustic impedances of the two media are not very different, part of the incident energy is transmitted into medium 2 and the hypothesis of perfect reflection does not apply. Thus, we have $a_1 > a_2$. This, for example, is the case of transmission at normal incidence from a liquid to a solid $(\varrho_2 c_2 > \varrho_1 c_1)$ or vice versa $(\varrho_1 c_1 > \varrho_2 c_2)$. The displacement equation becomes:

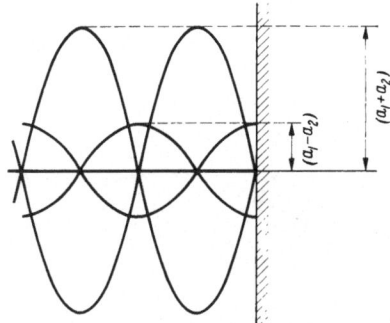

Fig. 51. Standing waves produced by a partially reflecting interface.

$$\xi = (a_1 + a_2) \sin \omega t \sin k x +$$
$$+ (a_1 - a_2) \cos \omega t \cos k x.$$

The system can now be regarded as composed of two standing waves, one of amplitude $(a_1 + a_2)$ as if the wave were reflected by a rigid interface, and the other of amplitude $(a_1 - a_2)$ as if the wave were reflected by a compliant interface. In either case, however, the interface absorbs energy.

The nodes and anti-nodes of the two standing waves do not coincide but are displaced respectively by $\Lambda/4$; moreover they cancel each other at two different moments separated by a time interval of $T/4$ $(T = 2\pi/\omega$, period).

In the resulting system, there are no amplitude nodes (zero amplitude), but in the corresponding positions there will be a minimum amplitude $a_1 - a_2$ (Fig. 51). The ratio between maximum and minimum amplitudes is called "the standing wave ratio" (S.W.R.):

$$\text{S.W. R.} = \frac{a_1 + a_2}{a_1 - a_2}. \tag{20.1}$$

It can be correlated with the coefficient of reflection, R, by the simple equation:

$$R = \left(\frac{a_2}{a_1} \right)^2 = \left(\frac{\text{S.W.R.} - 1}{\text{S.W.R.} + 1} \right)^2. \tag{20.2}$$

21. Modifications of field configuration. In order to modify the configuration of an ultrasonic field advantage is taken of the above-described characteristics of propagation in a variously composed medium.

For sound waves, in perfect analogy to optics, reflectors or lenses of a suitable shape and material can be constructed which are able to produce a particular distribution of sound energy. The corresponding equations of optics are applicable to acoustic systems. However, owing to the enormous difference between ultrasonic and optical wavelengths, diffraction phenomena will be much more evident. Moreover sound waves, otherwise than light waves, can, under the conditions explained, transform from longitudinal into shear waves and vice versa. Applications concern chiefly the concentration of sound energy and the formation of ultrasonic images.

Both lenses and concave reflectors concentrate ultrasonic energy on a small volume, transforming a beam of plane waves into spherical waves converging in the focus.

If the incident beam has a circular cross section of radius R, the distribution of sound intensity in the focal plane is given by[1]

$$I = A \pi^2 R^4 \left[\frac{2 J_1(\chi)}{\chi} \right]^2 \qquad (21.1)$$

where J_1 is BESSEL's function of first order of the argument

$$\chi = \frac{2 \pi r R}{f \Lambda},$$

where r is the radius of the diffraction picture and f is the focal distance. The factor A depends on the total acoustic power W distributed on the focal plane:

$$A = \frac{W}{(f \Lambda)^2}. \qquad (21.2)$$

The maximum of function (21.1) occurs when $r \to 0$, i.e. exactly in the focus. The radius r_0 of the central focal spot is

$$r_0 = 0.610 \Lambda f / R \qquad (21.3)$$

which is the value of r as determined from the first zero of the BESSEL's function in Eq. (21.1).

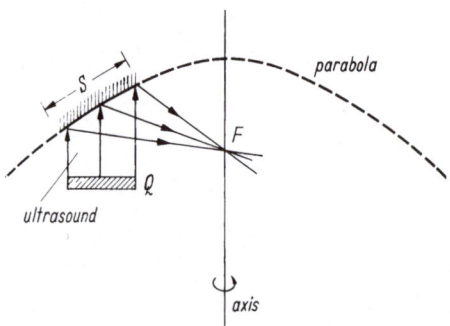

Fig. 52. Parabolic reflector.

$\alpha)$ *Reflectors.* The material employed in constructing a reflector must have a specific acoustic resistance very different from that of the fluid in which it is placed in order to avoid appreciable losses of sound energy on the reflecting surface. If the liquid is water, the most suitable materials are nickel and steel. The reflected energy is 89% for nickel and 88% for steel of the incident energy.

Fig. 52 shows the simplest apparatus by which to obtain a focalized beam. The transducer is a piezoelectric plate of radius R, very large compared to Λ, so that the condition for emission of plane waves can be regarded as satisfied. The reflector is a paraboloid sector S whose axis is parallel to the incident ultrasonic beam. The ultrasonic energy, therefore, converges in the focus F.

Various other solutions are possible to concentrate the energy in a region situated on the central axis of the transducer rather than sideways. A typical example[2] is shown in Fig. 53.

The focalizing device consists of two co-axial reflectors. The inner reflector is a cone of 90° vertex angle which transforms the axially incident plane waves into cylindrical ones. These, in turn, hit the second reflector which surrounds the cone. The reflecting surface of the latter is obtained by the revolution of a parabolic segment round an axis η perpendicular to the axis of the generating parabola and passing through its focus F. In this way, the cylindrical waves are converted into spherical ones and made to converge in F. The geometry of the apparatus

[1] F. E. FOX and V. GRIFFIN: J. Acoust. Soc. Amer. **21**, 352 (1949).
[2] A. BARONE: Acustica **2**, 221 (1952).

must not let the beam of emergent waves be impeded by the central conical reflector, a condition which is satisfied, if the generating parabola passes through the point with coordinates $y=H$ and $x=f-\dfrac{HR}{h}$, the symbols being those indicated in the figure.

These values, substituted in the parabola equation $y^2=2px$, determine the parameter p, as depending on the radius R of the incident ultrasonic beam and on the focal distance f of the concentrator.

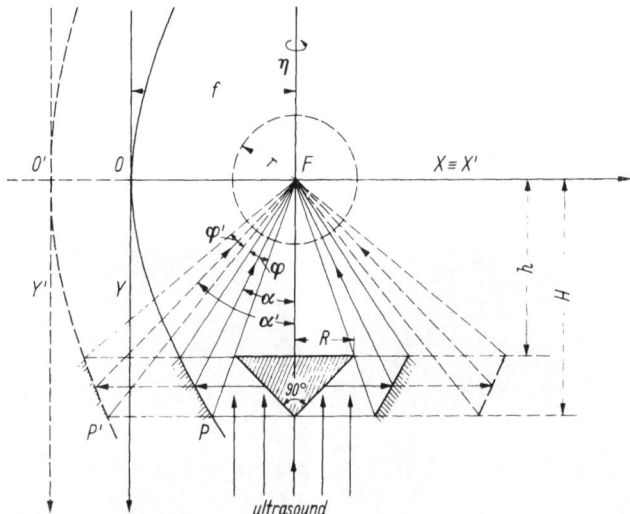

Fig. 53. Concentrator (after BARONE).

An observation on this concentrator relates to the gradient of intensity towards the focus. As a first approximation and at $\varphi<20°$ (which is effectively true in practice), we can write:

$$\frac{dI}{dr}=-\frac{(H+h)\,W}{\pi R \sin 2\alpha}\frac{1}{r^3} \tag{21.4}$$

where W is the total power radiated by the source.

As we see the gradient of the intensity diminishes as α increases up to a value $\alpha'=45°$.

Under these conditions, higher values of I or of dI/dr are attained at much smaller distances from the focus. Thus an increase of the paraboloid diameter causes a decrease in the specific power near the focus. Such a possibility may be useful in special applications.

A two-mirror concentrator of ultrasonic waves has also been developed by ROSENBERG[1] who used an inner convex mirror producing a homocentric diverging beam, surrounded by a second mirror in the shape of an ellipsoid of revolution.

β) Lenses. In order to have the best transmission through an acoustic lens, the material of which the lens is made must have a characteristic impedance ϱc very near to that of the fluid in which it is immersed. In this way almost the whole energy emitted by the transducer will reach the focal region.

On the other hand, in order to get small focal distances without giving the surface of the lens an excessive curvature, it is necessary that the propagation

---

[1] L. D. ROSENBERG: Dokl. Akad. Nauk SSSR. **91**, 1092 (1953).

velocities in the lens and the fluid are as different as possible. The materials which best meet these demands should be looked for among liquids[1] or plastic substances[2,3]. In the first case the lens consists of a thin-walled solid casing filled with a liquid of suitably chosen elastic constants.

Plastic lenses, however, have the advantage of being simpler to make. Fig. 54 taken from Sette's paper shows the behaviour of a Plexiglas concave plane-cylindrical lens in water.

In order to determine the focal length, the formula

$$f = \frac{R}{1 - \dfrac{c_l}{c_s}} \tag{21.5}$$

can be used, where f is the focal distance from the lens vertex, R is the curvature radius of the cylindrical or spherical interface, c_l the velocity of the sound in

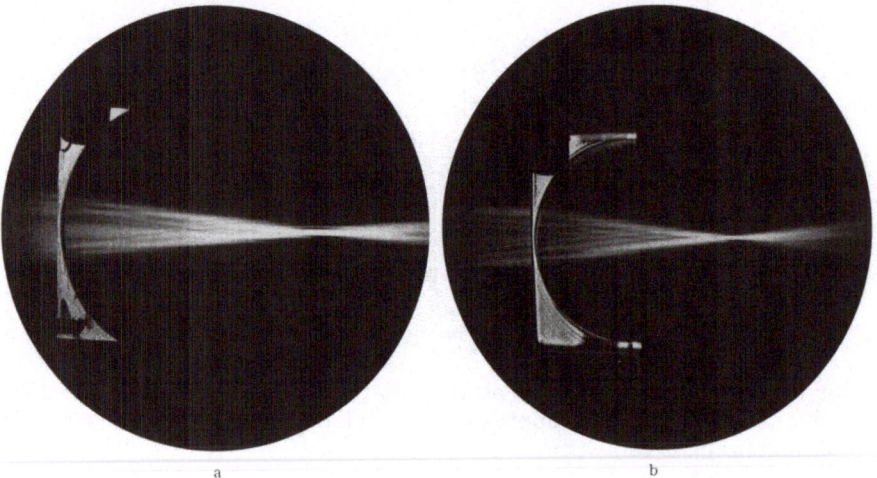

a   b

Fig. 54 a and b. Ultrasonic plexiglas lenses (after Sette). a radius = 25 mm; b radius = 20 mm; Frequency 8 MHz.

the liquid, and c_s the velocity of the longitudinal waves in the lens material. Since the velocity in the solid, c_s, is greater than c_l, the denominator of (21.5) is always positive, i.e. the focus is on the same side as the centre of curvature. The concave lens, therefore, will be convergent, the convex lens, divergent.

The face exposed to ultrasonic radiation is plane. With normal incidence, therefore, only a longitudinal wave is transmitted through the lens. On the curved interface, shear and longitudinal reflected waves will also form towards the inside of the lens, but the energy associated with them will be very small if the matching of the acoustic impedances between lens and fluid is good.

II. Detection of ultrasound.

The detection of ultrasound is effectuated by observation of physical effects produced by the waves. Such effects can be of a mechanical, thermal, or optical nature.

[1] A. Giacomini: Alta Frequ. **7**, 660 (1938).
[2] P. Ernst: J. Sci. Instrum. **22**, 238 (1945).
[3] D. Sette: Nuovo Cim. **6**, 135 (1949).

a) Mechanical effects.

Sound waves locally produce in a medium transversed a periodic variation of pressure (the so-called *acoustic pressure*) and a steady pressure (the so-called *radiation pressure*).

22. The acoustic pressure. The acoustic pressure can be detected by means of microphones similar in operation to those used in acoustics, but sufficiently sensitive also in the ultrasonic frequency range.

Ultrasonic generators usually emit waves of fixed frequency or, at least, of a frequency which varies but little within a restricted range. Microphones used to detect such waves can, therefore, consist of elements tuned at the wave frequency. Selectivity of the detector greatly increases sensitivity.

α) *Piezoelectric probes*. The simplest type of piezoelectric receiver may be a small quartz or titanate disk or element of another piezoelectric material, having a suitable cut. The electric output of such a transducer can be amplified and measured in the usual way.

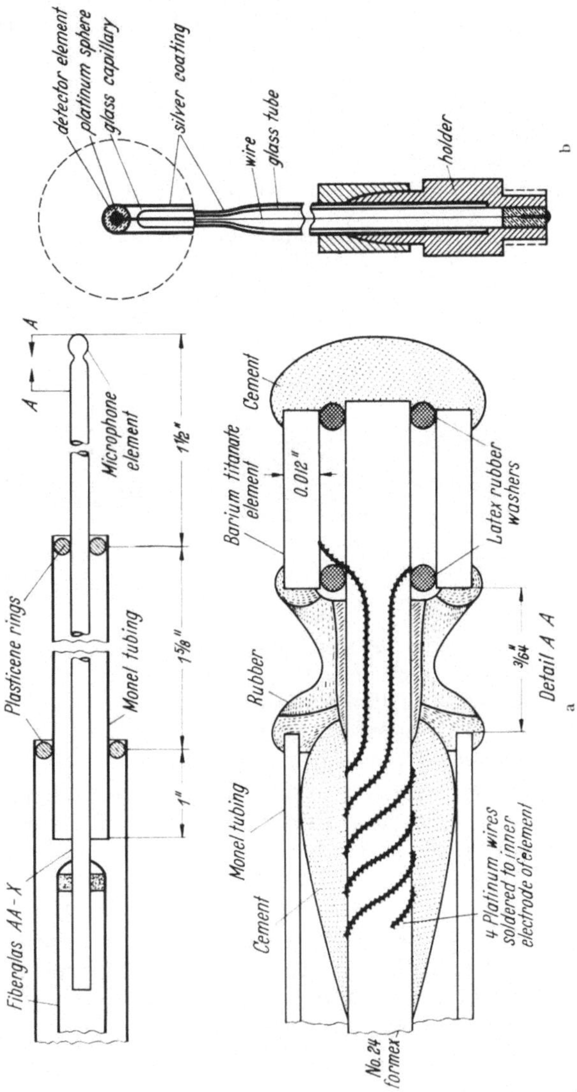

In order to study the energy distribution in an ultrasonic field, the microphone is bound to

Fig. 55a and b. Piezoelectric probe, a (after SCHILLING et al.), b (after ROMANENKO).

indicate the intensity existing at any point of the field and, at the same time, must not disturb the field. To this purpose, the detector must be sensitive and as small as possible. Such receivers are usually called *probes*.

Fig. 55a shows a typical example of a piezoelectric probe[1]. The sensitive element consists of a small barium titanate tube, about 1.5 mm long and having an external diameter also measuring about 1.5 mm. The wall thickness is 0.3 mm. Silver electrodes are fired onto the internal and external surface of the tube.

---

[1] H. SCHILLING: Atmospheric Physics and Sound Propagation. Penn. State Coll. Acoust. Lab. Final Rept. 1950.

The probe is designed to ensure mechanical insulation of the sensitive element from its support and to make it waterproof if used in a liquid.

A miniature piezoelectric probe which is very suitable for measurements of non-uniform ultrasonic fields has been developed by Romanenko[1]. It is shown schematically in Fig. 55b. The detector element is constructed in the form of a spherical layer of ceramic barium titanate which is approximately 0.05 mm thick; this is deposited on a very small platinum sphere which is soldered onto the end of a 0.05 mm diameter wire. The sphere is the inner electrode and the wire is the electrode lead.

The wire passes inside a glass capillary and the sphere is joined to the end-face of the capillary. The capillary itself is an extension of the glass tube which

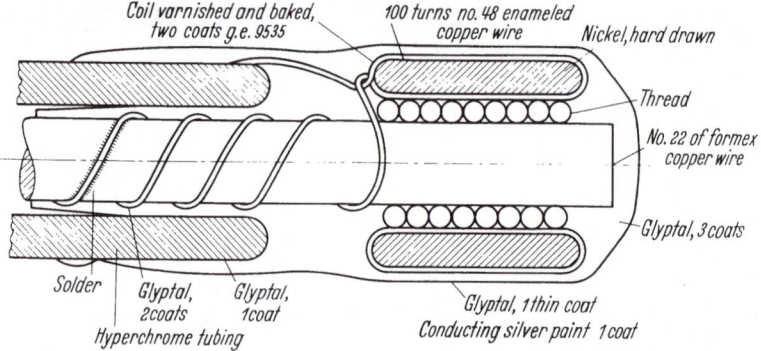

Fig. 56. Magnetostrictive probe (after Schilling et al.).

serves as a holder. A silver coating is fired onto the external surfaces of the tube, capillary and ceramic layer.

The maximum dimension of the sensitive element of the probe is approximately 0.2 mm. The response is good over an 1 to 10 MHz band and the directivity pattern is circular in a plane perpendicular to the axis of the probe.

β) *Magnetostrictive probes.* Magnetostrictive probes have also been constructed. Fig. 56 shows a typical example[1]. The sensitive element is a nickel tube with toroidal coil in which an e.m.f., appears, due to the inverse magnetostrictive effect. The magnetic polarization of the element is due to residual magnetism. The linear dimensions of the tube are of the order of a millimetre.

γ) *Electrostatic probes.* These work on the principle of the condenser microphone. The diaphragm exposed to the acoustic pressure can be tuned to the frequency of the ultrasonic wave. Capacitive probes of approximately 10 mm diameter have been constructed[2].

23. Radiation pressure. The separation surface of two media of different elastic properties experiences a force when hit by a beam of sound waves. This force depends on the intensity of the waves, on the elastic properties of the two media, on the orientation of the interface with regard to the direction of propagation, and on the area hit by the wave.

The origin of this force can be attributed to *radiation pressure* which is a physical quantity manifest in any type of wave propagation. E.g., the radiation pressure of light waves due to the electromagnetic momentum is well-known, and the same idea leads to introduce the concept of radiation pressure in the case

[1] E.V. Romanenko: Soviet Phys. Acoustics **3**, 364 (1957).
[2] G. Sacerdote: Alta Frequ. **2**, 516 (1933).

of sound waves, although here the situation is somewhat more complex. Its theoretical aspects have been thoroughly studied by RAYLEIGH[1], BRILLOUIN[2], LANGEVIN[3], SCHAEFER[4], BEYER[5] and others and more recently by BORGNIS[6].

In order to give a simple idea of radiation pressure we will follow a theory of first approximation suggested by HUETER and BOLT[7]. Let us consider a beam of plane sinusoidal waves propagating in a fluid. Then the acoustic pressure in the direction of propagation x, is $p = \varrho c u$ where $u = U \cos (\omega t - k x)$ is the vibration velocity of the particles.

If the medium were linear, no stationary pressure would be generated by the wave, but if the quantity ϱc varies periodically with the motion, the necessary pressure of a time independent component can easily be understood from an electric analog. An alternating current flowing in a resistance whose value varies periodically with the current, originates a direct current due to the non-linearity of the characteristics of such a circuit.

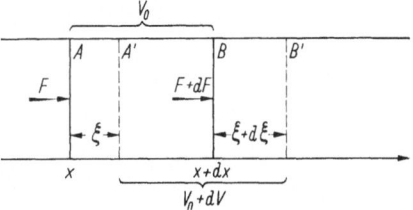

Fig. 57 shows a longitudinal section of a volume element V_0 in a fluid of density ϱ_0, crossed by a plane sound wave in x-direction. Let F and $F + dF$ be the forces

Fig. 57. Propagation of plane waves.

exerted at any particular moment, on the walls A and B of V_0. The corresponding displacements are ξ and $\xi + d\xi$, hence the volume will be changed by dV. The relative variation of density is

$$\frac{d\varrho}{\varrho_0} = \frac{dV}{V_0}$$

and, since during an adiabatic transformation:

$$\frac{dP}{P_0} = \gamma \frac{dV}{V_0},$$

(γ = ratio of specific heats and P_0 = static pressure), we obtain for the acoustic pressure:

$$dP = p = \gamma P_0 \frac{d\varrho}{\varrho_0}.$$

On the other hand, the actual density is:

$$\varrho = \varrho_0 + d\varrho = \varrho_0 \left(1 + \frac{p}{\gamma P_0}\right)$$

so that:

$$p = \varrho c u = \varrho_0 \left(1 + \frac{p}{\gamma P_0}\right) c u.$$

It is reasonable to approximate p in the second term of the bracket by its average value $\varrho_0 c u$ disregarding the correction $d\varrho$. Thus, if we only calculate a first order perturbation, we get:

$$p = \varrho_0 c U \cos (\omega t - k x) + \frac{\varrho_0^2 c^2 U^2 \cos^2 (\omega t - k x)}{\gamma P_0}. \tag{23.1}$$

[1] Lord RAYLEIGH: Phil. Mag. (6) **3** (1902).
[2] L. BRILLOUIN: Rev. d'Acoust. **5**, 99 (1936).
[3] P. LANGEVIN: See P. BIQUARD: Rev. d'Acoust. **1**, 93 (1932).
[4] C. SCHAEFER: Ann. Phys. Lpz. (5) **35**, 473 (1939).
[5] R. E. BEYER: Amer. J. Phys. **18**, 25 (1950).
[6] F. E. BORGNIS: Tech. Rept. 1, Calif. Inst. of Tech. 1951.
[7] F. T. HUETER and R. BOLT [3].

The mean value $\bar{p}$ of p is the direct component which we shall call the *radiation pressure* Π:

$$\Pi = \bar{p} = \frac{1}{2\pi} \int_0^{2\pi} p \, d(\omega t). \tag{23.2}$$

If we put (23.1) in (23.2), the integral of the first term is zero, and we obtain from the second term

$$\Pi = \frac{1}{2} \frac{\varrho_0^2 c^2 U^2}{\gamma P_0}, \tag{23.3}$$

or with

$$c^2 = \frac{\gamma P_0}{\varrho_0},$$

we get:

$$\Pi = \tfrac{1}{2} \varrho_0 U^2. \tag{23.4}$$

In the plane wave, the intensity I is defined as the energy contained in a cylinder of unit cross section and of length c; it is:

$$I = \tfrac{1}{2} \varrho_0 c U^2. \tag{23.5}$$

Hence:

$$\Pi = I/c = E_0 \tag{23.6}$$

where E_0 is the density of sound energy.

On the other hand, associated to the energy is the mechanical momentum of the particles in vibration. If in the ultrasonic beam we consider a surface of unit area parallel to the wave fronts, a mass element, dm, which is displaced across this surface, is contained in a volume $d\xi$, hence $dm = \varrho \, d\xi$. The flow of mass across unit area, therefore becomes

$$\frac{dm}{dt} = \varrho \frac{d\xi}{dt} = \varrho u.$$

Since the mean value of u is zero, there is no continuous flow of mass across the surface. Moreover the variation of the momentum, $d(mu)$, is equal to the impulse $F dt$ of the force on the unit area, i.e. the pressure p produced by the wave, or:

$$p = u \frac{dm}{dt} = \varrho U^2 \cos^2(\omega t - kx).$$

The mean value of p is not zero, but:

$$\bar{p} = \tfrac{1}{2} \varrho U^2 \tag{23.7}$$

which again represents the radiation pressure.

Thus, the radiation pressure can also be defined as the time average of the momentum carried by the plane wave across the unit area.

From the principle of conservation of momentum there follows that both, the source or an obstacle placed in the fluid before it, experience a force per unit area F/S directly proportional to the energy density existing in the fluid in a free field of sound. On the other hand, the force exerted on the source is directed opposite to the direction of propagation. As regards the obstacle, the elastic characteristics of the media separated by the interface determine the direction of the force. The proportionality coefficient is determined by the orientation of the interface and by the ratio m of the characteristic impedances of the two media separated by the interface.

Let us consider the general case of two media, *1* and *2* separated by an interface *Y*. Further, let the ultrasonic beam of progressive plane waves come from medium *1* and be directed to medium *2*, normally to *Y*. Let finally *I* be the intensity of the waves and $E_0 = I/c_1$ the corresponding density of energy which would exist in medium *1* in the absence of medium *2*. We then may consider the following cases:

1. Interface *Y* is completely absorbing. The whole of the acoustic energy is dissipated in *Y* and no waves reflected to *1* or transmitted to *2* subsist. The density of energy before the interface is not modified, hence: $F/S = E_0$.

2. The interface is not dissipative. In this case the action of the forces depends on the specific acoustic resistances of the two media and on the velocities of propagation. If *R* is the reflection coefficient, there will be in medium *1* an intensity:

$$I_1 = I(1 + R)$$

and in medium *2*

$$I_2 = I(1 - R).$$

The corresponding energy densities in the first and second medium,

$$E_1 = \frac{I}{c_1}(1 + R) = \frac{F_1}{S} \quad \text{and} \quad E_2 = \frac{I}{c_2}(1 - R) = \frac{F_2}{S},$$

represent the forces experienced by the unit area of *Y* in the respective media. The resulting force on the interface, from the side of medium *1*, will, therefore, be given by the equation:

$$\frac{F}{S} = \frac{(F_1 - F_2)}{S} = E_0\left[R\left(1 + \frac{c_1}{c_2}\right) + 1 - \frac{c_1}{c_2}\right]. \tag{23.8}$$

In the case of a perfectly reflecting interface ($m = 0$ or $m = \infty$ and $R = 1$) the pressure will be $F/S = 2E_0$, as is obvious since the energy density at the interface is doubled. In the case of a perfectly transparent interface ($m = 1$, $R = 0$) the whole energy is transmitted to medium *2* and the pressure will be:

$$\frac{F}{S} = \left(1 - \frac{c_1}{c_2}\right)E_0.$$

It should be noted, in this case, that if we have $c_1 > c_2$, the resulting force on the interface is directed in the opposite direction of propagation.

In the intermediate cases of partial reflection at the interface ($0 < R < 1$), the resulting force, as can be seen from Eq. (23.8), has the same direction as the propagation if

$$R > \frac{c_1 - c_2}{c_1 + c_2},$$

and it has the opposite direction if

$$R < \frac{c_1 - c_2}{c_1 + c_2}.$$

In this case we must have $c_1 > c_2$, otherwise we fall back on the preceding condition. Only if $R = 0$ can we have $R = (c_1 - c_2)/(c_1 + c_2)$, provided: $c_1 = c_2$. The two media are of equal specific acoustic resistance and equal velocity i.e. practically identical. We have $F/S = 0$ and no effect of radiation pressure is observed in the fluid; this was to be expected since we know that in a non-dissipative medium there is no transport of mass by the progressive wave.

A confirmation of these phenomena by experiments is illustrated in Fig. 58[1].

---

[1] G. HERTZ and H. MENDE: Z. Physik **114**, 354 (1939).

An ultrasound beam, directed vertically upwards hits the interface of two different fluids. In Fig. 58a the case of air above benzene is shown; $m = 3.8 \cdot 10^{-4}$, therefore $R \approx 1$. The interface can be regarded as perfectly reflecting and the forces tend upwards. Note the typical fountain produced by the ultrasonic beam.

In Fig. 58b water has been put above carbon tetrachloride; $m \approx 1$, therefore $R \approx 0$ and since $c_1 < c_2$ ($c_1 = 938$ m/sec; $c_2 = 1497$ m/sec) the forces are again directed upwards, but they have a value smaller than in the preceding case. Fig. 58c shows a layer of water above aniline; $m = 0.88$, therefore $R = 4 \cdot 10^{-3}$ and since $c_1 > c_2$ ($c_1 = 1656$ m/sec; $c_2 = 1497$ m/sec) paradoxically, as is seen, the forces are directed opposite to the way of propagation, the interface being pressed towards the source.

If the plane wave hits the interface at an angle Θ the component of the force experienced by the interface in the direction of propagation is $F_x = F \cos^2 \Theta$ and the component normal to the interface is $F_n = F \cos \Theta$.

a   b   c

Fig. 58a—c. Effects of radiation pressure on the interface of two fluids. a air above benzene; b water above carbon tetrachloride (after HERTZ and MENDE); c water above anilin (after HERTZ and MENDE)

In a homogeneous dissipative medium the intensity of a progressive plane wave decreases exponentially along the axis of propagation x.

$$I_x = I_0 e^{-2\alpha x}, \tag{23.9}$$

where I_0 is the intensity at $x = 0$ and α is the absorption coefficient for the vibration amplitude.

The gradient of the energy density along x is

$$\frac{dE}{dx} = \frac{1}{c} \frac{dI_x}{dx}$$

hence, in the same direction, the gradient of radiation pressure becomes

$$\frac{d\Pi}{dx} = -2\alpha E_x$$

with

$$E_x = I_x / c.$$

To this differential pressure, the origin of the hydrodynamic flow observed in viscous liquids inside the ultrasonic beam is attributed.

24. Detectors of radiation pressure. The radiation pressure can be determined by measuring the force exerted on a body of known shape, dimensions and acoustic properties, when hit by the sound wave.

As a measure of that force, one can, according to special circumstances, measure either the displacement of the body or the intensity of a restoring elastic force which is necessary to bring the detector back to its initial position. Such devices are quite generally called radiometers.

A very simple type of radiometer, often used in laboratories, is shown in Fig. 59. A small disk of suitable material is attached to a rigid support by means of a long and very light bifilar suspension. The disk is immersed in a liquid in which a plane ultrasonic wave hits the detector normally. By means of an optical comparator the displacement d of a point of the suspension at a distance l from the rigid support can be measured.

The force acting upon the detector will be:

$$F = w\,d/l, \qquad (24.1)$$

where w is the weight of the disk measured in the liquid. If the surface of the disk is completely absorbing, the ultrasonic intensity becomes

$$I = \frac{c\,w\,d}{S\,l}, \qquad (24.2a)$$

where c is the speed of sound in the liquid, and S is the detector surface hit by the waves. For frequencies of the order of a few MHz, cork, glass wool etc. can be used as absorbing materials.

If the surface of the disk is completely reflecting, we have:

$$I = \frac{c\,w\,d}{2\,S\,l}. \qquad (24.2b)$$

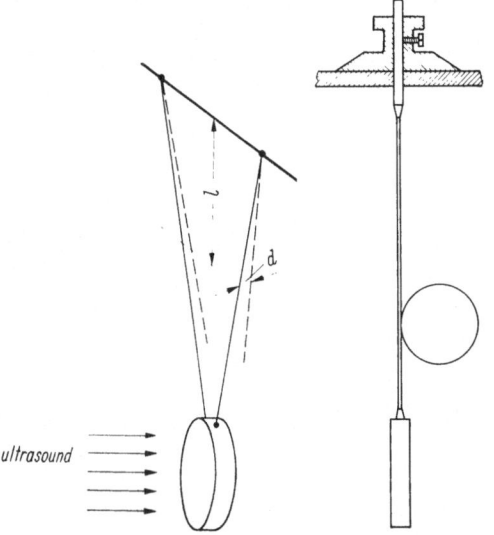

ultrasound

Fig. 59. Pendulum for measuring radiation pressure.

Fig. 60. Torsion pendulum (after BERGMANN).

For a partially reflecting detector we must take into account the specific acoustic resistances and introduce the reflection coefficient R according to Eq. (23.8).

In the case of normal incidence, reflection of the energy can affect the measurement because of formation of standing waves between transducer and detector. An oblique incidence is therefore preferable and may easily be obtained by giving to the detector a spherical, conical, or some other suitable shape.

A torsional pendulum has also been used as radiometer. The detecting disk is sideways attached to a rigid vertical wire, as indicated in Fig. 60. The torsion of the wire, due to the radiation pressure on the disk, is compensated by means of a counter torque produced by turning the suspension wire through a certain angle. If the dimension of the wire and its torsion modulus are known, the forces acting upon the wire can be determined.

The balance radiometer of Fig. 61 is intended for an ultrasonic beam that propagates vertically upwards. The detector is suspended at one of the plates. Weight p added to equilibrate the balance during the ultrasonic emission, represents the force acting on the detector which in the figure is a reflecting cone. Laterally reflected waves are absorbed by the material on the inner walls of the vessel. The ultrasonic intensity is:

$$I = c\,p/S\,(1 + R)\cos^2\alpha,$$

where S is the surface of the section hit by the ultrasonic beam, R is the reflection coefficient, and α is the aperture semi-angle of the cone. This device has become the basis for many commercial instruments used as dosimeters in ultrasonic therapy.

In the Technische Bundesanstalt of Brunswick[1], Germany, a floating radiometer is used, the essential parts of which are drawn in Fig. 62. The vessel A contains water and the internal vessel B contains carbon tetrachloride. A cylindrical body C is immersed in the water and kept in hydrostatic equilibrium because its graduated leg E is in the carbon tetrachloride. The ultrasonic beam emitted by the transducer Q pushes the float C downwards. The upper surface

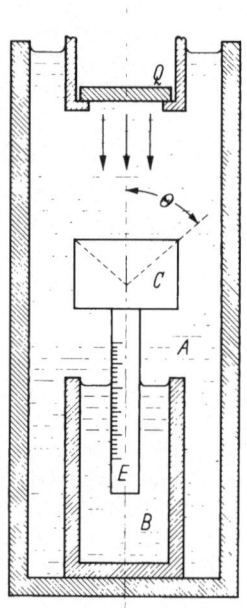

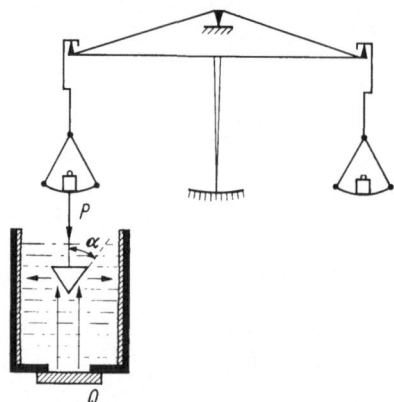

Fig. 61. Balance radiometer.   Fig. 62. Floating radiometer (after OBERST and RIECKMANN).

of C has a conic concave shape in order to prevent formation of standing waves and to help self-centering the float in the ultrasonic beam.

The apparatus must previously be calibrated with a known weight w. The ultrasound intensity is measured by reading off the lowering of the detector from the graduation on the leg. If S is the section hit by the beam, b_w the lowering of the float due to the calibration weight w and b_r the lowering due to radiation pressure, the measured intensity will be

$$I = \frac{\varkappa}{S} \frac{c\, b_r}{2 b_w \cos^2 \Theta} \, ,$$

where $\varkappa$ is a numerical constant depending on the units of measurement.

In the arrangements described as far, the hydrodynamic flow, caused by the gradient of radiation pressure, generally disturbs the measurements. Its effect is eliminated by placing a thin aluminium or collodium foil or plastic film of a few microns thickness in front of the detector which breaks the fluid flow, but transmits sound waves without appreciable attenuation.

In absolute measurements the attenuation due to the film can be taken account of.

[1] H. OBERST and P. RIECKMANN: Amtsblatt d. P.T.B. Braunschweig H. 3, 106 (1952).

A somewhat different method[1] of measurement has turned out to be useful in the case of the ultrasonic beam being modulated at low frequency with a given function. The detector is an ordinary microphone (of piezoelectric, electrostatic, or electromagnetic type) followed by an amplifier connected with an output meter. The amplitude of the signal detected is proportional to the variable component of the radiation pressure on the membrane of the microphone. Therefore, knowing the modulating function and the depth of modulation M it is easy to calculate the value E_0 which the ultrasonic energy density would have if $M = 0$. For example, if the modulation is sinusoidal, the output voltage indicated by a peak voltmeter will be:

$$V = \varkappa E_0 M \left(2 + \frac{M}{2} \right),$$

where $\varkappa$ is a proportionality constant depending on the sensitivity of the microphone, on the shape of its membrane and on the gain of the amplifier. When using this apparatus one has not necessarily to eliminate the fluid flow, since it operates, within reasonable limits, independently of the steady-state component of the pressure.

b) Thermal effects.

The transformation of mechanical energy into thermal energy is one of the causes of ultrasonic absorption in the propagation medium. The thermal energy produced depends on the ultrasound intensity and frequency. The absorption coefficient, quite independently of any molecular and structural phenomena, is proportional to the square of the frequency.

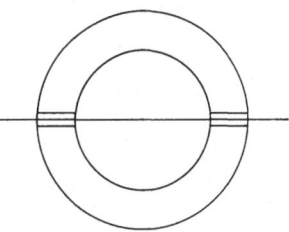

Ultrasonic intensities can, therefore, be measured by means of thermoelectric probes.

25. Thermoelectric probes. The junction of a thermocouple is covered with a small quantity of absorbing material and measures the increase of its equilibrium temperature when hit by an ultrasonic wave.

This kind of probe is usually employed for comparative measurements. Since the thermal equilibrium depends critically on the geometry and dimensions of the absorbing material, previous calibration in an ultrasonic field of known configuration is required to obtain absolute measurements of sound intensity.

FRY[2] designed a thermoelectric probe for absolute intensity measurement which eliminates the diffi-

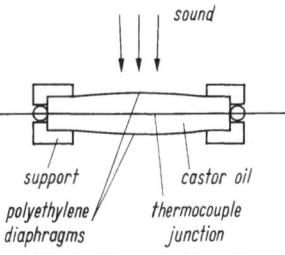

Fig. 63. Thermoelectric probe (after FRY).

culties of calibration in a known field. It does, however, require the ultrasonic radiation to be pulse-modulated and is unsuitable for analysing fields in which several frequencies co-exist.

Fig. 63 is a drawing of a Fry probe. The junction of the thermocouple is immersed in an absorbing liquid contained in a small cylindrical box whose walls, exposed to ultrasound, are formed by thin films of polyethylene of 0.075 mm thickness, so that they are practically transparent for ultrasonic waves. The wire of the thermocouple has 0.0125 mm diametre. The absorber contained in the probe box must have a specific acoustic resistance very near to that of the oute

[1] A. BARONE and M. NUOVO: Nuovo Cim., Suppl. No. 2 **7**, 359 (1950).
[2] W. J. FRY and R. BAUMAN FRY: J. Acoust. Soc. Amer. **26**, 311 (1954).

liquid, so that no appreciable part of the ultrasonic energy is reflected at the interface. If the outer liquid is water, and castor oil is used as absorber, the reflected energy is about 0.025% of the incident one. The probe is placed in such a way that the wire of the thermocouple is perpendicular to the direction of ultrasonic propagation.

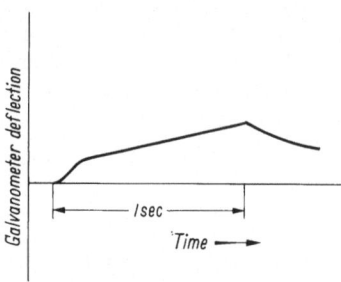

Fig. 64. Thermoelectric current due to ultrasonic pulse (after Fry).

The rise of temperature at the junction can be determined by recording on a photographic plate the deflection of the galvanometer measuring the thermoelectric current. Fig. 64 shows the record of the thermoelectric current due to a rectangular wave train of one second's duration. As soon as the pulse front arrives, the temperature increases rapidly for a short period, then it continues to rise almost linearly, but less steeply until the end of the pulse. Finally it decreases slowly to its original value. The rapid initial increase is caused by viscous forces acting between the wire and the liquid in immediate contact. The subsequent linear increase is due to ultrasound absorption in the surrounding liquid.

The ultrasonic intensity in the immediate neighbourhood of the thermocouple is given by the equation:

$$I = \frac{\varrho C}{2\alpha} \left(\frac{dT}{dt} \right)_0,$$

where α is the absorption coefficient for the amplitude, ϱC is the thermal capacity per unit volume of the liquid contained in the probe and $(dT/dt)_0$ indicates the variation in time of the temperature of the liquid at the beginning of the disturbance. This last quantity can be determined from the slope of the linear part of the record in Fig. 64. Knowing the absorption coefficient α and the thickness of the liquid traversed in the probe, one can immediately find the absolute value of the intensity of the field at the point where the junction of the thermocouple is placed.

c) Optical effects of ultrasound.

26. Diffraction of light. In 1932, Debye and Sears[1] in U.S.A., Lucas and Biquard[2] in France, found that a transparent medium, traversed by an ultra-

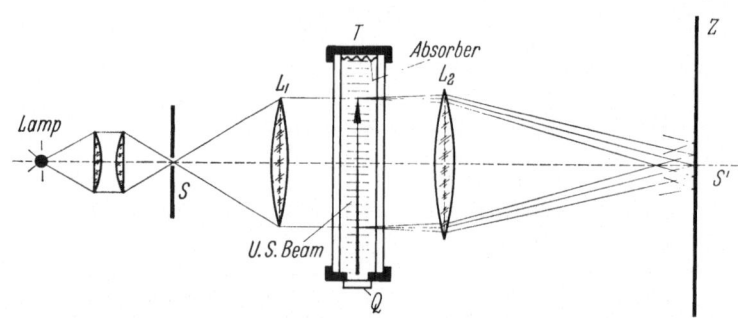

Fig. 65. Optical arrangement for observing light diffraction produced by ultrasound.

sonic wave, diffracts light. This effect is due to periodic changes of the refraction index which, in their turn, depend on the sound pressure. It is the more conspic-

[1] P. Debye and F.W. Sears: Proc. Nat. Acad. Sci. **18**, 409 (1932).
[2] R. Lucas and P. Biquard: C. R. Acad. Sci. Paris **194**, 2132 (1932); **195**, 121 (1932).

uous and easy to observe, the greater the variations of the refraction index are and the closer they succeed each other at short distance. These conditions most readily apply in the case of liquids and within the range of ultrasonic frequencies to which very small wavelength correspond. The effect, of course, occurs also in gases and transparent solids, but there it is more difficult to demonstrate it.

A simple way of demonstrating this kind of light diffraction is outlined in Fig. 65. The light source (in the figure, a slit S illuminated by the lamp) is placed in the focus of lens L_1 from which a beam of parallel light emerges. This penetrates the vessel containing a transparent liquid and transversed by a beam of plane ultrasonic waves. The direction of the light rays is perpendicular to the direction of the ultrasound wave. The latter is emitted by a transducer Q driven by a high-frequency oscillator.

The walls of the vessel are plane parallel glass plates. A second lens L_2 projects the real image S' of the source on a screen Z coinciding with the focal plane. If ultrasound is present in the liquid, two series of diffraction images of the source appear on both sides of S'. Fig. 66 shows a photo of such diffraction spectra produced by a beam of 8 MHz ultrasonic plane waves, the light source being an incandescent lamp.

Fig. 66. Diffraction spectra produced by 8 MHz ultrasound (light source: incandescent lamp).

The light of the smallest wavelength is the least deviated, hence in any spectral order, as in the case of optical gratings, red colour will be found at the greatest distance from the central (zero order) fringe.

In order to understand this phenomenon we have to assume that the ultrasonic wave produces variations of the refraction index in the liquid so that on the exit plane from the ultrasonic beam the phase of the light wave varies from point to point because of the different velocity with which the light has propagated through regions having a different refraction index.

In a sinusoidal wave, propagating in z-direction, the pressure is:

$$P = P_0 + \Delta P \cos \frac{2\pi}{T} \left(t - \frac{z}{c}\right), \qquad (26.1)$$

where T is the period, P_0 the average pressure in the medium, and ΔP the amplitude of pressure variation. Corresponding to such a distribution of pressure, the refraction index will be:

$$\mu = \mu_0 + \Delta \mu \cos \frac{2\pi}{T} \left(t - \frac{z}{c}\right).$$

On the other hand, since the velocity of light is so much greater than that of sound, we can assume that, during the whole of the time in which the interaction between light and elastic wave occurs, the refractive index maintains a distribution of the type:

$$\mu = \mu_0 + \Delta \mu \cos \frac{2\pi z}{\Lambda} \qquad (26.2)$$

independent of time.

The propagation of light waves in a direction normal to z (e.g., y) in the disturbed medium, should take place in such a way as to satisfy a differential equation of the type:

$$\frac{\partial^2 h}{\partial y^2} = \left(\mu_0 + \Delta\mu \cos\frac{2\pi z}{\Lambda}\right)^2 \frac{1}{c_{\text{light}}^2} \frac{\partial^2 h}{\partial t^2},$$

(26.3)

where h is one of the characteristic parameters of the light propagation, e.g. the instantaneous value of the electric field.

The solution of an equation of this type can give the spatial distribution of light intensity. The mathematical problem is, however, rather complicated and we prefer to follow the simpler procedure, also suggested by LUCAS and BIQUARD[1], which consists of determining the paths of the light rays in the medium traversed by the ultrasonic wave.

The light rays which hit the ultrasonic beam in a direction parallel to the front of the sound wave will be bent because of the variation of the refractive index—a phenomenon similar to what happens in the case of mirage. The radius R of curvature of every single ray is given by the equation:

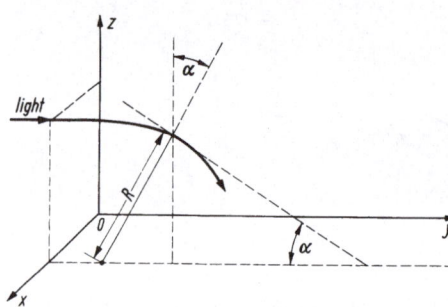

$$\frac{1}{R} = \frac{1}{\mu}\frac{d\mu}{dr},$$

(26.4)

where $d\mu/dr$ represents the variation of the refractive index in a direction r normal to the light ray. Following Fig. 67, in which the directions of the ultrasound wave and of the light rays are as stated above, the curvature of the light path will be given at any point by:

Fig. 67. Bending of light rays.

$$\frac{1}{R} = \frac{1}{\mu}\frac{d\mu}{dz}\cos\alpha$$

(26.5)

where α is the angle between the tangent to the ray and the y axis. Since

$$\tan\alpha = \frac{dz}{dy} \equiv \dot{z},$$

we obtain from Eq. (26.5)

$$\frac{1}{R} = \frac{1}{\mu}\frac{d\mu}{dz}\frac{1}{(1+\dot{z}^2)^{\frac{1}{2}}}$$

and, since the curvature is

$$\frac{1}{R} = \frac{\ddot{z}}{(1+\dot{z}^2)^{\frac{3}{2}}},$$

we can write the differential equation for the light rays:

$$\frac{\ddot{z}}{1+\dot{z}^2} = \frac{1}{\mu}\frac{d\mu}{dz}$$

(26.6)

or also:

$$\frac{d\mu}{\mu} = \frac{1}{2}\frac{d(1+\dot{z}^2)}{1+\dot{z}^2}.$$

By integration we find

$$\log(1+\dot{z}^2) = \log\mu^2 + C$$

which we can as well write in the form

$$A\mu = (1+\dot{z}^2)^{\frac{1}{2}}$$

(26.7)

with A being an integration constant depending on the initial conditions. For the initial height $z = z_1$ we have $\dot{z} = 0$ since the ray penetrates into the ultrasonic

[1] R. LUCAS and P. BIQUARD: J. Phys. Radium 10, 464 (1932).

beam parallel to y, so that:

$$A = \frac{1}{\mu} = \frac{1}{\mu_0 + \Delta\mu \cos \frac{2\pi z_1}{\Lambda}} .$$

Eq. (26.7) then becomes:

$$1 + \dot{z}^2 = \left(\frac{\mu_0 + \Delta\mu \cos^2 \frac{2\pi z}{\Lambda}}{\mu_0 + \Delta\mu \cos^2 \frac{2\pi z_1}{\Lambda}} \right)^2 . \tag{26.8}$$

Since the variation $\Delta\mu$ is very small also for considerable ultrasonic intensities (e.g. in water, for a pressure excess of 10^5 Nt/m², the relative variation of μ is approximately $2 \cdot 10^{-5}$), we can write instead of Eq. (26.8):

$$\frac{dz}{dy} = \left[\frac{2\Delta\mu}{\mu_0} \left(\cos \frac{2\pi z}{\Lambda} - \cos \frac{2\pi z_1}{\Lambda} \right) \right]^{\frac{1}{2}}$$

and obtain by another integration:

$$y = \int_{z_1}^{z} dy = \int_{z_1}^{z} \frac{dz}{\left[\frac{2\Delta\mu}{\mu_0} \left(\cos \frac{2\pi z}{\Lambda} - \cos \frac{2\pi z_1}{\Lambda} \right) \right]^{\frac{1}{2}}} . \tag{26.9}$$

This integral defines z as an elliptic function of y. If, indeed, we put:

$$h = \sin \frac{\pi z_1}{\Lambda} \quad \text{and} \quad t = \frac{1}{h} \sin \frac{\pi z}{\Lambda} , \tag{26.10}$$

then Eq. (26.9) can be rewritten in the form

$$F(h, t) = K(h) + \frac{2\pi}{\Lambda} y \left(\frac{\Delta\mu}{\mu_0} \right)^{\frac{1}{2}} \tag{26.11}$$

where:

$$F(h, t) = \int_0^t \frac{dt}{[(1 - t^2)(1 - h^2 t^2)]^{\frac{1}{2}}} \tag{26.12}$$

and

$$K(h) = \int_0^1 \frac{dt}{[(1 - t^2)(1 - h^2 t^2)]^{\frac{1}{2}}} \tag{26.13}$$

are respectively, LEGENDRE's incomplete and complete elliptic integrals of the first kind. Defining in the usual way by the identity

$$t = \operatorname{sn} \int_0^t \frac{dt}{[(1 - t^2)(1 - h^2 t^2)]^{\frac{1}{2}}} \tag{26.14}$$

the elliptic function "sinus amplitude", sn, we finally obtain

$$\sin \frac{\pi z}{\Lambda} = h \operatorname{sn} \left[K(h) + \frac{2\pi y}{\Lambda} \left(\frac{\Delta\mu}{\mu_0} \right)^{\frac{1}{2}} \right] . \tag{26.15}$$

From this equation it is easy to construct the trajectories $z = z(y)$ of light rays crossing the ultrasonic beam. Using LEGENDRE's tables, we first obtain the values of $K(h)$ for each individual h—i.e., corresponding to every point of incidence—and then we may, for different values of y, determine the values of z from Eq. (26.15).

Using Eq. (26.15) in this way, LUCAS and BIQUARD have drawn the paths, shown in Fig. 68, of 21 light rays corresponding to as many incident rays on a total tract of the ultrasonic beam equal to one wavelength.

On the ordinate axis the values $z' = \dfrac{\pi z}{\Lambda}$ are indicated which are equal to half the phase angle of the wave, and on the abscissa axis there are shown the values:

$$y' = \frac{2\pi}{\Lambda}\, y \left(\frac{\Delta\mu}{\mu_0}\right)^{\frac{1}{2}}.$$

The light rays normally penetrating into the ultrasonic beam, therefore, converge in the pressure maxima of the wave, in a kind of focal region distant about

$$y = \frac{\Lambda}{4\,(\Delta\mu/\mu_0)^{\frac{1}{2}}}$$

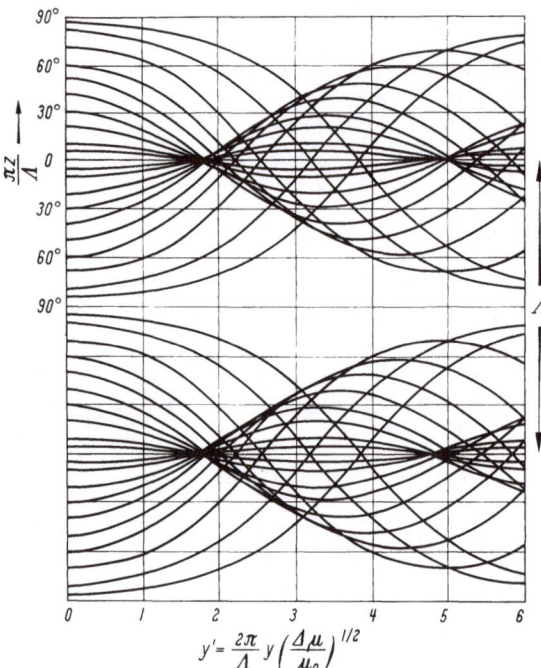

(i.e. $y' \approx \pi/2$) from the point of incidence. They then diverge, concentrate again, and so on.

Inside the ultrasonic beam, in the direction of sound propagation, we observe a periodical distribution of light intensity, a period being equal to one wavelength Λ of the ultrasound. The distribution law is not simple, but the light leaving the ultrasonic beam will appear to come from regions of maximum and minimum intensity which alternate at the distance $\Lambda/2$ as if it had been generated by a series of equidistant sources of more or less large dimension.

If the distance between these sources, i.e. Λ, is sufficiently small, the light waves emitted, by interfering with each other, give rise to a phenomenon of diffraction similar to that pro-

Fig. 68. Patterns of light rays in an ultrasonic beam according to LUCAS and BIQUARD.

duced by a grating. The angle φ of deviation of the diffracted light, from the direction of incidence, is determined by an equation similar to that of optical gratings:

$$\sin \varphi = n\, \frac{\lambda}{\Lambda}, \tag{26.16}$$

where λ is the wavelength of the light, deviated according to the angle φ, and n is the order of interference.

The light wave emerging from the ultrasonic beam will thus be composed of an ensemble of plane wave beams propagating in the direction resulting from Eq. (26.16) symmetrically to the direction of incidence. If therefore the light leaving the ultrasonic beam is made to converge by a lens, as in Fig. 65, we shall observe in the focal plane diffraction spectra of various order. Each spectrum will be composed of as many coloured images of the source as there are wavelengths present in the light radiation.

We owe the calculation of the distribution of light intensities in diffraction spectra to RAMAN and NAGENDRA NATH[1] who used RAYLEIGH's theory of the diffraction of a plane wave normally incident on a periodically corrugated surface.

[1] C.V. RAMAN and N.S. NAGENDRA NATH: Proc. Ind. Acad. Sci. **2**, 406, 413 (1935); **3**, 75, 119, 459 (1936).

RAMAN and NATH do not consider the curvature of the light rays, but they only take account of the variation of light velocity in passing through the compression and expansion regions of the ultrasonic beam. It follows from this hypothesis, that, on the plane of emergence from the sound wave, the amplitude of light vibration is everywhere constant while its phase is represented by a periodical function of the position having a period equal to the ultrasonic wavelength. In penetrating the ultrasonic beam, the phase plane of the incident light wave is transformed into a surface corrugated according a sinusoidal law. The sound wave behaves, therefore, as a phase grating. This hypothesis is compatible with the presence of diffraction spectra of higher order than the first, since phase gratings represented in optics by echelon grantings (as opposed to rule grantings) produce spectra of a higher order even if the distribution of the phases is sinusoidal. On the hypothesis we have made, if the incidence of the light rays is normal to the ultrasonic beam, the intensity of the light in the zero order fringe I_0—assumed to be equal to one in the absence of ultrasound—is given by the square of BESSEL's function of zero order:

$$I_0 = J_0^2(a). \tag{26.17a}$$

The argument

$$a = 2\pi l \, \Delta\mu/\lambda \tag{26.17b}$$

represents the phase lag of the light wave; l is the length of the optical path in the ultrasonic beam, $\Delta\mu$ is the maximum variation of the refractive index produced by the ultrasonic wave, and λ the wavelength of the light in vacuo. Similarly the relative intensity of light in each spectrum of n-th order is given by:

$$I_n = J_n^2(a), \tag{26.18}$$

where n will be regarded as positive for the right-hand side spectra and negative for those on the left-hand side. Since

$$J_{+n}^2(a) = J_{-n}^2(a),$$

left- and right-hand side spectra of the same order have the same intensity.

Moreover, for BESSEL's functions there holds the sum rule

$$J_0^2(a) + 2\sum_{n=1}^{\infty} J_n^2(a) = 1,$$

so that, whatever be the value of a, the sum of the light intensities over all spectra is constant and equal to unity, i.e. equal to the intensity of the incident light. Fig. 69 shows the relative intensities of the light in several orders of diffraction for 24 successive values of the parameter a, calculated using the formulae of RAMAN and NATH.

In Fig. 70 we see how the relative intensity of the light varies as a function of the parameter a in the central fringe and in the first five orders of diffraction. The broken curves represent the theoretical values and the unbroken curves indicate experimental results according to SANDERS[1]. It can readily be seen that the results very definitely confirm the prediction of RAMAN and NATH.

RAMAN and NATH also studied the light diffraction in the case of standing waves, and Fig. 71 shows a comparison between the theoretical prediction and the experimental curves obtained by SANDERS.

---

[1] F.H. SANDERS: Canad. J. Res. **14**, 158 (1936).

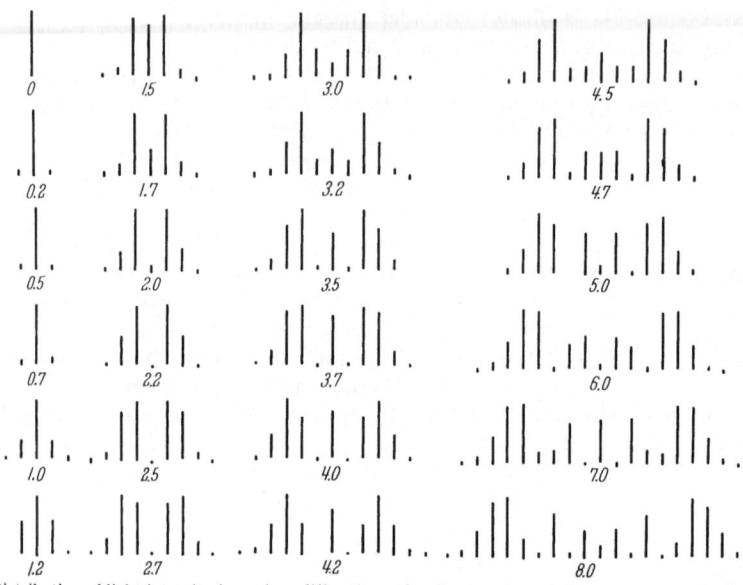

Fig. 69. Distribution of light intensity in various diffraction orders for 24 values of the parameter $a = 2\pi l \, \varDelta\mu/\lambda$ (after RAMAN and NAGENDRA NATH).

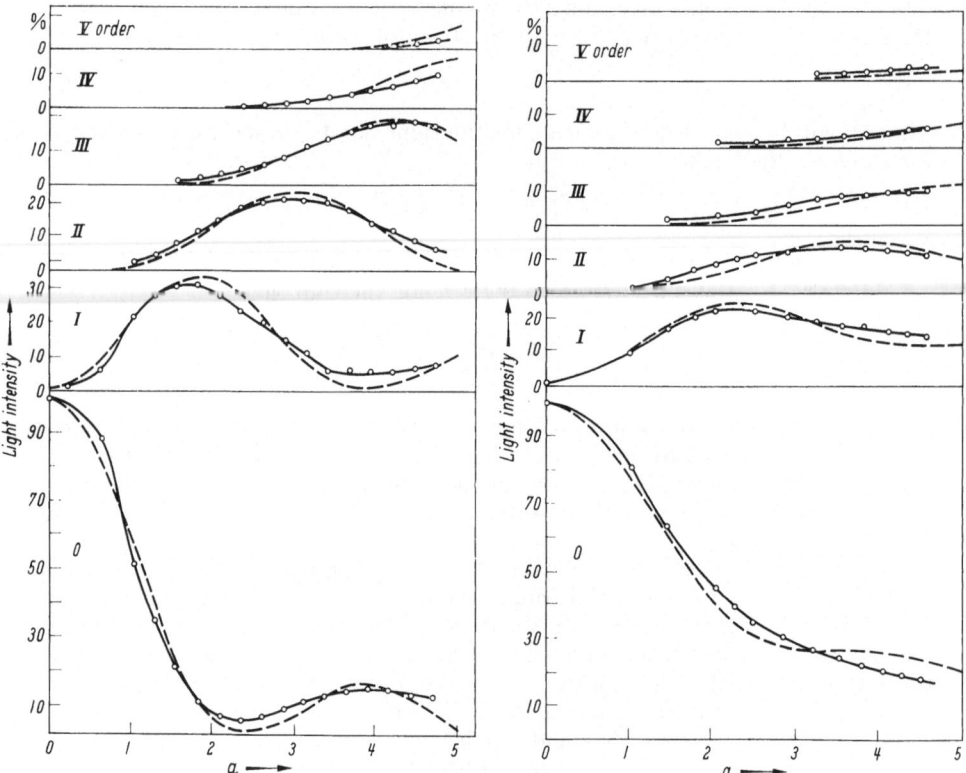

Fig. 70. Light intensity in the central fringe and in the first 5 orders depending on the parameter a, for progressive waves (after SANDERS).

Fig. 71. Light intensity in the central fringe and in the first 5 orders depending on the parameter a, for standing waves (after SANDERS).

Light diffraction due to standing waves differs from that due to progressive waves by the simple fact that it is produced by an ultrasonic grating which, during each period of acoustic vibration, is twice cancelled by interference. At these moments the light falls all into the central fringe so that, on the average, complete cancellation of undiffracted light can never occur. The light intensity in the spectra will therefore be smaller for standing waves.

27. Diffraction of light due to oblique incidence. When the incidence of the light on the ultrasonic beam is not normal, then the relative light intensity in a fringe of order n, according to the theory of RAMAN and NATH, is:

$$I_n = J_n^2 \left(a \sec \Phi \, \frac{\sin N \pi}{N \pi} \right) \tag{27.1}$$

where Φ is the angle of incidence and $N = \frac{1}{\Lambda} \tan \Phi$. N represents, therefore, the number of wavelengths encountered by the light ray obliquely crossing the ultrasonic beam.

Eqs. (26.17), (26.18) and (27.1) generally yield good approximations, but under certain conditions, experimental results do not agree with them. In order to eliminate these discrepancies it is necessary to investigate the validity limits of the theory of RAMAN and NATH.

This theory is based on the hypothesis that, in consequences of the sinusoidal variation of the refractive index, the ultrasonic beam of plane waves behaves like a sinusoidal phase grating.

However, we know that light rays are not rectilinear, but bent toward the regions of higher refractive index. As LUCAS and BIQUARD have shown[1], this leads to a concentration of light intensity in periodic distances of an ultrasonic wavelength Λ in direction of the sound propagation. Consequently the light leaving the ultrasonic beam can, under certain conditions, presents itself as periodically altered in both, phase and amplitude. The result therefore is a combined phase and amplitude grating producing a light distribution in the diffraction spectra which does not follow exactly the laws expressed by the formulae of RAMAN and NATH.

According to the theory of LUCAS and BIQUARD the bent rays first meet at a distance l_1 from the point of incidence on the ultrasonic beam, so that $y' \approx \pi/2$ as seen in Fig. 68. An ultrasonic beam wider than l_1 (i.e. $y' \geq \pi/2$) possesses, in considerable measure, the characteristics of an amplitude grating. If, however, $y' \leq \pi/2$, then the amplitude grating property correspondingly decreases and the phase grating characteristics preponderate. If $y' \leq \pi/2$, WILLARD ascertained[2] that the experimental data agree with the formulae of RAMAN and NATH with a maximum deviation of 10%.

WILLARD therefore suggests that the criterion for the correct use of RAMAN and NATH's formulae should be established by the relation:

$$\frac{4 \pi^2 l^2 \Delta \mu}{\Lambda^2 \mu_0} \leq \frac{\pi^2}{4}, \tag{27.2}$$

where l is the width of the ultrasonic beam, and that the diffraction occurring under this condition should be called "normal" to distinguish it from the "anomalous diffraction" to be discussed later.

[1] R. LUCAS and P. BIQUARD: See footnote 1 on p. 126.
[2] G.W. WILLARD: J. Acoust. Soc. Amer. **21**, 101 (1949).

α) *Normal diffraction.* The criterion expressed by (27.2) may be put more simply:

$$f^2 l q \le \frac{c^2 \mu_0}{16 \lambda} \quad \text{(where } q = a/2\pi)$$

or, expressing the frequency f in MHz:

$$f^2 l q \le \frac{10^{-12} \mu_0 c^2}{16 \lambda}.$$

If the medium of propagation is water $(c = 1.5 \cdot 10^5 \text{ cm/sec}, \mu_0 = 1.33)$ and the green mercury light $(\lambda = 5.46 \cdot 10^{-5} \text{ cm})$ is used, we have:

$$f^2 l q \le 34.4 \text{ MHz}^2 \times \text{cm}.$$

The width, frequency or intensity of the ultrasonic beam, therefore, cannot be indefinitely increased without passing beyond the validity limits of normal diffraction and the applicability of the corresponding formulae.

β) *Anomalous diffraction.* The criterion of normal diffraction is not sufficient to guarantee the validity of Eq. (27.1) for oblique incidence. From this equation, in fact, it follows that spectra of the same order, both on the left and on the right, have the same intensity, but in experiments carried out by BHAGAVANTAM and RAO[1] at frequencies of 50, 100 and 180 MHz, it was observed that the first order spectra on the left and right did not appear simultaneously when the angle of incidence was varied.

The complete image of diffraction, obtained e.g. with a device such as shown in Fig. 65, is therefore dissymmetrical with respect to the central fringe. This fact can be attributed to a diffraction phenomenon of BRAGG's type. It is supposed that each light ray obliquely incident on the front of an ultrasonic wave is partly scattered and partly proceeds unchanged and behaves again in the same way at the front of the next ultrasonic wave, and so on, until it leaves the beam.

The rays scattered in the direction of specular reflection from every point of the different ultrasonic wave fronts, will all be in phase on a plane perpendicular to this direction if the angle of incidence satisfies BRAGG's condition:

$$n \lambda = 2 \Lambda \sin \varphi, \tag{27.3}$$

where n is an integer and λ is the wavelength of the light in the medium. In such a direction a considerable increase of light intensity will occur. Thus, e.g., if the angle of incidence is such as to satisfy BRAGG's condition for the first order on the right (or left) the corresponding spectrum of the first order in the right (or left) will appear by itself with a sharp intensity peak.

WILLARD indicates the conditions for BRAGG's diffraction. They depend on the factor $f^2 l$, on the number N of ultrasonic wavelengths traversed by the non-diffracted light ray and on the order of the spectrum. N must be equal at least to 1; for higher values, the angle of incidence satsfying BRAGG's condition becomes more and more critical; for lower values, there is no diffraction of this kind.

The validity of Eq. (27.3) for a given spectrum implies that the diffraction order be equal to n.

Since $\tan \varphi = N \Lambda / l$, for small angles of incidence, we can replace the tangent by the sine and obtain instead of Eq. (27.3):

$$n \lambda / \Lambda = 2 N \Lambda / l.$$

[1] S. BHAGAVANTAN and B. R. RAO: Nature, Lond. **158**, 484 (1946); **159**, 267 (1947); **161**, 927 (1948).

Putting λ_0 for the wavelength of the light in vacuo and μ_0 for the refractive index of the medium traversed by the ultrasound wave, we can write:

$$l/\Lambda^2 = 2N \mu_0/n \lambda_0$$

or, when expressing the frequencies in MHz,

$$f^2 l = 2 \cdot 10^{-12} \mu_0 c N/n \lambda.$$

under the assumption $N \geq 1$. If the medium is water and we use the green mercury light, we get:

$$f^2 l = 1100 N/n.$$

The condition for BRAGG'S reflection occurring in the first order of diffraction will therefore be:

$$f^2 l = 1100 N,$$

for the second order:

$$f^2 l = 550 N,$$

and so on.

Contrary to what has been established for normal diffraction, the criterion for BRAGG'S diffraction is independent of the ultrasound vibration amplitude.

28. Consequences of the Doppler effect. The frequency of the light diffracted by the ultrasonic field is altered by Doppler effect. The variation is very small since the speed of light, c_l, is much greater than that of sound, c. However, the effect can lead to considerable interference phenomena between the light of the different orders of the spectrum.

Let ν_0 and λ_0 be the frequency and the wavelength of the light incident normally upon a beam of progressive plane ultrasonic waves. If $\pm \varphi_n$ is the angle by which the original light beam has been diffracted into the symmetric fringes of n-th order, the wavelength of the light will become, for the Doppler effect, $\lambda_n = \lambda_0 (c_l \pm c \sin \varphi_n)/c_e$ and the corresponding frequency ν_n will be:

$$\nu_n = \frac{c_l}{c_l \mp c \sin \varphi_n} \nu_0;$$

but since

$$\sin \varphi_n = n \lambda_n/\Lambda,$$

the frequency of the ultrasound being $f = c/\Lambda$, it follows that:

$$\nu_n = \nu_0 \pm n f.$$

Consequently the light diffracted into the fringes on the same side as the direction of propagation of the ultrasound wave has the frequency $\nu_{n+} = \nu_0 + nf$; in the fringes on the opposite side, the frequency is: $\nu_{n-} = \nu_0 - nf$.

However, the diffraction in the ultrasonic beam is complex and it should be borne in mind that, as the light penetrates into the ultrasonic field, the single light beams undergo a further diffraction and part of the light already sent into the direction of any particular order n is in turn deviated to other orders and so on, until it leaves the ultrasonic beam. To each deviated light beam there belongs its own Doppler effect, but, as can easily be seen in Fig. 72, only light of frequency $\nu_0 + nf$ and $\nu_0 - nf$ appears in the respective orders.

The Doppler effect can be confirmed experimentally by letting the light rays of the first order on the right and left interfere with one another on a screen[1]. The resulting light will be modulated with a frequency: $(\nu_0 + f) - (\nu_0 - f) = 2f$.

---

[1] P. DEBYE, H. SACK and F. COULON: C. R. Acad. Sci. Paris **198**, 922 (1934).

Through every point of the screen, $2f$ interference fringes will thus pass per second. If we then interrupt the incident light periodically with the same frequency $2f$, we get a stroboscopic image of the interference picture, i.e. an indirect image of the ultrasonic waves.

Repeating the experiment with light of any pair of n-th order spectra we obtain the light modulated with the frequency $2nf$ and the stroboscopic image will present an interfringe n-times smaller than the preceding.

In the case of standing waves the above argument becomes more complicated by the fact that another ultrasonic wave propagates with the same velocity but in the opposite

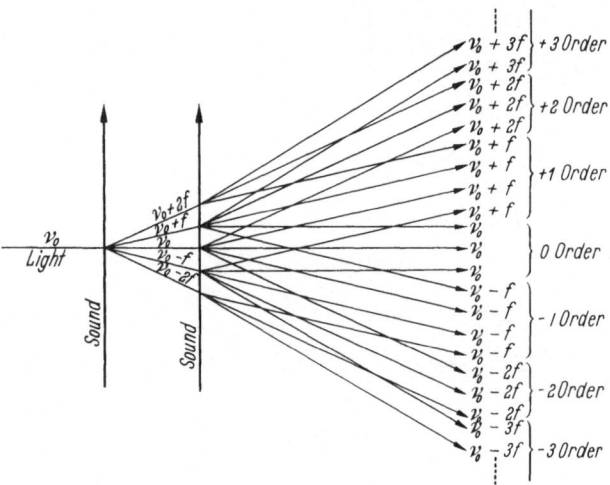

Fig. 72. Doppler frequencies in the light diffracted by progressive plane ultrasonic waves.

direction. In each single diffraction fringe of order n, frequencies $\nu_0 \pm nf$ are certainly present, but they are not the only components appearing in the fringe. In fact, if we consider the diffraction mechanism already explained in the case of progressive waves, we realize the multiplicity of the light components having Doppler frequency which are present in each spectral order. This fact is illustrated in Fig. 73.

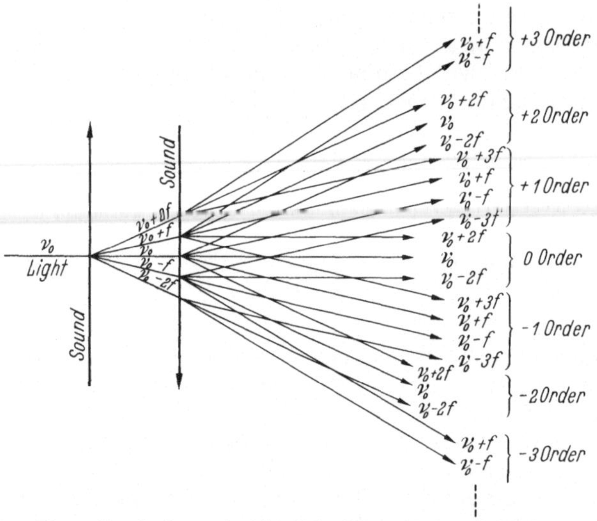

A more complete list of frequencies of the light components in the various orders of diffraction is given in the table[1]. The intensity of these components varies with the intensity of the ultrasonic wave. Frequency spectra and relative intensities for five values of the parameter a, defined in Eq. (26.17a), are given in Fig. 74.

Fig. 73. Doppler frequencies in the light diffracted by standing plane ultrasonic waves.

Repeating, with standing waves the interference experiments, we note that it is not necessary to periodically interrupt the light in order to obtain a stable picture of the interference fringes. Fig. 75 shows three photographs of fringes obtained by BÄR[2]; (a) is the image of interference fringes obtained using only the

---

[1] L. BERGMANN [1].

[2] R. BÄR: Helv. phys. Acta **8**, 591 (1936).

Table. *Doppler frequencies of light diffracted by standing waves in the various diffraction orders*
(v_0 = frequency of the incident light)

Order n	Frequencies							
0	v_0		$v_0 \pm 2f$		$v_0 \pm 4f$		$v_0 \pm 6f$	$\cdots$
± 1		$v_0 \pm f$		$v_0 \pm 3f$		$v_0 \pm 5f$		$v_0 \pm 7f$ $\cdots$
± 2	v_0		$v_0 \pm 2f$		$v_0 \pm 4f$		$v_0 \pm 6f$	$\cdots$
± 3		$v_0 \pm f$		$v_0 \pm 3f$		$v_0 \pm 5f$		$v_0 \pm 7f$ $\cdots$
± 4	v_0		$v_0 \pm 2f$		$v_0 \pm 4f$		$v_0 \pm 6f$	$\cdots$
± 5		$v_0 \pm f$		$v_0 \pm 3f$		$v_0 \pm 5f$		$v_0 \pm 7f$ $\cdots$

light of the two diffraction spectra of the first order; in (b) only the light of the two spectra of the second order is used and in (c) only the light of the third order spectra.

Reuniting by superposition the entire light diffracted, an interference picture similar to that of Fig. 75 a is obtained, but it is much clearer and sharper. It is

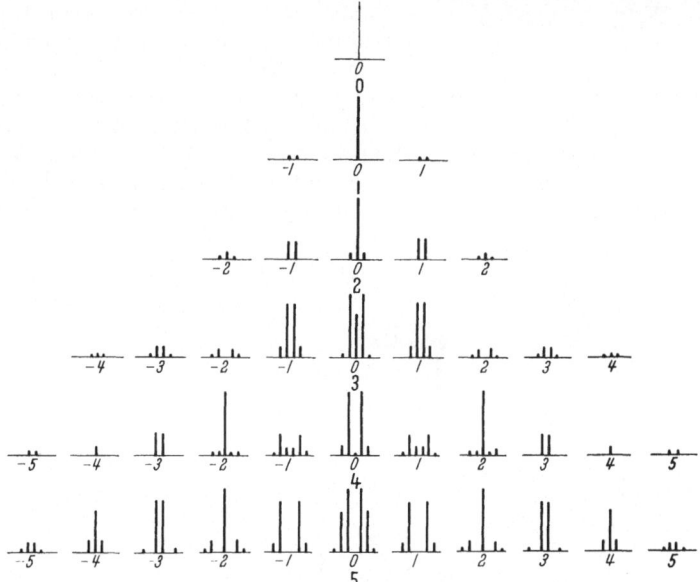

Fig. 74. Frequency spectra and relative intensity of light diffracted by standing waves, depending on the parameter a (for a = 0, 1, 2, 3, 4, 5) (after RAMAN and NAGENDRA NATH).

a direct picture of standing waves in the fluid. BÄR[1] also called attention to the fact that, as regards optical effects, the standing-wave field is equivalent to two separate beams of progressive waves moving in opposite directions, crossed successively by the same light beam.

It also can be seen from the table, that the light of the central fringe (zero order) includes components which by their interference give rise to a periodical variation (in time) of the intensity of light, their fundamental frequency being $2f$. The

---

[1] R. BÄR: Helv. phys. Acta **9**, 678 (1936).

variation is usually not sinusoidal owing to the multiplicity of the harmonic components present. The depth of modulation depends on the parameter a, i.e. on the ultrasound intensity, as can also be concluded from Fig. 74.

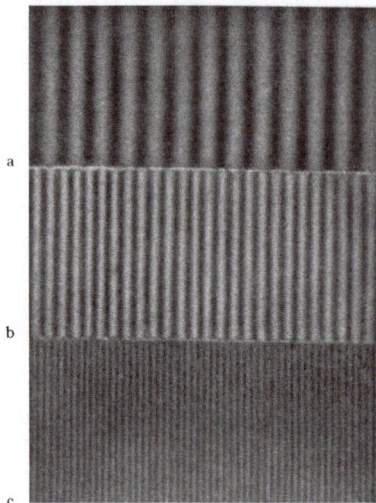

a

b

c

Fig. 75 a—c. Interference fringes obtained using only: a light diffracted in the first order; b light diffracted in the second order; c light diffracted in the third order (after BÄR).

We may get an idea of the origin of this modulation if we bear in mind that the stationary grating is twice cancelled out in each period of the ultrasonic wave; at these moments there is no optical diffraction and all the light falls into the zero order. At any other time, the light is partially distributed through the higher diffraction orders. Hence, the entire diffracted light in its turn, is modulated with the same fundamental frequency $2f$, but with a phase difference of $\pi/2$.

d) Visualization of the ultrasonic field.

29. Method of secondary interferences. Using the device shown in Fig. 76, we can obtain[1], on a projection screen Z, an image of the standing waves produced in a transparent liquid by means of the quartz Q and the plane reflector R facing it. The vessel is similar to that in Fig. 64. The distance from Q to R must be $N\Lambda/2$ with N being an integer in order to obtain a pure field of standing waves. The light source S is in the focus of lens L_1 and is of dimensions small enough to let a beam of parallel light be sent across the ultrasonic field. Another lens L_2 focuses a part of the ultrasonic

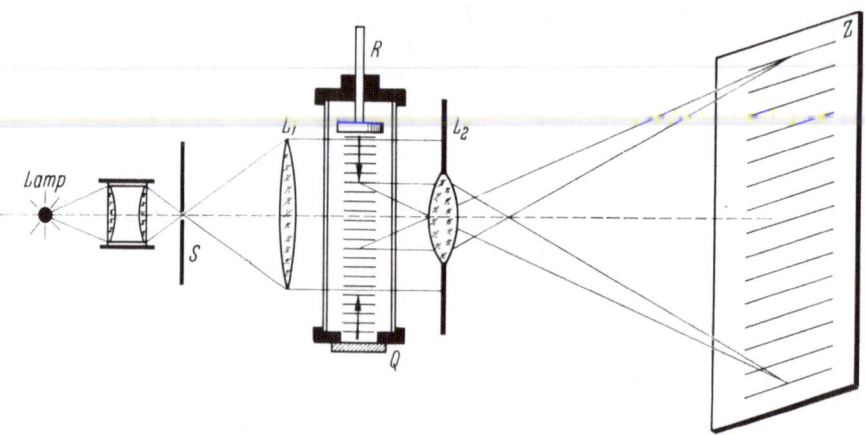

Fig. 76. Optical arrangement for visualization of ultrasonic waves.

field on the screen Z where a series of parallel bright lines will appear. Fig. 77 is a photographic picture of this kind obtained by HIEDEMANN and ASBACH. For every ultrasonic wavelength there are two luminous lines, at a distance from one another corresponding to half an ultrasonic wavelength and depending

---

[1] C. BACHEN, E. HIEDEMANN and H.R. ASBACH: Z. Physik **87**, 734 (1934). — Nature, Lond. **133**, 176 (1934).

on the optical enlargement produced by lens L_2. The origin of these lines can be interpreted in two ways, viz.

(1) As an image of interferences due to the superposition of light diffracted in the spectra of all orders including zero order; it is thus a consequence of the

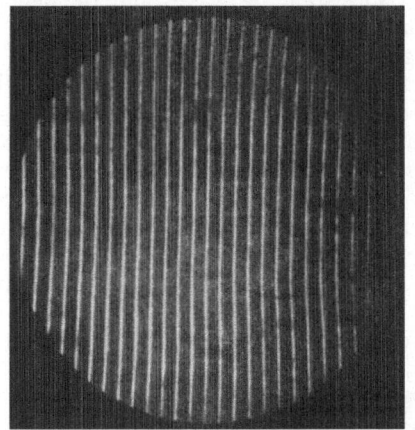

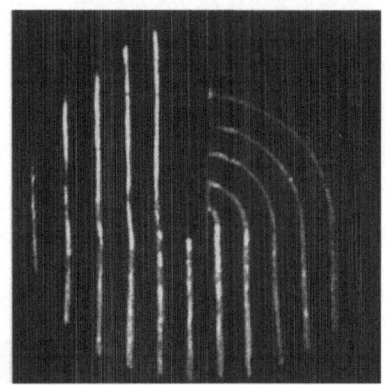

Fig. 77. Photographic image of standing waves (after HIEDEMANN and ASBACH).

Fig. 78. Photographic image of progressive waves (after HIEDEMANN and ASBACH).

multiplicity of Doppler frequencies in the light emerging from the ultrasonic field, as has been shown when discussing BÄR's experiments (p. 134);

(2) as an image of the lines where the light rays converge according to the theory of LUCAS and BIQUARD (p. 128). For the standing waves, however, the position of these lines of convergence oscillates, in the direction of light propagation, between the infinite and a position which is the nearer to the entrance plane of the light into the ultrasonic field, the greater is the intensity of the ultrasounds.

During each period the rays emerging from the beam become in fact parallel in the two successive moments when the standing ultrasonic grating disappears.

Replacing the reflector R by a layer of absorbing material (e.g. glass wool or cork) a beam of progressive waves is obtained in the vessel. One can then get a fixed image of the ultrasonic wave fronts only by stroboscopically modulating the light, e.g. with a Kerr cell, at a

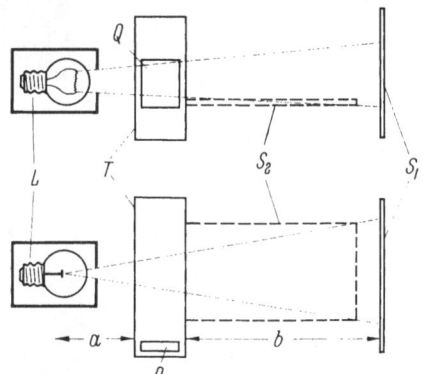

Fig. 79. Optical arrangement for observing ultrasonic waves with divergent light (after BERGMANN and GOEDLICH).

frequency equal to that of the ultrasound. Fig. 78 shows an image of this kind: a metal plate partially interrupts the beam and gives rise to an acoustic diffraction phenomenon clearly visible near the lower edge of the obstacle. Compared with the image of the standing waves, the distance between the luminous lines is twice as great because it corresponds to an entire wavelength.

BERGMANN and GOEDLICH[1] have shown that one can make the waves visible also by illuminating the ultrasonic field with divergent, rather than parallel, light.

[1] L. BERGMANN and H. J. GOEDLICH: Phys. Z. **38**, 9 (1937).

Fig. 79 shows the device used. The source of light is the rectilinear filament of a lamp L, placed parallel to the ultrasonic waves, S_1 is a photographic plate lighted frontally and S_2 is another photographic plate receiving grazing light. The

Fig. 80. Image of standing waves obtained on S_1 plate.

distances a and b are 15 and 50 cm, respectively. Using standing waves and frequencies of the order of a few MHz, one can obtain on S_1 an image as in Fig. 80, and on S_2 an image as in Fig. 81. If the frequency is increased, the whole series of lines more and more decreases in number, until, for values of the order of 10 MHz, only one bright line is left, as shown in the photograph of Fig. 82 taken on plate S_2.

With progressive waves a clear picture is obtainable only at the highest frequencies (above 10 MHz), but it is again only one distinct line, as shown in the photographs of Fig. 83 a, b. One easily observes in Fig. 83 a, or in Fig. 83 b where the direction of ultrasound is slightly inclined with respect to the symmetry

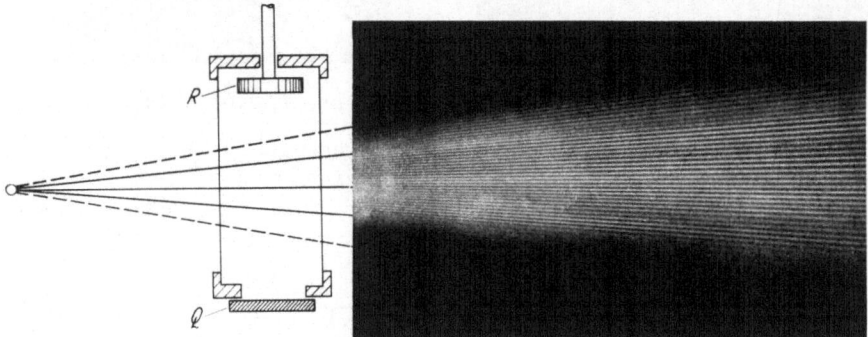

Fig. 81. Image of standing waves obtained on S_2 plate.

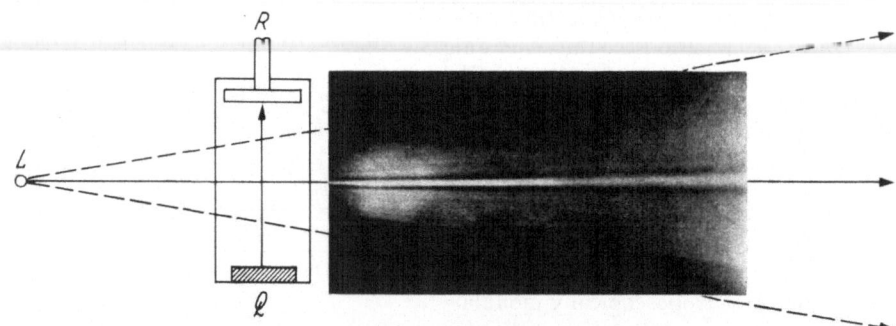

Fig. 82. Image of high frequency (10 MHz) standing waves on S_2 plate.

plane of the light beam, that the luminous line is always directed parallel to the ultrasonic wave fronts. This fact can be explained by the refraction of the light according to the theory of Lucas and Biquard. Indeed, light convergency phenomena are possible only in the directions of the ultrasonic wave fronts; in the other directions the light waves, especially at high frequencies, undergo irregular refraction which leads to a more uniform distribution of the light energy. At low frequencies the divergence of the light beam is obviously less important and we obtain an image containing a greater number of lines. The presence of the line depends,—also in the case of progressive waves—on the fact that, in the prescribed direction, the light exhibits always a maximum of convergence.

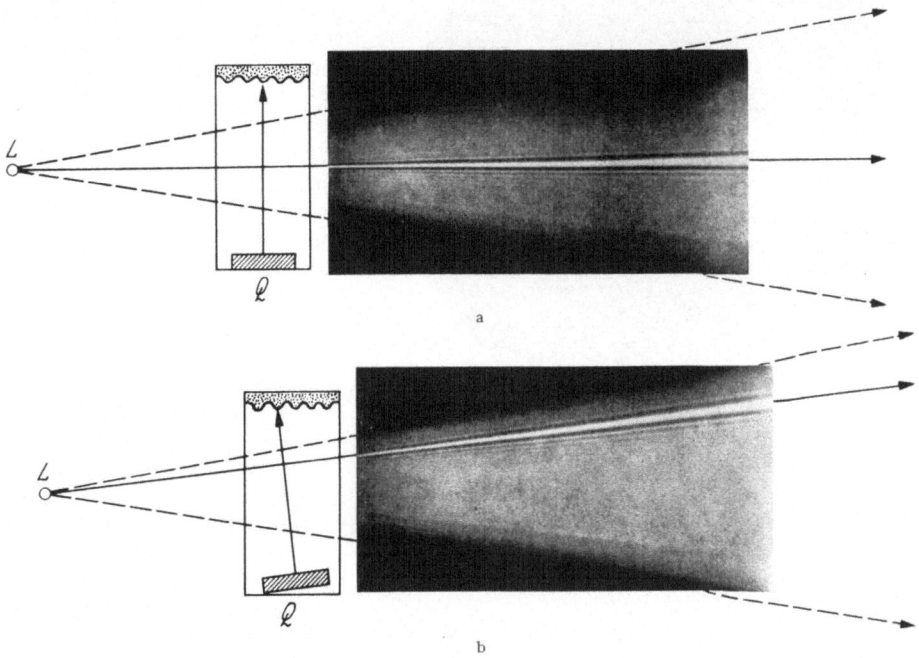

Fig. 83 a and b. Image of progressive waves obtained on S_2 plate.

30. Schlieren method. By this method an image of the configuration of an ultrasonic field is obtained, no matter whether it is formed by progressive or

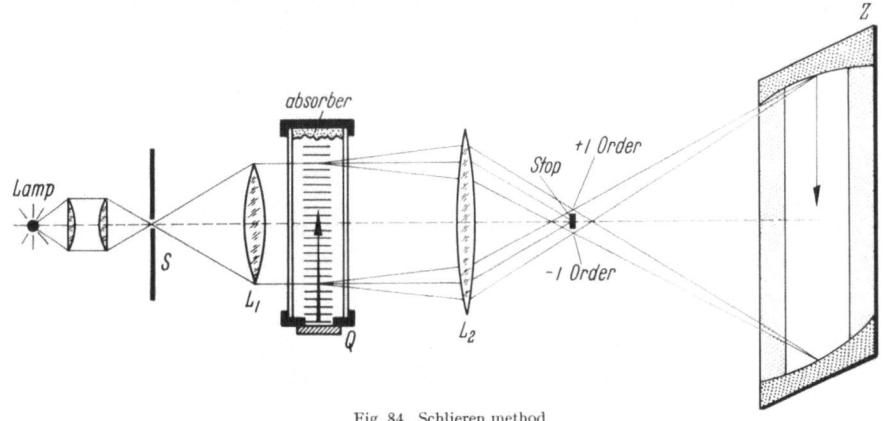

Fig. 84. Schlieren method.

by standing waves. Under certain conditions, the method also enables us to know the distribution of the ultrasonic intensities in the field.

The experimental device is shown in Fig. 84. The optical system resembles that of Fig. 76, except that there is a small opaque obstacle similar in shape to the source in the focus of lens L_2 (which, in this case, has a much greater focal distance). In the absence of ultrasound no light reaches the screen which remains

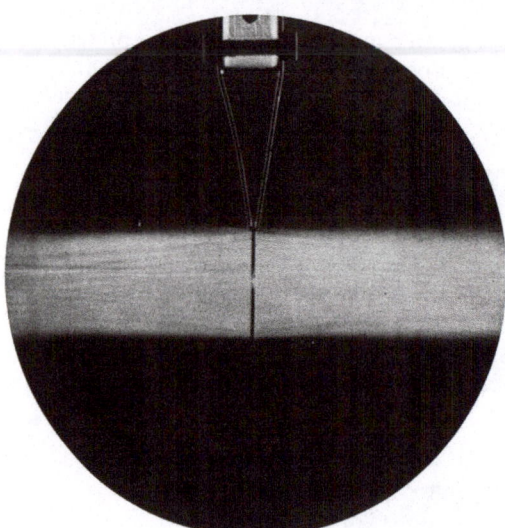

Fig. 85. Schlieren image on an ultrasonic beam produced by a bilateraly radiating quartz.

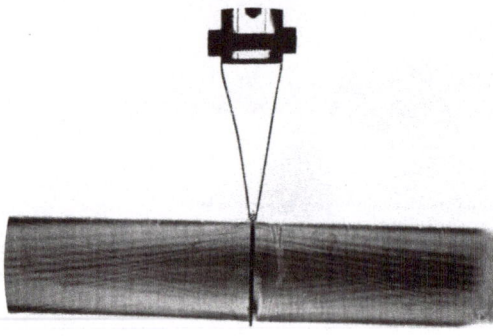

dark. But when the ultrasonic grating forms in the fluid, the light diffracted by it, for the most part is distributed in the spectra on the focal plane of L_2 at the sides of the principal optical axis. This light is not stopped by the obstacle and reaches the screen where it reconstructs the exact configuration of the ultrasonic field.

In order to make visible fields constituted by several ultrasonic beams, however directed, but always normal to the light beam, it is best to use a point source and a disk-shaped stop, since diffraction occurs according to the direction of the ultrasonic beams.

When all ultrasonic beams have the same direction, the light source can also be a rectilinear incandescent filament or an illuminated slit, provided they are arranged at a right angle to the direction of sound propagation. In this case the stop will be stripe-shaped.

Fig. 85 as well as Figs. 46 to 49 and 54 are all examples of images of an ultrasonic field obtained by the schlieren method.

If, instead of stopping the zero-order light, the remaining diffracted light is stopped by means of a diaphragm or a slit, then the image obtained on the screen Z will be produced by subtracting the diffracted light from the zero-order light. This image will be dark on light ground, as shown in Fig. 86.

In any case, within the limits to which the theory by Raman and Nath applies, the intensity of the light in each element of the field image, depends upon the parameters of the field according to the equations given in Sect. 26 and upon the diffraction orders used.

Fig. 86. Image obtained using only zero order light.

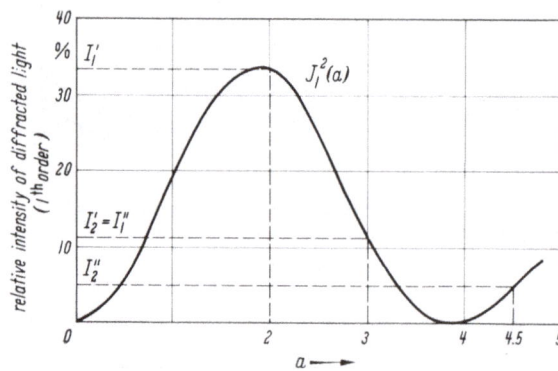

relative intensity of diffracted light (1th order)

Fig. 87. Relative intensity of first order diffraction light.

If the image is formed only by a part of the diffracted light—e.g., if a suitable diaphragm or slit is placed in the focal plane of L_2—the image will show several colours if the light of the source S is not monochromatic. In order to explain this fact, let us suppose that we use only the spectrum of first order and that

Fig. 88. Coloured image of a complex ultrasonic field (after BARONE). (The beam radiated downwards strikes a wedge-shaped reflector giving rise to two reflected beams which come to interfere with the radiation from the other side of the quartz plate.)

the light has only two components of wavelength λ_1 (e.g. red light) and λ_2 (e.g. green light), both having the same amplitude. Let the ultrasound in a certain area A' of the field have such an intensity that the parameter $a\,(=2\pi\,\varDelta\mu\,l/\lambda)$ be $a_1' = 2$ for the light component of wavelength λ_1 and $a_2' = 3$ for the other component. The relative intensity of the two light components, as deduced from the first order curve in Fig. 87, will then become

$$I_1' \approx 33\% \quad \text{and} \quad I_2' \approx 11\%.$$

If we now consider another area A'' of the field where the sound intensity is greater, so that $a_1'' \approx 3$ for λ_1, we shall, consequently, have $a_2'' \approx 4.5$ for λ_2. The intensities of the light components will now become

$$I_1'' \approx 11\% \quad \text{and} \quad I_2'' \approx 5\%.$$

Hence the intensity ratio of the two coloured components will differ according to the intensity of the ultrasonic field, and the resulting image will be thus differently coloured. This effect is shown in Fig. 88 where the ultrasonic field is due to the interference of beams propagating in different directions. The light source radiates white light, and only the first orders of diffraction have been used.

III. Methods of measuring the propagation constants.

a) Continuous wave methods.

Two categories of methods must be distinguished: continuous wave methods and non-stationary methods.

In the first category we will consider interferometers and resonance methods. Interferometers, in which electrical or optical detection may be used, are suitably employed only for measurements in fluids. Resonance methods are chiefly used in the study of solids.

31. Electrical detection interferometers. The first device of this kind was PIERCE's[1]. His model has later been modified in many ways[2,3] but its principle can be described as follows:

The fluid column between the piezoelectric plate Q and the reflector R exactly parallel to it, is in resonance at the driving frequency if the distance l between the radiating surface of Q and the reflecting surface of R is equal to a whole number N of half wavelengths of the ultrasound in the fluid:

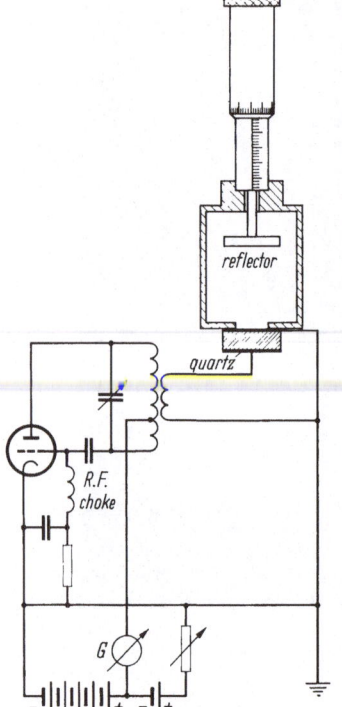

$$l = N \Lambda/2.$$

These are the optimal conditions for the establishment of a system of standing waves. When l varies, the load impedance Z_l of the quartz plate varies periodically. Supposing R reflects perfectly, we can write:

$$Z_l = A \operatorname{Tan}\left[\alpha l + j\left(\frac{\omega}{c} l + \frac{\pi}{2}\right)\right] \quad (31.1)$$

Fig. 89. Basic circuit of electrical detection interferometer.

where α is the absorption coefficient for the amplitude and A is a constant which is reduced to the characteristic impedance ϱc of the medium in the ideal case in which the fluid is wanting absorption $(\alpha=0)$.

If the transducer is coupled to its driving oscillator as shown in Fig. 89, the variations of the load impedance are transferred to the anodic circuit of the

[1] G.W. PIERCE: Proc. Amer. Acad. Arts Sci. **60**, 271 (1925).
[2] J.C. HUBBARD: Phys. Rev. **38**, 1011 (1931); **41**, 523 (1932).
[3] F.E. FOX: Phys. Rev. **52**, 973 (1937).

vacuum tube and, for each maximum of Z_l which, as indicated in (31.1), will occur every time when:

$$l = N\,\pi\,c/\omega = N\,\Lambda/2,$$

there will be a minimum of the anodic current i, and vice versa. The intensity of current in the galvanometer G, plotted versus the positions of the reflector is shown in Fig. 90. The reflector is displaced by means of the micrometer screw M and the distance, measured between two successive maxima of i, is $\Lambda/2$. If the frequency of the oscillator is known, the velocity can be obtained. In practice, one measures the displacement L corresponding to a sufficiently large number N of peaks so as to increase the precision of the measurements. Therefore:

$$c = 2L\,f/N. \qquad (31.2)$$

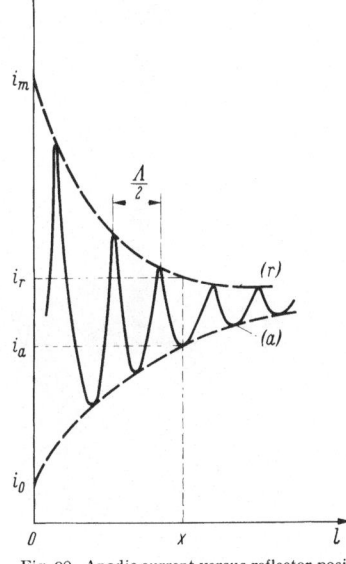

The maxima and minima of the current i determine, on the $i\,l$ plane, two curves r and a from which the absorption coefficient α can be calculated. If, indeed, i_m and i_0 are the extrapolated maximum and minimum values of the current and i_r, i_a are the maximum and minimum values of the current corresponding to a value x of l, we can show that the equation

$$\mathrm{Tan}\,\alpha x = \left[\frac{\left(\dfrac{i_m}{i_r}-1\right)\left(\dfrac{i_m}{i_0}-\dfrac{i_m}{i_a}\right)}{\left(\dfrac{i_m}{i_0}-1\right)\left(\dfrac{i_m}{i_a}-1\right)}\right]^{\frac{1}{2}}$$

applies[1], and permits us to determine the value of α.

Fig. 90. Anodic current versus reflector position (after Hueter and Bolt).

Another way of using the interferometer is by keeping the distance L fixed, but so as to have a system of standing waves, and to explore the space between the piezoelectric plate and the reflector with a probe. By moving the probe along the direction of sound propagation, a series of maxima and minima is found which succeed each other at distances $\Lambda/4$. The velocity and absorption can also be measured by this method.

A fixed path interferometer was proposed by Matta and Richardson[2] for the measurement of vapours propagation constants. The detector is a hot-wire probe consisting of a very thin wire through which a current passes. When the wire is hit by a sound wave (a standing wave in this case) its resistance varies proportionally to the velocity of the fluid particles in contact with it. If the wire is directed parallel to the wave fronts and displaced along the ultrasonic beam, the variation of its resistance measured by means of a suitable bridge gives an indication of the maxima and minima of vibration velocity, i.e. of the antinodes and nodes of the standing wave.

In these methods the precision can reach 10^{-3} in the case of velocity measurements, and a few percent in the case of absorption measurements.

32. Optical detection interferometers. These interferometers are primarily intended for measuring the propagation velocity of ultrasound in liquids. The precision may reach values of the order of 10^{-4}. In a transparent liquid a system

[1] T. Hueter and R. Bolt [3].

[2] K. Matta and E. G. Richardson: J. Acoust. Soc. Amer. **23**, 58 (1951).

of standing waves is produced by means of a piezoelectric plate and a reflector, arranged in the usual way. The vessel containing the liquid has glass walls, so that a parallel light beam can hit the ultrasonic field at a right angle to the sound propagation. Using the method of secondary interference (see Sect. 29) an enlarged image of the standing waves is projected on a screen and these waves, of course, will appear as a series of parallel luminous lines. If the vessel is now displaced by means of a screw, along the direction of sound propagation, the image of the waves, too, will undergo a parallel displacement on the screen. If, for a total displacement L of the vessel, N is the number of lines passing a reference line drawn on the screen, it will be easy to obtain the velocity by using the simple Eq. (31.2), the frequency f being known.

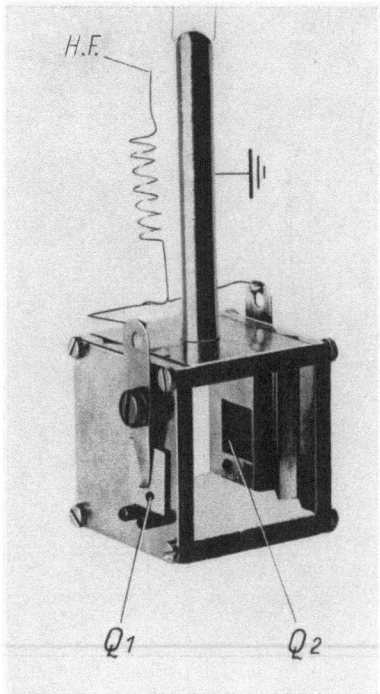

Fig. 91. Interferometer vessel with two opposed quartzes (after GIACOMINI).

A useful modification of this method was contributed by GIACOMINI[1] who substituted for the system of standing waves two beams of progressive waves of the same frequency, radiated by two separate quartz plates. The two beams propagate in opposite direction and their axes are parallel, but do not coincide. The light passes successively the two beams but, as BÄR has shown, the image produced on the screen is identical with that one would get if the two beams were superposed as in standing waves. The advantage of this modification consists in the elimination of the reflector and in the fact that the image of the standing waves is always very good even if the velocity is changed.

Fig. 91 shows the interferometric vessel with two opposed quarzes Q_1 and Q_2.

Optical methods generally proved very helpful, even in the least promising cases. For example, NOZDREW[2] succeeded in measuring velocities in the critical region of a liquid-vapour system, notwithstanding the strong opalescence of the fluid in this region. Of course, the precision lowers to a few percent.

An optical detection interferometer capable of velocity measurements also in opaque liquids has been proposed by BARONE[3]. A quartz plate (Fig. 92) radiates two plane wave beams from its two opposed faces, one into the upper liquid under examination, and the other into an insulating and transparent auxiliary lower liquid. A reflector R is placed in the upper vessel and sends the incident supersonic beam, through the quartz plate, down into to the auxiliary liquid. No appreciable attenuation results from crossing the quartz plate since this plate is obviously in resonance with the wave frequency. There are thus two ultrasonic beams in the auxiliary liquid and they interfere with each other. Since the surface of the reflector is slightly inclined with respect to the quartz plate, the two interfering beams, as shown in Fig. 93, give rise to inter-

[1] A GIACOMINI: Rend. Accad. naz. Lincei **2**, 791 (1947).
[2] V. F. NOZDREW: Soviet Phys. Acoustics **1**, 249 (1955).
[3] A. BARONE: Nuovo Cim. **5**, 717 (1957).

ference lines along the bisector of angle α formed between the directions of the two beams. Through two windows in the wall of the auxiliary vessel, a beam of parallel light passes across the ultrasonic field in a direction which is normal to the plane of angle α and, by means of the schlieren method, an image of the interference lines is projected on a screen (Fig. 94).

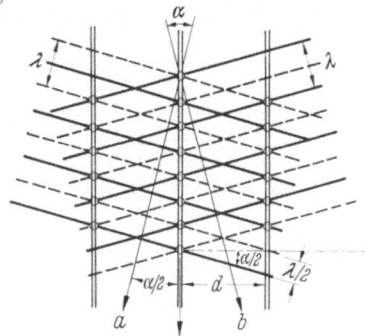

Fig. 93. Interference between two beams of progressive plane waves.

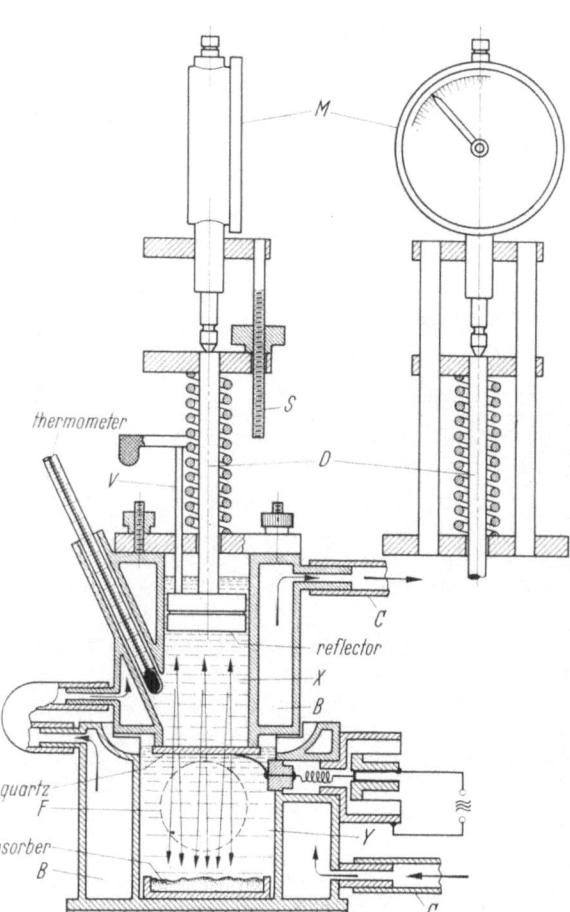

Fig. 92. Interferometer with auxiliary liquid (after BARONE).

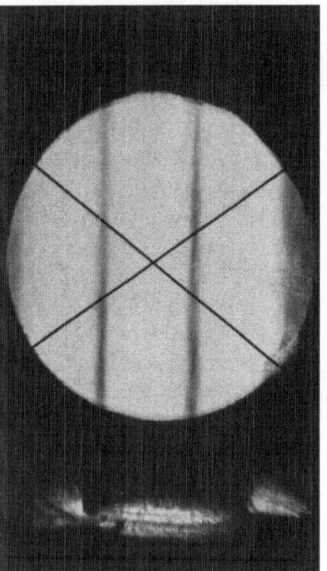

Fig. 94. Interference lines obtained in the auxiliary liquid of the interferometer.

When the reflector is displaced in the direction of the incident beam, the phase relation between the two beams changes in the auxiliary liquid and each inter-ference line is displaced parallel to itself.

For a displacement of the reflector equal to half a wavelength of the ultra-sound in the test liquid, the phase variation is 2π and, in the interference picture, the lines re-take their previous positions.

The displacement L of the reflector (measured by the centesimal comparator M) which is necessary to cause a certain number N of interference lines to pass across a reference line, drawn on the screen, renders the wavelength Λ and hence, if the frequency is known, the velocity can be calculated using Eq. (31.2).

If the angle of incidence on the reflector is small, there are large distances between the interference lines, even at the highest frequencies (10 to 15 MHz). This gives a considerable precision of approximately $2 \cdot 10^{-4}$ to this method. If the incidence angle α lies above 1° or 2°, simple geometrical considerations on the reflected beam show that the velocity measured (c^*) must be corrected, using the equation:

$$c = \frac{1 + \cos 2\alpha}{2} c^*,$$

c being the corrected value.

33. Measurement of the absorption of progressive waves. In a strictly plane progressive wave, the intensity varies according to Eq. (23.9):

$$I_x = I_0 e^{-2\alpha x},$$

where α is the absorption coefficient for the amplitude. Any method of measuring ultrasonic intensity can therefore be used to calculate the coefficient α.

Using a pendulum (as e.g. Biquard[1]) one can measure the radiation pressure Π_x at different distances x from the source and draw $\log \Pi_x$ versus x. The straight line thus obtained follows the equation

$$\log \Pi_x = \text{const} - 2\alpha x \tag{33.1}$$

and from its slope, α is readily determined.

Measurements are reliable only if no standing waves are allowed to form between the detector and the sound source. This is obtained, for example, if the interface of the detector is completely absorbing or if it is so shaped as to reflect away the incident wave far off the source. The disturbing action of the hydrodynamic flow in the ultrasonic beam can be removed by using a suitable screen (see Sect. 24).

Optical methods of detection are widely used in this kind of measurements, if the liquids to be examined are transparent. The experimental equipment then is similar to that of Fig. 65. In this case monochromatic light is used. The light beam crossing the ultrasonic grating has a small cross-section because the relative intensity of the field must be measured at several points. On the focal plane of lens L_2 a screen is arranged which, through a slit, permits only zero-order light to reach a photoelectric cell. If the ultrasonic power is sufficiently low (no second order present), the photoelectric current depends linearly on the sound intensity (see Fig. 70).

Displacing the vessel in the direction of the propagation in order to analyse successive parts of the ultrasonic beam, we can draw, for the photoelectric current, a diagram similar to that of Eq. (33.1) and calculate the absorption coefficient α. Instead of working with zero-order light, we can use first-order diffraction light[2] whose intensity is proportional to the ultrasonic energy within the same range of sound intensities.

Precisions are of the order of a few percent.

34. Resonance methods. The sample to be examined, in the shape of a rod or a plate, is caused to vibrate (in extensional, torsional, or flexural oscillation) by means of an alternative force suitably applied. Measurements are carried out by determining the frequencies of the driving force to which the resonance corresponds on one of the vibration modes of the sample.

---

[1] P. Biquard: C. R. Acad. Sci. Paris **193**, 226 (1931); **197**, 309 (1933).
[2] D. Sette: Nuovo Cim. **7**, 55 (1950).

The shape of the sample and the particular manner in which the driving force is applied to it, determine the type of oscillation obtained; the dimensions and resonance frequencies allow the velocities of propagation of the corresponding waves to be calculated. From these latter, if the density is known, the elastic constants of the material are easily determined. The attenuation of the various types of elastic waves is usually calculated by means of the Q-factor of the vibrating element deduced from the resonance curve.

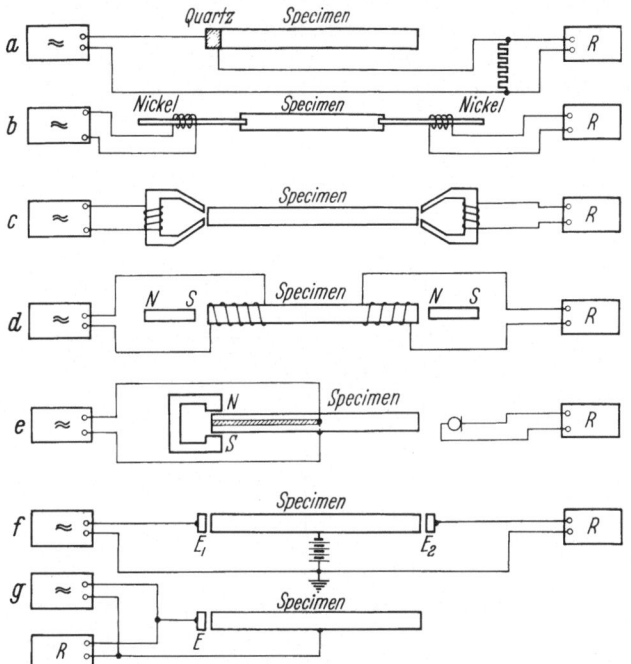

Fig. 95 a—g. Some methods of exciting extensional vibrations in a rod (after GIACOMINI).

In order to increase accuracy, the sample, if possible, is placed in a vacuum.

In Fig. 95 some ways of exciting extensional oscillations in a rod are indicated[1]. The method shown in Fig. 95a illustrates how the vibromotive force is obtained by means of a piezoelectric crystal[2,3]. The frequency of the driving voltage is caused to vary until the rod vibrates at resonance in the mode desired. The crystal is connected with the detector which enables the conditions of resonance to be ascertained.

In Fig. 95b the driving force and the detection are ensured by means of two magnetostrictive transducers connected with the rod[4].

The method shown in Fig. 95c can be used only in the case of ferromagnetic solids. Excitation and detection are obtained electromagnetically[5].

In Fig. 95d the driving force is obtained by interaction between an external magnetic field and eddy currents generated at one end of the rod by means of a suitable coil[6].

[1] A. GIACOMINI: Accad. Lincei: Recenti progressi della scienza delle costruzioni. 1956.
[2] S.L. QUIMBY: Phys. Rev. 25, 252 (1925).
[3] L. BALAMUTH: Phys. Rev. 45, 715 (1934).
[4] E.G. STANFORD: Nuovo Cim. 7, Suppl. No. 2, 333 (1950).
[5] R.L. WEGEL and H. WALTER: Physics 6, 141 (1935).
[6] C. ZENER, F.C. ROSE and R.H. RANDALL: Phys. Rev. 56, 343 (1939).

In Fig. 95e an application of the electromagnetic excitation discussed in Sect. 4 is shown. The alternating current circulates in a thin metallic layer on one end of the bar and interacts with an external magnetic field normal to it. In the equipment shown here, the detection is obtained by means of sound radiation from the other end of the bar.

In Fig. 95f the excitation is obtained by electrostatic attraction between an electrode E_1 and one end of the metal rod[1]. The detection system is also electrostatic, based on the periodical variation of capacitance between the other end of the vibrating rod and a second electrode E_2.

As seen in Fig. 95g the electrostatic method can be simplified considerably by using a single electrode[2]. In this case the variations of the capacitance C between

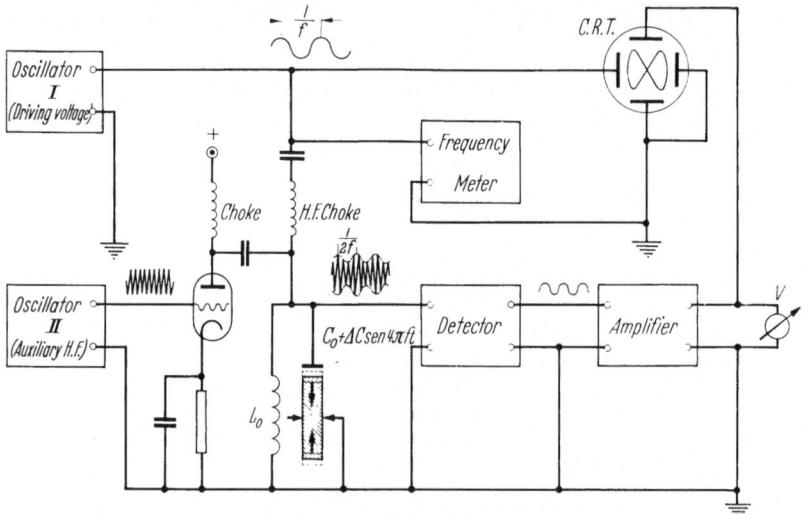

Fig. 96. Electrostatic method using a single electrode (after Bordoni and Nuovo).

the electrode and the rod are detected by the effect which they exert on an auxiliary high frequency oscillation. The capacity C, in fact, is a part of the tank circuit of an auxiliary oscillator. In this way the vibrations of the rod produce a frequency modulation and thus they are detected.

In a variant of this method[3], the periodic component of capacity C causes variations in the tuning of a resonant circuit, inserted between the auxiliary oscillator and the detecting circuit.

Under static conditions, the frequency f_a of the auxiliary oscillator corresponds to one of the two points of maximum slope of the resonance curve of the tuned circuit so that, when the sample starts to vibrate, the high frequency voltage across this circuit is modulated according to the mechanical oscillation. A block diagram of the apparatus is shown in Fig. 96. The driving oscillator I sends to the electrode E a voltage of frequency f which, by electrostatic action, communicates to the rod a vibromotive force of frequency $2f$. If this frequency corresponds to the mechanical resonance of the rod, the capacitance of the electrode, under the above conditions, modulates the amplitude of the electrical wave produced by the oscillator II, whose frequency is much higher than $2f$. The amplitude

[1] R.B. Jacobs and D. Bancroft: Rev. Sci. Instrum. **9**, 279 (1938).
[2] P.G. Bordoni: Nuovo Cim. **4**, 177 (1947).
[3] P.G. Bordoni and M. Nuovo: Private communication 1961.

of the modulating signal is measured by means of a voltmeter and compared in a cathode-ray tube with the driving voltage. As the frequency f varies, the voltmeter readings allow the resonance curve of the rod to be determined.

Fig. 98 shows how different types of oscillation can be produced in the sample by means of electrostatic excitation.

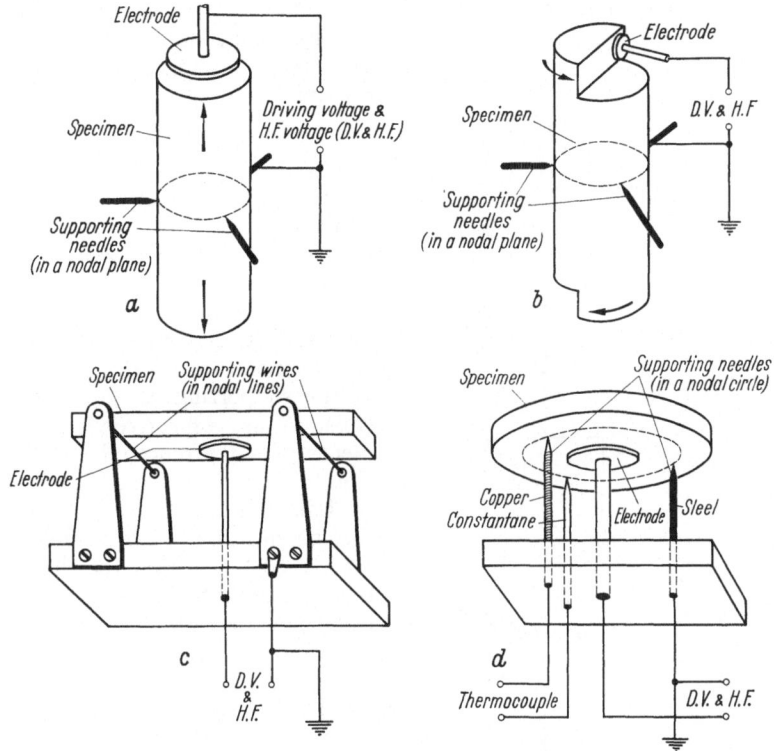

Fig. 97 a—d. Electrostatic methods of exciting various types of vibrations. a Extensional vibrations (after BORDONI); b torsional vibrations (after BORDONI and NUOVO); c flexural vibrations in a bar (after BARDUCCI and PASQUALINI); d flexural vibrations in a circular plate (after BORDONI, NUOVO and VERDINI).

Using the method shown in Fig. 97a, the sample, a cylindric rod, is excited according to its *extensional* vibration modes. The frequency of the lowest mode is:

$$f = \frac{1}{2l}\left(\frac{E}{\varrho}\right)^{\frac{1}{2}},$$

where l is the length of the sample.

In Fig. 97b, owing to the dissymmetry of the driving force with respect to the axis of the sample, *torsional* waves are generated[1]. The frequency of the fundamental mode is

$$f = \frac{1}{2l}\left(\frac{\mu}{\varrho}\right)^{\frac{1}{2}}.$$

To obtain *flexural waves in a prismatic rod* the arrangement shown in Fig. 97c is used[2]. The rod is fastened by two wires in correspondence with the nodal lines.

---

[1] P. G. BORDONI and M. NUOVO: Acustica **4**, 184 (1954).
[2] I. BARDUCCI and G. PASQUALINI: Nuovo Cim. **5**, 416 (1948).

The frequency of the vibration mode with two nodes is:

$$f = \frac{d}{0.975\, l^2} \left(\frac{E}{\varrho}\right)^{\frac{1}{2}},$$

where d is the thickness.

In Fig. 97d a *circular plate* is shown *in flexural vibration*, supported by three needle points in a nodal circle[1]. For a thin plate having thickness d and radius R, the frequency of the vibration mode with m nodal circles and n nodal diameters is:

$$f = a_{m,n} \frac{d}{R^2} \left[\frac{E}{\varrho\,(1-\sigma^2)}\right]^{\frac{1}{2}},$$

where $a_{m,n}$ is an eigenvalue which depends on m and n. For the first symmetrical mode ($m=1$, $n=0$) we have: $a_{1,0}=0.4170$ and $r_{1,0}=0.678R$, r being the radius of the nodal circle. A useful device is also shown here for controlling the temperature of the sample. Two of the supporting needles (copper and constantane, respectively) are the elements of a thermocouple whose junction is the metal of the plate.

Using all these methods, absorption can be measured with a precision of 10^{-3} to 10^{-2}. On the other hand, since the precision attainable in the resonance frequency measurement is very high, provided low dissipations occur, sound velocity measurements may reach an accuracy of 10^{-5}.

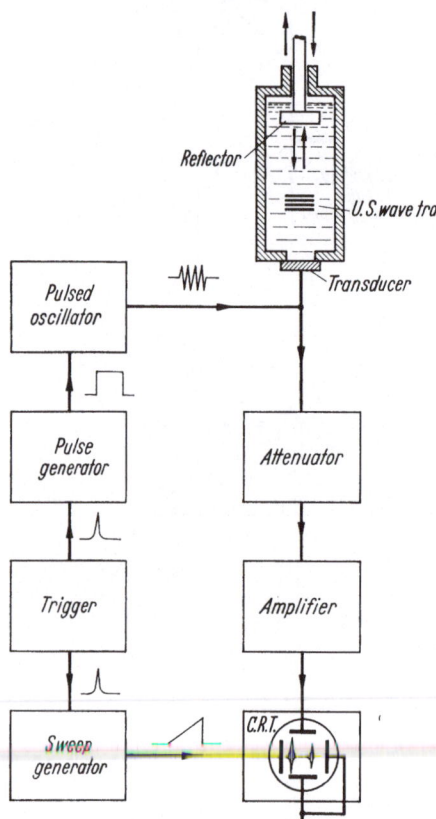

Fig. 98. Pulse method.

b) Non-stationary methods.

35. Pulse method. A short ultrasonic wave train radiated into the material under examination is detected by a receiver after having travelled a distance of known length. Measurement of the time taken, enables the propagation velocity to be determined. On the other hand, if the intensity of the received pulse is measured for different values of the distance, the absorption coefficient in the medium can be deduced.

Various types of circuit were used to carry out this technique[2-4]. Essentially, the method may be described as in Fig. 98 which is designed for measurements in liquids.

At the same moment at which a radiofrequency pulse is applied to the crystal transducer, horizontal displacement of the luminous spot on the screen of a cathode-ray tube, begins. The ultrasonic wave train, generated by the crystal, propagates in the liquid and after reflection returns to the transducer which transforms it

[1] P. G. Bordoni, M. Nuovo and L. Verdini: Nuovo Cim. **14**, 273 (1959).
[2] J. R. Pellam and J. H. Galt: Chem. Phys. **14**, 608 (1946).
[3] J. M. Pinkerton: Proc. Phys. Soc. Lond. **62**, 162, 286 (1949).
[4] B. Chick, G. Anderson and R. Truell: J. Acoust. Soc. Amer. **32**, 186 (1960).

again into an electric signal. This signal is amplified and recorded by the cathode-ray tube with a delay equal to the time taken by the ultrasonic pulse to cover the entire distance in the liquid.

If the sweep velocity in the oscillagraph is known and constant, the measurement of the delay is easily carried out. For a greater precision it is preferable to vary the ultrasonic path by a known distance l and to bring back the oscillographic signal to the original position by means of a controlled delay τ of the sweep oscillation. The measured velocity is obviously:

$$c = l/\tau.$$

In order to determine the absorption coefficient one measures the variation of amplitude of the received pulse produced by a known displacement of the reflector. The measurement is made easier by using a calibrated attenuator, which, after the displacement is accomplished, brings back the height of the pulse to its initial value.

The precision of this method, which is very suitable in the high frequency range, is of the order of a few percent. In special cases of velocity measurements, the precision may reach 10^{-3}.

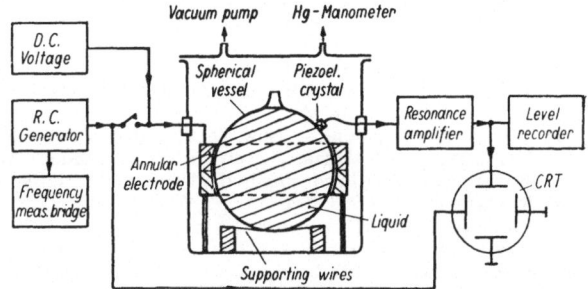

Fig. 99. Decay method for measuring sound absorption in liquids (after KURTZE and TAMM).

36. Decay methods. A last category of methods of measurement of absorption coefficient, which belongs partly to the continuous wave methods and partly to the pulse method, must at least briefly be mentioned here. It applies well to solids and liquids, provided that internal losses of the material are low.

By means of a suitable driving source the specimen is excited to one of its modes of vibration and then the source is suddenly turned off. The exponential decay of the vibration amplitude is presented by a logarithmic recorder as a rectilinear plot whose slope enables the logarithmic damping decrement δ and, in turn, the absorption coefficient α to be calculated:

$$\alpha = \delta/\Lambda.$$

For solid specimens, extensional, torsional and flexural vibrations can be analyzed with this technique. Devices which produce very low losses in exciting and detecting vibrations, must be employed. The electrostatic method is often preferred. The working frequency is generally low in order to obtain a small decay of vibrations and therefore a higher accuracy.

In the case of liquids, a spherical glass vessel filled with the liquid to be investigated is excited electrostatically in a radial mode at fundamental resonance[1]. For this purpose one electrode is painted with a conductive varnish on a ring-shaped section of the spherical vessel; the other is an external ring electrode. Detection is attained by means of a piezoelectric pick-up cemented to the vessel. Fig. 99 shows a block diagram of this apparatus.

The precision of absolute measurements of absorption is low because it is difficult to calculate exactly the damping due to the wall of the resonator and to

---

[1] G. KURTZE and K. TAMM: Acustica **3**, 33 (1953).

the supporting device. However, comparative measurements with a standard liquid can be carried out with a good accuracy. The useful frequency range extends at several kilohertz.

Another method, called *reverberation method*, similar in principle to the one just described, has been developed by MEYER and SKUDRZYK[1]. A vessel of low degree of symmetry filled with the liquid is excited by a frequency band to a large number of normal modes of oscillation, so that a diffuse sound field is formed. If the normal modes are nearly uniformly influenced by the boundary losses, an approximately exponential decay of the sound energy results. This is partly due to the effect of the walls, but, comparing the decay constant with that found for a standard liquid, it is easy to deduce the damping due to the liquid investigated.

General references.

[1] BERGMANN, L.: Der Ultraschall. Stuttgart: S. Hirzel 1954.
[2] HIEDEMANN, E.: Ultraschallforschung. Berlin: W. de Gruyter Co. 1939.
[3] HUETER, T.F., and R.H. BOLT: Sonics. New York: J. Wiley & Sons 1955.
[4] CADY, W.G.: Piezoelectricity. New York, London: McGraw-Hill Book Co. 1946.
[5] MASON, W.P.: Piezoelectric crystals and their application to ultrasonics. New York: D. Van Nostrand Co. 1950.
[6] CARLIN, B.: Ultrasonics. New York: McGraw-Hill Book Co. 1960.
[7] CRAWFORD, A.E.: Ultrasonic engineering. London: Butterworths Sci. Publ. 1955.
[8] MATAUSHEK, J.: Einführung in die Ultraschalltechnik. Berlin: VEB Verlag Techn. 1957.
[9] RICHARDSON, E.G.: Ultrasonic physics. London: Elsevier Publ. Co. 1952.
[10] VIGOUREUX, P.: Ultrasonics. London: Chapman & Hall Ltd. 1950.
[11] HERFORTH, L., u. H.M. WINTER: Ultraschall. Leipzig: B. Teubner 1958.
[12] WOOD, A.B.: A Text-book of sound. London: G. Bell & S. Ltd. 1944.

---

[1] E. MEYER and E. SKUDRZYK: Unpublished report 1943. See Acustica **3**, 33 (1953).

High Frequency Ultrasonic Stress Waves in Solids.

Rohn Truell and Charles Elbaum.

With 49 Figures.

1. Introduction. The propagation behavior of high frequency stress waves in solids is determined by the measurement of the attenuation and velocity of the stress waves as a function of whatever variables are of interest. Such measurements permit one to study the influence, on the propagation behavior, of any property of a solid that is sufficiently well coupled to the lattice. Various types of defects in the lattice are, for example, closely coupled with the lattice and a change in the type or density of defects will, in such cases, change the propagation behavior of a stress wave. Altogether there are ten or more known "interactions" between stress waves and various properties of a solid. Electrons are coupled to the lattice and interact with stress waves in at least two distinct ways. Lattice waves interact with other stress waves, and nuclear interaction with stress waves has been studied. Radiation induced defects of several types have been investigated, and paramagnetic resonance effects can be influenced by stress waves of the proper frequency.

Before going further with a discussion of such interactions and their consequences, a brief review of certain features of elastic wave propagation will be given.

I. Stress waves.

2. Stress, strain, and displacement relations. Study of the propagation of stress waves [1] to [5] in an elastic solid yields information about the velocities of the stress waves and about the attenuation or energy loss of these waves. The elastic coefficients of a solid may be obtained from the velocities and the density, and these elastic coefficients are in turn connected with interatomic binding forces and compressibility. Deviations from purely elastic behavior and the connection of such behavior with loss mechanisms or non-linear effects will be introduced later.

In an isotropic elastic solid material, which occupies the entire space, there are two types of elastic waves that can be propagated through the material. One type of wave is variously called compressional, longitudinal, or dilatational where the particle motion in a plane wave in the solid is in the direction of propagation. The second type of wave has the names transverse, shear, or distortional where the particle motion in a plane wave is transverse to the direction of the propagation[1]. When the solid is finite in extent, hence bounded by surfaces, surface waves may also be propagated.

The *stress tensor* σ_{ij} represents the components of force per unit area acting on an element of area in the solid; in this discussion the first subscript denotes the direction of the normal to the plane on which the stress component acts and the second subscript denotes the direction of the stress component. Because of symmetry conditions it makes no difference if the indices are interchanged.

---

[1] In general, however, waves are neither longitudinal nor transverse, as pointed out later.

The *strain tensor* ε_{ij} defines the deformation for small deformation, in the neighborhood of a point specified in the undeformed medium by the position vector $\boldsymbol{x}\,(x_1,\,x_2,\,x_3)$. The strain tensor may be written in terms of the displacement gradients $\partial s_i/\partial x_j$ where s_i are the components of the displacement vector $\boldsymbol{s}\,(s_1,\,s_2,\,s_3)$

$$\varepsilon_{ij} = \frac{1}{2}\left(\frac{\partial s_i}{\partial x_i} + \frac{\partial s_i}{\partial x_j} + \frac{\partial s_k}{\partial x_i}\,\frac{\partial s_k}{\partial x_j}\right). \tag{2.1}$$

With small deformations and displacement gradients the products are usually neglected as second order terms so that

$$\varepsilon_{ij} = \frac{1}{2}\left(\frac{\partial s_j}{\partial x_i} + \frac{\partial s_i}{\partial x_j}\right) = \frac{1}{2}\,(s_{j,i} + s_{i,j}) \tag{2.2}$$

where the comma notation for differentiation has been used.

The ε_{ij} defined by (2.1) have a simple geometrical meaning such that $\varepsilon_{ij}\,(i=j)$ is the change in length per unit length of a straight line segment originally parallel to the x_j axis. $\varepsilon_{ij}\,(j \neq i)$ is twice the change in an angle whose sides were originally parallel to the x_j and x_i axes.

The displacement $\boldsymbol{s}\,(s_1,\,s_2,\,s_3)$ may consist of three parts [5]

$$\boldsymbol{s} = \boldsymbol{s}_0 + \boldsymbol{s}' + \boldsymbol{s}'' \tag{2.3}$$

where $\boldsymbol{s}_0$ is a rigid body translation, $\boldsymbol{s}'$ is a rigid body rotation, and $\boldsymbol{s}''$ is the local deformation. If we consider only the last term $\boldsymbol{s}''$, the components of this may be written

$$s_i'' = \varepsilon_{ij}\,x_j \quad (i = 1,\,2,\,3). \tag{2.4}$$

The assumption that strain is proportional to stress, for small strains, is *Hooke's law*. The elastic coefficients are defined in terms of a linear stress strain relation[1]

$$\sigma_{ik} = c_{ikjl}\,\varepsilon_{jl}. \tag{2.5}$$

There are nine equations with nine coefficients each, or 81 coefficients altogether.

This expression is the most general linear stress strain law yielding zero stress for zero strain. It follows from the symmetry of the stress and strain tensors (no body moments) that

$$c_{ikjl} = c_{kijl} = c_{iklj}. \tag{2.6}$$

These symmetry conditions reduce the number of coefficients from 81 to 36, and the most general linear stress-strain relation will therefore depend on 36 coefficients. In addition the condition that there exist an elastic potential (i.e. that the medium be elastic) imposes the further symmetry condition

$$c_{ikjl} = c_{jlik}. \tag{2.7}$$

The last condition amounts to having the strain energy be a function of state and independent of the path by which the state is reached. This condition for the existence of an elastic potential further reduces the number of independent elastic coefficients from 36 to 21 in number.

The total number of independent coefficients c_{ijkl} depends on crystal symmetry and is greatest for crystals of low symmetry. The lowest symmetry is that for the triclinic system where 21 constants are required.

---

[1] The summation convention has been used throughout this paper: Subscripts occurring twice are summation dummies. In Eq. (2.5), e.g., summation is extended over j and l.

A sufficiently detailed discussion of the elastic coefficients may be found, for example, in HEARMON [14.1], MASON [1], and NYE [14.2] and for numerical values in HUNTINGTON [14.3]. It must be noted that the elastic coefficients in the tensor notation c_{ijkl} are usually written in a condensed or matrix notation where the following correspondence of indices is used.

$$\begin{array}{ccccccc}
\text{tensor notation} & 11 & 22 & 33 & 23, 32 & 13, 31 & 12, 21 \\
\text{matrix notation} & 1 & 2 & 3 & 4 & 5 & 6
\end{array}$$

thus c_{1231} in tensor notation would be written c_{65} in the matrix notation.

In an *isotropic elastic solid* the coefficients must be independent of the particular set of rectangular coordinate axes chosen. This restriction leads to only two independent constants

$$\left.\begin{aligned}
c_{12} &= c_{13} = c_{23} = \lambda, \\
c_{44} &= c_{55} = c_{66} = \mu, \\
c_{11} &= c_{22} = c_{33} = \lambda + 2\mu.
\end{aligned}\right\} \tag{2.8}$$

λ and μ are known as *Lamé's elastic coefficients* and they completely define the elastic properties of an isotropic solid.

3. Equations of motion and solutions. One obtains the equations of motion for an elastic medium by considering the forces acting on an element of volume in the medium. In particular if one considers the forces on opposite pairs of faces on a small rectangular parallelopiped, then by taking the differences of these pairs of forces, one arrives at the components of the resultant force (neglecting body forces) acting on the volume considered. On equating the force components to the acceleration components for a medium of density ϱ one obtains the following equations of motion:

$$\sigma_{ij,j} = \varrho \ddot{s}_i \quad (i = 1, 2, 3) \tag{3.1}$$

where $\mathbf{s}(s_1, s_2, s_3)$ is the displacement vector, and $\mathbf{x}(x_1, x_2, x_3)$ is the position vector.

Elastic waves can be propagated in any direction in a crystalline medium regardless of whether or not it is isotropic.

There are, for each direction in a crystal, three independent waves (each with a distinct velocity) whose displacements are mutually orthogonal. In general no one of these displacements coincides either with the normal to the wave front or a direction at right angles to the wave front normal. In other words, these waves are in general neither compressional waves nor transverse waves. There are, however, special directions in any crystal (depending on its type) for which the wave front normal coincides with the displacement vector for the compressional wave. When it happens that one of the three orthogonal displacement vectors coincides with the wave front normal, the other two displacement vectors lie in the plane of the wave front and are therefore two transverse waves each with its own velocity.

Returning to the stress-strain relation, Eq. (2.5),

$$\sigma_{ij} = c_{ijkl}\,\varepsilon_{kl} \quad (i, j = 1, 2, 3)$$

together with the equations of motion (3.1), one has

$$c_{ijkl}\,\varepsilon_{kl,j} = \varrho \ddot{s}_i \tag{3.2}$$

where

$$\varepsilon_{kl,j} = \tfrac{1}{2}\,(s_{l,kj} + s_{k,lj});$$

hence,

$$c_{ijkl}(s_{l,kj} + s_{k,lj}) = 2\varrho\,\ddot{s}_i$$

and the fact that $c_{ijlk} = c_{ijkl}$ makes the following more compact form possible

$$c_{ijkl}\,s_{l,kj} = \varrho\,\ddot{s}_i \qquad (i = 1, 2, 3). \tag{3.3}$$

Consider a plane wave travelling in a direction given by the propagation or wave vector $\boldsymbol{k}\,(k_1, k_2, k_3)$ which is normal to planes of constant phase. The unit vector $\boldsymbol{n}\,(n_1, n_2, n_3)$ is also normal to the phase or wave front and

$$\boldsymbol{k} = \frac{\omega}{v}\,\boldsymbol{n} = \frac{2\pi}{\lambda}\,\boldsymbol{n}.$$

$\boldsymbol{s}$ is the particle displacement which is in general not parallel to the wave vector $\boldsymbol{k}$. $\boldsymbol{s} = \boldsymbol{s}\,(s_1, s_2, s_3)$. One can write the expression for the components of any one of the plane waves as

$$s_l = s_{0l}\,e^{i(\omega t - \boldsymbol{k}\cdot\boldsymbol{r})} \qquad (l = 1, 2, 3). \tag{3.4}$$

Then

$$s_{l,kj} = -\,n_k\,n_j\,s_{0l}\left(\frac{\omega^2}{v^2}\right)e^{i(\omega t - \boldsymbol{k}\cdot\boldsymbol{r})},$$

and the equation of motion becomes

$$c_{ijkl}\,s_{0l}\,n_k\,n_j = \varrho\,v^2\,s_{0i} \qquad (i = 1, 2, 3). \tag{3.5}$$

4. Propagation directions and velocities. There are, for any direction of propagation, three real values of velocity with mutually orthogonal displacements. In the particular case of *cubic symmetry*,

$$\left.\begin{aligned} c_{11} &= c_{22} = c_{33}, \\ c_{12} &= c_{21} = c_{13} = c_{31} = c_{23} = c_{32}, \\ c_{44} &= c_{55} = c_{66} \end{aligned}\right\} \tag{4.1}$$

and all other values are zero. The special form of equations for cubic symmetry is

$$\left.\begin{aligned} (c_{11} - c_{44})\,s_{01}\,n_1^2 + (c_{12} + c_{44})\,s_{02}\,n_1\,n_2 + (c_{12} + c_{44})\,s_{03}\,n_1\,n_3 &= (\varrho\,v^2 - c_{44})\,s_{01}; \\ (c_{12} + c_{44})\,s_{01}\,n_2\,n_1 + (c_{11} - c_{44})\,s_{02}\,n_2^2 + (c_{12} + c_{44})\,s_{03}\,n_2\,n_3 &= (\varrho\,v^2 - c_{44})\,s_{02}; \\ (c_{12} + c_{44})\,s_{01}\,n_3\,n_1 + (c_{12} + c_{44})\,s_{02}\,n_3\,n_2 + (c_{11} - c_{44})\,s_{03}\,n_3^2 &= (\varrho\,v^2 - c_{44})\,s_{03}. \end{aligned}\right\} \tag{4.2}$$

The compatibility condition for the solution of Eqs. (4.2) requires that the determinant of the coefficients of the displacement components s_{01}, s_{02}, s_{03} be zero. A cubic equation in v^2 results, and the three roots of this cubic equation are the allowed solutions which make Eqs. (4.2) compatible. Since there are three solutions or velocities, regardless of the direction of propagation, it is clear that there are only three distinct modes of propagation. Along certain specific directions in the crystalline medium two of these modes are transverse and one is compressional [6]. It is in these special directions that the wave front normal coincides with the displacement component for the compressional wave. It can be shown that the three independent wave displacements are mutually orthogonal, hence that, in the special circumstances just mentioned, two of these waves are pure transverse waves and one of them is a pure compressional wave.

If Eqs. (4.2) are solved for $\boldsymbol{s}\,(s_{01}, s_{02}, s_{03})$, it can be shown [7] that

$$(s_{01}, s_{02}, s_{03}) = \frac{n_1}{\lambda + C\,n_1^2},\ \frac{n_2}{\lambda + C\,n_2^2},\ \frac{n_3}{\lambda + C\,n_3^2},$$

where

$$\left.\vphantom{\begin{aligned}&\\&\end{aligned}}\right\} \tag{4.3}$$

$$\lambda = \left(\frac{\varrho\,v^2 - c_{44}}{c_{44}}\right) \quad \text{and} \quad C = \frac{c_{12} + 2c_{44} - c_{11}}{c_{44}}.$$

The expression for C determines the anisotropy in a cubic crystal since, when $c_{11} - c_{12} = 2c_{44}$, the crystal is elastically isotropic. The factor $\dfrac{2c_{44}}{c_{11} - c_{12}}$ is sometimes called the *anisotropy factor*. The value of this factor is unity for elastic isotropy; it may, however, be less than or greater than unity, depending on the material.

Since the use of stress waves for ultrasonic measurements in single crystals is usually confined to the use of compressional or transverse waves propagated in principal directions in the crystals, such waves will be discussed. In particular this will be done for cubic crystals.

The condition that a pure longitudinal wave be propagated is that the wave front normal coincide with the displacement vector **s**. It has been pointed out above that since there are three distinct velocities[1], and since only one of the three waves can be a compressional mode when the wave front normal coincides with the propagation direction, the remaining two waves must in this case be transverse waves. The condition that the displacement vector of the compressional wave be parallel to the wave front normal is

$$
\left.
\begin{aligned}
\mathbf{s} \times \mathbf{n} &= 0; \\
s_{02}\, n_2 - s_{03}\, n_3 &= 0, \\
s_{03}\, n_1 - s_{01}\, n_3 &= 0, \\
s_{01}\, n_2 - s_{02}\, n_1 &= 0
\end{aligned}
\right\}
\tag{4.4a}
$$

or

$$
n_1 : n_2 : n_3 = s_{01} : s_{02} : s_{03} \tag{4.4b}
$$

and, from Eq. (4.3),

$$
s_{01} : s_{02} : s_{03} = \frac{n_1}{\lambda + C n_1^2} : \frac{n_2}{\lambda + C n_2^2} : \frac{n_3}{\lambda + C n_3^2}. \tag{4.5}
$$

Eqs. (4.4b) and (4.5) lead to

$$
n_1 : n_2 : n_3 = \frac{n_1}{\lambda + C n_1^2} : \frac{n_2}{\lambda + C n_2^2} : \frac{n_3}{\lambda + C n_3^2}. \tag{4.6}
$$

The conditions under which these ratios are satisfied are as follows (if $C \neq 0$)[2]:

$$
\left.
\begin{array}{ll}
n_1 = n_2 = n_3 & \langle 111 \rangle \\
n_1 = n_2 = n_3 = 0 & \langle 110 \rangle \\
n_1 = n_3, \quad n_2 = 0 & \langle 101 \rangle \\
n_2 = n_3, \quad n_1 = 0 & \langle 011 \rangle
\end{array}
\right\}
\tag{4.7}
$$

$$
\text{equivalent}
\left\{
\begin{array}{ll}
n_1 \neq 0, \quad n_2 = n_3 = 0 & \langle 100 \rangle \\
n_2 \neq 0, \quad n_1 = n_3 = 0 & \langle 010 \rangle \\
n_3 \neq 0, \quad n_1 = n_2 = 0 & \langle 001 \rangle.
\end{array}
\right.
$$

From Eqs. (4.2) one can find the corresponding values of ϱv^2 for each of the conditions of (4.7). For example, the $\langle 110 \rangle$ direction compressional wave would require that

$$
n_1 = n_2, \quad n_3 = 0
$$

---

[1] Consequently there are three distinct displacement vectors which might be labeled $\mathbf{s}^1, \mathbf{s}^2, \mathbf{s}^3$ and each of these vectors has three component displacements such as those appearing in Eqs. (4.2). It was not considered necessary for this discussion to make use of the notation to identify the three separate waves.

[2] When $C = 0$ (i.e., $c_{11} - c_{12} = 2c_{44}$) the medium is elastically isotropic, it is seen from Eq. (4.6) that $\mathbf{n}$ is completely arbitrary—as it should be.

and, since these are direction cosines,

$$n_1^2 + n_2^2 = 1 \quad \text{or} \quad n_1 = n_2 = \frac{1}{\sqrt{2}}.$$

These values of n_1 and n_2 used in Eqs. (4.2) yield the following determinant of the coefficients (compatibility) to be set equal to zero.

$$\begin{vmatrix} \left(\frac{1}{2}(c_{11} + c_{44}) - \varrho v^2\right) & \frac{1}{2}(c_{12} + c_{44}) \\ \frac{1}{2}(c_{12} + c_{44}) & \left(\frac{1}{2}(c_{11} + c_{44}) - \varrho v^2\right) \end{vmatrix} = 0$$

and this yields a value

$$\varrho v_{\langle 110 \rangle}^2 = \frac{1}{2}(c_{11} + c_{12} + 2c_{44}).$$

In the same way one finds two other values of v and altogether [1], [7]

$$\left. \begin{aligned} \varrho v_{\langle 100 \rangle}^2 &= c_{11}, \\ \varrho v_{\langle 110 \rangle}^2 &= \tfrac{1}{2}(c_{11} + c_{12} + 2c_{44}), \\ \varrho v_{\langle 111 \rangle}^2 &= \tfrac{1}{3}(c_{11} + 2c_{12} + 4c_{44}) \end{aligned} \right\} \tag{4.8}$$

for compressional waves propagating in each case in the directions given by the subscripts on the velocity.

The condition that pure transverse waves be propagated is that the displacement vector s be perpendicular to the wave front normal, namely that $s \cdot n = 0$, hence that

$$s_{01} n_1 + s_{02} n_2 + s_{03} n_3 = 0. \tag{4.9}$$

Eq. (4.9) together with (4.5) leads to

$$\frac{n_1^2}{\lambda + C n_1^2} + \frac{n_2^2}{\lambda + C n_2^2} + \frac{n_3^2}{\lambda + C n_3^2} = 0 \tag{4.10}$$

or

$$\lambda^2 + 2\lambda C (n_1^2 n_2^2 + n_2^2 n_3^2 + n_3^2 n_1^2) + 3 C^2 n_1^2 n_2^2 n_3^2 = 0.$$

Values of $\lambda = \left(\dfrac{\varrho v^2 - c_{44}}{c_{44}}\right)$ obtained from Eq. (4.10) are substituted in Eqs. (4.2), and one then finds that the directions which satisfy the conditions for transverse waves are the same conditions as those for compressional waves. In addition it becomes necessary to pay attention to the plane of polarization when the propagation is in the $\langle 110 \rangle$ direction.

The values of ϱv^2 for transverse waves are [1], [7]

$$\left. \begin{aligned} \varrho v_{\langle 100 \rangle}^2 &= c_{44} \\ \varrho v_{\langle 110 \rangle}^2 &= c_{44} & \text{(polarized in } \langle 001 \rangle \text{ direction)} \\ \varrho v_{\langle 110 \rangle}^2 &= \tfrac{1}{2}(c_{11} - c_{12}) & \text{(polarized in } \langle 110 \rangle \text{ direction)} \\ \varrho v_{\langle 111 \rangle}^2 &= \tfrac{1}{3}(c_{11} + c_{44} - c_{12}). \end{aligned} \right\} \tag{4.11}$$

It becomes necessary in some cases to calculate the energy flux associated with elastic waves. Such calculations are essential, for example, in determining scattering cross sections, and for calculating internal conical refraction behavior in anisotropic crystals.

5. Energy and energy flux. At this point the purpose is that of showing how the energy flux may be calculated from stresses and displacements.

The elastic energy W in a given field taken as the sum of kinetic and potential energy is as follows

$$W = \frac{1}{2} \int \varrho \frac{\partial s_i}{\partial t} \frac{\partial s_i}{\partial t} d\tau + \frac{1}{2} \int \sigma_{ij} \varepsilon_{ij} d\tau \qquad (5.1)$$

with [cf. Eq. (2.2)]

$$\varepsilon_{ij} = \frac{1}{2} (s_{i,j} + s_{j,i}).$$

The time derivative of W will yield the energy flux into or out of the field:

$$\frac{\partial W}{\partial t} = \int \varrho \frac{\partial s_i}{\partial t} \frac{\partial^2 s_i}{\partial t^2} d\tau + \frac{1}{2} \int \frac{\partial}{\partial t} (\sigma_{ij} \varepsilon_{ij}) d\tau.$$

Now

$$\frac{\partial}{\partial t} (\sigma_{ij} \varepsilon_{ij}) = \frac{\partial}{\partial t} (c_{ijkl} \varepsilon_{kl} \varepsilon_{ij}) = 2 c_{ijkl} \varepsilon_{kl} \frac{\partial \varepsilon_{ij}}{\partial t} = 2 \sigma_{ij} \frac{\partial \varepsilon_{ij}}{\partial t},$$

hence

$$\frac{\partial W}{\partial t} = \int_\tau \left\{ \varrho \frac{\partial s_i}{\partial t} \frac{\partial^2 s_i}{\partial t^2} + \sigma_{ij} \frac{\partial \varepsilon_{ij}}{\partial t} \right\} d\tau \qquad (5.2)$$

and (5.2) together with the differential equations of motion (3.1) becomes

$$\frac{\partial W}{\partial t} = \int_\tau \left\{ \frac{\partial s_i}{\partial t} \sigma_{ij,j} + \sigma_{ij} \frac{\partial \varepsilon_{ij}}{\partial t} \right\} d\tau. \qquad (5.3)$$

This volume integral can be changed to a surface integral over the surface S_c bounding the volume τ. Using GREEN's theorem on the first term of the volume integral

$$\int_\tau \left(\frac{\partial s_i}{\partial t} \sigma_{ij,j} + \sigma_{ij} \frac{\partial s_{i,j}}{\partial t} \right) d\tau = \int_{S_e} \sigma_{ij} \frac{\partial s_i}{\partial t} n_j d\sigma,$$

and the fact that

$$\frac{\partial \varepsilon_{ij}}{\partial t} = \frac{\partial}{\partial t} (s_{i,j})$$

yields the result that

$$\frac{\partial W}{\partial t} = \int_\tau \frac{\partial}{\partial x_j} \left(\sigma_{ij} \frac{\partial s_i}{\partial t} \right) d\tau = \int_{S_e} \sigma_{ij} \frac{\partial s_i}{\partial t} n_j d\sigma. \qquad (5.4)$$

The surface integral just derived must represent the flow of energy into the volume τ through the closed surface S_c. Consequently, the component of outward flux of energy f_j is

$$f_j = - \sigma_{ij} \frac{\partial s_i}{\partial t}. \qquad (5.5)$$

6. Scattering relations. In the particular case of the calculation of the energy scattered by a spherical obstacle one can consider a sphere of radius b larger than the radius a of the scattering obstacle. The time rate at which energy is being carried away through the spherical surface of radius b by the scattered wave is given by the tensor whose components have just been derived. Integrating over the sphere of radius b [2], [8]

$$\dot{F}_s = \int_{S_e} \left(\sigma_{xr} \frac{\partial s_x}{\partial t} + \sigma_{yr} \frac{\partial s_y}{\partial t} + \sigma_{zr} \frac{\partial s_z}{\partial t} \right)_{\substack{\text{scattered} \\ \text{wave}}} n_r d\sigma \qquad (6.1)$$

where σ_{xr}, σ_{yr}, and σ_{zr} are stress components[1] acting in x, y, z directions on a surface normal to the radial vector r. σ_{xr}, σ_{yr}, σ_{zr}, s_x, s_y, s_z of the scattered wave

---

[1] Both stress and displacement components are assumed to have harmonic time dependence $\sigma_{xr} = \sigma_{xrs} e^{i\omega t}$ etc.

are in general complex and only the real parts will be used; hence $\sigma_{xr} = \frac{1}{2}(\sigma_{xr} + \sigma_{xr}^*)$ and $s_x = \frac{1}{2}(s_x + s_x^*)$, etc., will be substituted. Then the integral (6.1) taken over the sphere of radius b surrounding the spherical scatterer, and averaged timewise, yields

$$\dot{F}_s = \frac{i\,\omega}{2} \int_{S_e} \{(\sigma_{xrs}\,s_{xs}^* + \sigma_{yrs}\,s_{ys}^* + \sigma_{zrs}\,s_{zs}^*) - (\sigma_{xrs}^*\,s_{xs} + \sigma_{yrs}^*\,s_{ys} + \sigma_{zrs}^*\,s_{zs})\}\,n_r\,d\sigma, \quad (6.2)$$

or in spherical components

$$\left.\begin{aligned}
\dot{F}_s = \frac{i\,\omega}{2} \int_0^\pi \{(\sigma_{rrs}\,s_{rs}^* + \sigma_{\vartheta rs}\,s_{\vartheta s}^* + \sigma_{\varphi rs}\,s_{\varphi s}^*) - \\
- (\sigma_{rrs}^*\,s_{rs} + \sigma_{\vartheta rs}^*\,s_{\vartheta s} + \sigma_{\varphi rs}^*\,s_{\varphi s})\}_{r=b}\,2\pi\,b^2\,\sin\vartheta\,d\vartheta.
\end{aligned}\right\} \quad (6.3)$$

The total scattering cross section of a scatterer is usually defined as the ratio of the total energy scattered per unit time to the energy per unit area normal to propagation direction per unit time in the incident wave. The energy flux incident may be calculated from relation (5.5)

$$\dot{F}_i = \frac{i\,\omega}{2}\{(\sigma_{xzi}\,s_{xi}^* + \sigma_{yzi}\,s_{yi}^* + \sigma_{zzi}\,s_{zi}^*) - (\sigma_{xzi}^*\,s_{xi} + \sigma_{yzi}^*\,s_{yi} + \sigma_{zzi}^*\,s_{zi})\} = \frac{\varrho_1\,\omega^3}{k_1}. \quad (6.4)$$

In this calculation of the incident flux for a plane wave

$$s_{xi} = s_{yi} = 0, \qquad s_{zi} = e^{-ik_1 z}, \qquad \sigma_{xzi} = \sigma_{yzi} = 0, \qquad \sigma_{zzi} = -i\,\frac{\varrho_1\,\omega^3}{k_1}\,e^{-ik_1 z}.$$

The total scattering cross section is then

$$\gamma = \frac{\dot{F}_s}{\dot{F}_i}. \quad (6.5)$$

Further details are given in Sect. 11.

It is frequently desirable or necessary to propagate waves in a direction not quite along a principal axis of the crystal. In such cases pure transverse or pure compressional modes are not possible and the amount of deviation from pure modes depends on the amount of deviation from a principal or pure mode axis. A description of the perturbed or quasi-pure modes resulting from small mis-orientations of the propagation direction with respect to the principal axes of a crystal has been developed by Waterman [9]. The effect of such misorientations on the phase velocity and on the elastic displacement vectors is determined to-gether with the energy flux of pure and quasi-pure modes.

These considerations are important because of their influence on measurements, but the matter will not be pursued further here.

7. Non-linear or anharmonic effects. The discussion has thus far contained the assumption that the propagation of waves in the lattice involves only a linear relation between stress and strain. In other words the assumption is made that the lattice is linearly elastic[1]. A characteristic property of linearly elastic waves is that any wave may be expressed as a linear combination of sepa-rate monochromatic waves. Each monochromatic wave may be propagated independently and such waves do not interact with one another. Under these circumstances there would be no losses, and the attenuation of a stress wave in infinite elastic medium would be zero. In order that absorption be present the medium must possess non linear or anharmonic features, in which case the equa-

---

[1] Linearly elastic as a special case of elastic where elastic is taken to mean that the total strain energy depends only on the state of deformation and is independent of the path by which this state is reached.

tions of motion are non linear, and they do not admit simple harmonic solutions as in the purely elastic case. Non linear behavior in a solid can be related to higher order coefficients in the strain energy function and for this reason the higher order coefficients have become of interest in megacycle ultrasonic wave propagation measurements in crystalline solids. The second order moduli are those related to the velocities of a stress wave in a purely elastic solid, but they are not related to any loss mechanisms; the loss mechanisms are related only to the coefficients of higher order than the second. Consequently, those velocity *changes*, associated directly with particular loss mechanisms, are also related only to the higher order coefficients. The first of such higher order coefficients are those of the third order, and they form a tensor, c_{ijklmn}, of the sixth order.

The strain function, the energy of deformation, may be written in the form

$$E = E_0 + g\, c_{ij}\, \varepsilon_{ij} + \tfrac{1}{2}\, c_{ijkl}\, \varepsilon_{ij}\, \varepsilon_{kl} + c_{ijklmn}\, \varepsilon_{ij}\, \varepsilon_{kl}\, \varepsilon_{mn} + \cdots,$$

where the ε_{ij} are the complete strain components defined early in this article and the c's are the coefficients characteristic of the material under investigation. If the strain energy is zero before deformation E_0 is zero. At the same time the second term is a potential energy which can be set equal to zero since the reference level does not matter.

$$E = \tfrac{1}{2}\, c_{ijkl}\, \varepsilon_{ij}\, \varepsilon_{kl} + c_{ijklmn}\, \varepsilon_{ij}\, \varepsilon_{kl}\, \varepsilon_{mn} + \cdots.$$

The c_{ijkl} are the elastic coefficients discussed earlier in this article. The c_{ijklmn} are the third order coefficients in the expansion of the strain energy; they form a sixth order tensor, and in general there are 729 components. An isotropic material requires three independent third order coefficients, and a cubic crystal may require six or eight third order coefficients depending on the symmetry. Other crystal systems require more independent coefficients, and the triclinic system requires 56 coefficients. A specific example of a strain energy expression for cubic crystals is given in Refs. [15] and [16].

Since the non linear effects in a crystal lattice are connected with third order terms or higher in the strain energy function, the coefficients of these terms must contain or represent the physical quantities which determine the non linear behavior.

In the consideration of anharmonic effects arising from third order terms in the strain energy function there are at least two ways of handling the deformation formalism necessary for the discussion of the non linear or anharmonic effects of the third order.

One of these methods is that of simply writing the strain energy function as above and continuing with the standard elasticity theory as discussed, for example, in "Theory of Elasticity" by LANDAU and LIFSHITZ [10]. The second method is that of finite deformation elasticity as given by MURNAGHAN [11] or RIVLIN [12]. In any case the results lead to discussion of third order elastic constants and related topics.

The determination of third order elastic constants has been discussed by BIRCH [13] for anisotropic single crystals. The available data (up to 1956) have been summarized by HEARMON [14].

Using the finite strain method of MURNAGHAN, SEEGER and BUCK [15] have given a detailed discussion of the experimental determination of the third order elastic constants together with the necessary relations for evaluating experimental results. A discussion of the available data on third order constants is also given.

A detailed account of the evaluation of the six third order elastic constants of germanium has been given by BATEMAN, MASON, and McSKIMIN [16].

As pointed out above, the third order elastic constants are of interest in ultrasonic work because of the dependence of the lattice vibration spectrum on strains and because they provide a means of describing the coupling between ultrasonic waves and lattice vibration waves (phonon-phonon interaction). One type of non linear behavior is that in which the medium is non linear to provide the intermode coupling, but elastic to insure that there are no absorption loss mechanisms present. In other words, one assumes that one has only non linear elasticity with which to deal.

A second type of non linear behavior is that in which absorption losses are present and the material is no longer elastic.

Both types of non linear behavior, with and without absorption losses, may be present. It is clear, however, that even the first type of non linear, but elastic, effect may couple energy out of the ultrasonic beam and so appear as a loss as far as attenuation measurement is concerned.

8. Loss interactions, terms and relations. Before discussing the various loss mechanisms in a crystalline solid, it is necessary to point out that several points of view may be taken regarding what one calls a loss. If one measures the attenuation of the ultrasonic beam, then any energy deviated from the beam by scattering or any energy absorbed within the boundaries of the beam contribute to the attenuation measured[1]. Any absorption loss process must by definition involve a non linear relation between stress and strain, and ultimately all of the ultrasonic energy becomes heat. On the other hand, the measurement of the loss in the beam either by absorption within the beam or scattering from the beam provides a means of studying certain fundamental properties of the solid.

The energy dissipation mechanisms or the scattering losses from the beam (or both) are expressed in terms of the attenuation factor α either in decibels per unit length of path or in decibels per microsecond. These relations are shown by using a plane stress wave

$$\sigma = \sigma_0 e^{-\alpha x} \sin(\omega t - kx) \tag{8.1}$$

where ω is the angular frequency, k is the propagation factor, and α is the attenuation per unit length of path. Since the attenuation is given by the envelope of the high frequency wave one can use

$$\sigma(x) = \sigma_0 e^{-\alpha x} \tag{8.2}$$

to specify the attenuation α. It is assumed that α is not dependent on x although in some cases the problem can be handled when α is a function of x. Hence

$$\alpha = -\frac{\left(\dfrac{d\sigma(x)}{dx}\right)}{\sigma(x)} = -\frac{d}{dx}\left(\log_e \sigma(x)\right),$$

so that for two different points x_1 and x_2

$$\alpha = \frac{1}{x_2 - x_1} \log_e \frac{\sigma(x_1)}{\sigma(x_2)} \qquad (x_1 < x_2)$$

and since any ratio of two amplitudes such as $\sigma(x_1)$ and $\sigma(x_2)$ must, in order to be expressed in decibels, be written

$$20 \log_{10}\left(\frac{\sigma(x_1)}{\sigma(x_2)}\right) \text{ (decibels)}$$

---

[1] Discussion of losses due to geometry, diffraction effects, bonds and so on have been omitted because no attempt is made here to discuss measurement techniques.

or in units of nepers

$$\log_e\left(\frac{\sigma(x_1)}{\sigma(x_2)}\right) \text{(nepers)}.$$

Then

$$\alpha = \frac{20}{x_2 - x_1} \log_{10}\left(\frac{\sigma(x_1)}{\sigma(x_2)}\right) \text{decibels/unit length}, \tag{8.3}$$

or

$$\alpha = \frac{1}{x_2 - x_1} \log_e\left(\frac{\sigma(x_1)}{\sigma(x_2)}\right) \text{nepers/unit length}, \tag{8.4}$$

hence

$$\alpha \text{ (db/unit length)} = 8.686\alpha \text{ (nepers/unit length)}. \tag{8.5}$$

Another expression for energy loss is that of logarithmic decrement δ which is defined for a harmonically oscillating system as $\delta \equiv (W/2E)$ where W is the energy loss per cycle in the specimen, and E is the total vibrational energy stored in the specimen. It turns out that

$$\delta \text{ (nepers)} = \alpha \text{ (nepers/cm)} \cdot \lambda \text{ (cm)} = \frac{\alpha \text{ (nepers/cm)}}{\nu \text{ (sec}^{-1})} v \text{ (cm/sec)}. \tag{8.6}$$

Finally

$$\alpha \text{ (db/}\mu\text{sec)} = 8.68 \times 10^{-6} v \text{ (cm/sec)} \cdot \alpha \text{ (nepers/cm)} \tag{8.7}$$

or

$$\alpha \text{ (db/}\mu\text{sec)} = 8.68 \times 10^{-6} \nu \text{ (sec}^{-1}) \delta \text{ (nepers)}. \tag{8.8}$$

There is, of course, the equivalent measure of dissipation called the Q of a system which is defined as

$$Q \equiv 2\pi \frac{E}{W}, \tag{8.9}$$

where E and W are as defined above. From the definition of Q and δ it is seen that

$$Q \delta = \pi. \tag{8.10}$$

9. Types of interactions. Most, if not all, of the known interaction effects between elastic stress waves and some property of the crystal lattice can be found in the following list:

1. Direct scattering by defects,
2. Thermoelastic or heating effects,
3. Dislocation damping,
4. Stress wave interaction with conduction electrons in metals,
5. Magneto-elastic loss effects in ferromagnetic materials, (a) magnetic domain wall motion, (b) spin wave—phonon effects,
6. Stress wave (ultrasonic) interaction with thermal elastic waves (lattice vibrations),
7. Stress wave interaction with nuclear spin systems,
8. Stress wave interaction with electron spins in paramagnetic centers,
9. Stress wave interaction with charge carriers in piezoelectric materials,
10. Acoustoelectric effect in semi-conductors.

These interactions will be described in what follows; some of the effects will be discussed in more detail than others, either because there is more understanding or information available or because they seem to be of relatively greater interest or importance at present.

II. Scattering.

10. Statement of the problem. Scattering of stress waves in a solid is brought about by the presence of gradients in the elastic properties in the solid. Gradients may, of course, be very sharp; they may in fact be infinite as at boundaries in

11*

the solid where reflection and refraction may occur in a regular or an irregular manner, thereby producing coherent or incoherent scattering.

The importance of scattering in ultrasonic measurements arises from the fact that defects in solids may cause both scattering from and absorption in the ultrasonic beam. In many cases it is necessary to separate the two effects and possibly others as well. This is especially true at the high frequencies, now available, where the ultrasonic wavelengths in the solid may be equal to or less than those of visible light. Even in single crystals defects may be present which cause scattering of ultrasonic stress waves. Such defects may be induced in single crystals, for example, by irradiation, by the formation of precipitates, and by strain effects.

The general problem of expressing analytically the attenuation by scattering for an ultrasonic beam propagating through a medium containing scatterers with any size, shape, distribution of sizes, density of scatterers, wavelength relative to scatterer size, and so on, is one of extreme complexity even as to the statement of the problem. In what follows a review is given of most of what has been accomplished thus far in experimental and analytical work in the scattering of high frequency stress waves in solid materials.

Scattering of an acoustic wave by an obstacle in a fluid medium [17] to [24] has been studied rather extensively since the time of Lord Rayleigh. Scattering of an acoustic wave by an elastic obstacle in an elastic medium [25] to [30] has been studied in detail relatively less extensively and only for simple geometries such as spheres and cylinders. The calculation of the scattering cross section for a single elastic scatterer in an elastic medium has been done in a general way for spherical scatterers for both compressional waves [8] and for transverse waves [31]. Such calculations have also been made for the scattering cross section of a spherical cavity, a rigid sphere, a fluid filled cavity in an elastic medium.

11. Scattering cross section and attenuation. The scattering cross section is the ratio of the total energy scattered per unit time to the energy per unit area per unit time in the incident wave front normal to the direction of propagation. The cross section is given explicitly in Eq. (6.5). The detailed calculation of the stresses and displacements is given in Refs. [8] and [31]. The result for compressional waves and spherical scatterers [8] in an isotropic medium is:

$$\gamma = 4\pi \sum_{m=0}^{\infty} \frac{1}{2m+1} \left[|A_m|^2 + m(m+1) \frac{k_1}{\varkappa_1} |B_m|^2 \right] \tag{11.1}$$

where m is integer and A_m and B_m are two sets of coefficients that must be determined from the elastic boundary conditions at the boundary between scatterer and medium. There are four equations arising from these boundary conditions, and they are given in Sect. 13. The quantities $k_1 = \omega/V_{c_1}$ and $\varkappa_1 = \omega/V_{t_1}$ are the propagation factors for compressional and transverse waves. V_{c_1} is the compressional wave velocity and V_{t_1} the transverse wave velocity. γ has the dimensions of length squared, and it is convenient for certain purposes to normalize γ to the geometrical cross section πa^2 so that the normalized cross section is

$$\gamma_N = \frac{\gamma}{\pi a^2}. \tag{11.2}$$

The medium is considered to be isotropic. If the scattering cross section for a single scatterer is known, and if the individual scatterers can be regarded as independent of each other, the amount of energy scattered is γI per single scatterer and $n_0 \gamma I$ per unit volume at any point x where the intensity is I. n_0 is the density of scatterers. One of the difficult points associated with scattering is that of determining when the scatterers can be regarded as independent and when they cannot be so regarded.

In the preceding discussion α was defined in terms of the stress amplitude. In terms of intensity and the scatterer density, the loss dI in intensity at a point x is

$$dI = - n_0 \gamma I = - 2\alpha I$$

or the attenuation

$$\alpha = \tfrac{1}{2} n_0 \gamma, \tag{11.3}$$

as long as the scatterers are independent. When the scatterers are not independent, multiple scattering effects are present and the problem of obtaining a valid and useful expression for α is much more complex. The scattering cross section γ as calculated [8], [31] for an elastic scatterer in an elastic medium contains the frequency dependence. The frequency dependence of γ may vary from the fourth power of the frequency, as with the Rayleigh approximation, to any value between the fourth power and the first power of the frequency under circumstances where the Rayleigh conditions for scattering are not satisfied.

The coefficients A_{em}, B_{em}, C_{em}, and D_{em} can be calculated at least numerically by solving the four simultaneous equations given in Sect. 13. Such numerical calculations have been carried out for a few cases. For the case of Rayleigh scattering with $la \ll 1$ where l is either k_1, k_2, $\varkappa_1$, or $\varkappa_2$ and a is the radius of the spherical scatterer, the coefficients A_{em}, B_{em}, C_{em}, and D_{em} can be explicitly evaluated as closely as desired (see Sect. 13). With terms of order $(la)^2$ and higher neglected in comparison with unity, the following approximation (RAYLEIGH) is obtained:

$$\gamma_{\text{elastic sphere}} = \frac{4\pi}{9} g_e k_1^4 a^6 \tag{11.4}$$

or

$$\gamma_N = \frac{\gamma}{\pi a^2} = \frac{4}{9} g_e k_1^4 a^4, \tag{11.5}$$

where

$$g_e = \left\{ \left[\frac{3\left(\frac{\varkappa_1}{k_1}\right)^2}{3\left[\left(\frac{\varkappa_2}{k_2}\right)^2 - 4\right]\frac{\mu_2}{\mu_1} + 4} - 1 \right]^2 + \frac{1}{3}\left[1 + 2\left(\frac{\varkappa_1}{k_1}\right)^3\right]\left[\left(\frac{\varkappa_2}{\varkappa_1}\right)^2 \frac{\mu_2}{\mu_1} - 1\right]^2 + \right.$$
$$\left. + 40\left[2 + 3\left(\frac{\varkappa_1}{k_1}\right)^5\right]\left[\frac{\left(\frac{\mu_2}{\mu_1} - 1\right)}{2\left[3\left(\frac{\varkappa_1}{k_1}\right)^2 + 2\right]\frac{\mu_2}{\mu_1} + \left(9\left(\frac{\varkappa_1}{k_1}\right)^2 - 4\right)}\right]^2 \right\}, \tag{11.6}$$

where

$$k = \frac{\omega}{V_c} = \frac{\omega}{\left(\frac{\lambda + 2\mu}{\varrho}\right)^{\frac{1}{2}}} \quad \text{and} \quad \varkappa = \frac{\omega}{V_t} = \frac{\omega}{\left(\frac{\mu}{\varrho}\right)^{\frac{1}{2}}}$$

the ratio

$$\frac{\varkappa}{k} = \frac{V_c}{V_t} \quad \text{and} \quad \frac{\varkappa_1}{\varkappa_2} = \frac{V_{t_2}}{V_{t_1}}.$$

Consequently, the only data required to evaluate g_e are compressional and transverse velocities for both scatterer and matrix media.

An alternative form for g_e is that involving the *bulk modulus*

$$K = \lambda + \tfrac{2}{3}\mu$$

where λ and μ are the Lamé constants and $G=\mu$ is the *shear modulus*:

$$g_e = \frac{1}{\left(1+\frac{4}{3}\frac{G_1}{K_2}\right)^2}\frac{1}{\left(\frac{K_2}{K_1}\right)^2}\left(\frac{K_2-K_1}{K_1}\right)^2 + \frac{1}{3}\left[1+2\left(\frac{K_1}{G_1}+\frac{4}{3}\right)^{\frac{5}{3}}\right]\left(\frac{\varrho_2-\varrho_1}{\varrho_1}\right)^2 +$$
$$+ 40\left[2+3\left(\frac{K_1}{G_1}+\frac{4}{3}\right)^{\frac{5}{3}}\right]\left\{\frac{1}{\left[3\left(2\frac{G_2}{G_1}+3\right)\frac{K_1}{G_1}+4\left(3\frac{G_2}{G_1}+2\right)\right]^2}\right\}\left(\frac{G_2-G_1}{G_1}\right)^2.$$
(11.7)

g_e is put in this form because it allows comparison with certain relations, largely empirical, found in the literature, relations involving

$$\left\langle\frac{\Delta K}{K}\right\rangle^2, \quad \left\langle\frac{\Delta \varrho}{\varrho}\right\rangle^2, \quad \text{and} \quad \left\langle\frac{\Delta G}{G}\right\rangle^2.$$

The cross section γ, discussed thus far, is that for a compressional wave. Separate solutions are required when the incident wave is transverse, and this case [31] has also been worked out for an elastic sphere in an elastic medium as well as for a spherical cavity in an elastic medium, for a fluid filled cavity in an elastic medium, and for a rigid sphere in an elastic medium. The case of an elastic sphere in an elastic medium is generally of most interest and it will be discussed here. In this case, for an incident transverse wave, the boundary conditions require six linear equations as compared with four for the compressional wave case. The result of evaluating Eq. (6.5) for the transverse wave case [31] is

$$\gamma_N = \frac{\gamma}{\pi a^2} = 2\sum_{n=1}^{\infty}(2n+1)\left\{\frac{1}{(\varkappa_1 a)^2}(a_n a_n^* + b_n b_n^*) + \frac{\varkappa_1 a}{n(n+1)(k_1 a)^3}d_n d_n^*\right\} \quad (11.8)$$

where again γ_N is the scattering cross section normalized to the geometrical cross section πa^2 of the spherical scatterer. The manner of calculating a_n, b_n, and d_n is discussed in Sect. 14.

In the Rayleigh approximation, for which $la \ll 1$, a limiting form of the scattering cross section can be calculated for transverse elastic waves for an elastic sphere in an elastic medium.

$$\gamma_N = \frac{8}{3}\left[1+\frac{1}{2}\frac{k_1^3}{\varkappa_1^6}\right]\left\{\frac{\left[3\frac{\varkappa_1^2}{\varkappa_2^2}-3\frac{\varkappa_2^2}{\varkappa_1^2}-4\frac{k_2^2\varkappa_1^2}{\varkappa_2^4}+10\frac{k_2^2}{\varkappa_2^2}-6\frac{k_2^2}{\varkappa_1^2}\right]}{\left[1-10\frac{\varkappa_1}{k_1}+6\frac{k_2}{k_1}-6\frac{k_2}{k_1}\frac{\varkappa_2^2}{\varkappa_1^2}+9\frac{\varkappa_2^2}{\varkappa_1^2}\right]^2}\right\}(\varkappa_1 a)^4. \quad (11.9)$$

Again, as with the compressional wave case, the fourth power frequency dependence is present; this fourth power frequency dependence is also present in the Rayleigh limit for the cavity and rigid sphere for transverse waves.

12. Multiple scattering and scattering density. Thus far this discussion has been concerned with the calculation of scattering cross sections for independent scatterers for which the expression $\alpha=\frac{1}{2}n_0\gamma$ would be valid. It is clear that even in this case γ might have any frequency dependence equal to or less than the fourth power of the frequency depending on how well or how poorly the Rayleigh conditions are satisfied.

At some stage, where the scatterers become sufficiently dense and closely spaced and where the wavelength becomes comparable with the scatterer size, not only will the Rayleigh approximation not be valid, but the scatterers will not be independent. Multiple scattering will occur and the expression $\alpha=\frac{1}{2}n_0\gamma$ will not apply.

The approach to multiple scattering involves a method [32], primarily due to Foldy [33], which consists of averaging a joint probability distribution for a given configuration of scatterers, considered as one state of an ensemble, over all configurations of scatterers or states of the ensemble. This "configurational average" technique is used to calculate what is called the exciting field ψ^E in a medium having N identical scatterers located at $r_1, r_2, \ldots, r_N$.

An incident wave ψ^I plus the scattering from all other scatterers except the j-th produces the exciting wave ψ^E which in turn causes a scattered wave ψ^S to be generated by the j-th scatterer. ψ^S is obtained from ψ^E at the j-th scatterer by an operator method involving what is called a scattering operator. A configurational average value $\langle \psi^E \rangle$ is obtained, and from this an averaged total field is obtained. The total field at any point is the sum of the incident and all scattered waves. The details of this calculation are complicated and only the result is given here. The reader is referred to Ref. [32] for details as well as for a discussion of the literature on this subject.

The multiple scattering field calculation yields a complex propagation constant β in the relation

$$\left(\frac{\beta}{k_1}\right)^2 = \left[1 + \frac{2\pi n_0 f(0)}{k_1^2}\right]^2 - \left[\frac{2\pi n_0 f(\pi)}{k_1^2}\right]^2. \tag{12.1}$$

$k_1 = \omega/v$ is the propagation constant (real) for the incident wave; n_0 is the density or number of scatterers per unit volume; $\beta = \left(\frac{\omega}{V'} + i\,\alpha\right)$ where V' is the velocity with scattering present, and α is the attenuation in nepers/cm

$$\left.\begin{aligned} f(0) &= -\sum_{m=0}^{\infty} (-i)^m A_{em}^*, \\ f(\pi) &= -\sum_{m=0}^{\infty} (-i)^m (-i)^m A_{em}^*, \end{aligned}\right\} \tag{12.2}$$

where in general A_{em} must be found from the solution of the four equations in Sect. 13 previously mentioned in connection with the evaluation of γ for a single scatterer with a compressional wave.

The special case for Rayleigh scattering $la \ll 1$ allows an approximation for the A_{em} as given in Sect. 13.

On rewriting the expression for β as

$$\beta = k_1 \left[1 + \frac{4\pi n_0 f(0)}{k_1^2} + \frac{4\pi^2 n_0^2}{k_1^4} \{f(0)^2 - f(\pi)^2\}\right]^{\frac{1}{2}}, \tag{12.3}$$

one sees that if the scatterers were isotropic, i.e., $f(0) = f(\pi)$, the third term would disappear and one would have

$$\beta = k_1 \left[1 + \frac{4\pi n_0 f(0)}{k_1^2}\right]^{\frac{1}{2}}. \tag{12.4}$$

This simplification will not apply in most cases of interest in high frequency work in solids. One can see this by calculating $f(0)$ and $f(\pi)$ for some particular case, or one can see qualitatively that at low frequencies and long wavelengths a scatterer of given size would scatter more uniformly in all directions than it would at higher frequencies.

A different type of approximation is possible under conditions of weak scattering density. Scattering density is defined as $\left(\frac{n_0 f(\theta)}{k_1^2}\right)$, and when $\left(\frac{n_0 f(\theta)}{k_1^2}\right) \ll 1$ the so-called condition of weak scattering density is satisfied. This situation arises physically under circumstances where the number of scatterers per unit volume is sufficiently small or where the individual scatterer is weak, as, for example, where the material of the scatterer has elastic properties very similar to those of the matrix medium or where the gradient of the elastic properties is small.

In Eq. (12.3) for β the elastic scattering density appears linearly and quadratic-ally, and if $\left(\dfrac{n_0 f(\vartheta)}{k_1^2}\right) \ll 1$, the quadratic terms may be neglected in comparison with the linear term in the scattering density. Then

$$\beta = k_1 \left[1 + \frac{4\pi\, n_0 f(0)}{k_1^2}\right]^{\frac{1}{2}} \approx k_1 \left[1 + \frac{2\pi\, n_0 f(0)}{k_1^2}\right]. \tag{12.5}$$

On looking at this weak scattering approximation and recalling that $\beta = \left(\dfrac{\omega}{V'} + i\alpha\right)$ and that k_1 is real,

$$\left(\frac{\omega}{V'} + i\alpha\right) = k_1 + \frac{2\pi\, n_0 f(0)}{k_1},$$

consequently

$$\frac{\omega}{V'} = k_1 \left[1 + \frac{2\pi\, n_0}{k_1^2}\, \mathrm{Re}\, f(0)\right] \tag{12.6}$$

$$\alpha = k_1 \left[\frac{2\pi\, n_0}{k_1^2}\, \mathrm{Im}\, f(0)\right] = \frac{1}{2}\, n_0\, \gamma \tag{12.7}$$

where

$$\gamma = \frac{4\pi}{k_1}\, \mathrm{Im}\, f(0);$$

hence with this weak scattering approximation the modified phase velocity V' is determined by the real part of $f(0)$ and the attenuation is determined by the imaginary part of $f(0)$. This agrees with the results of the independent scatterer relation (11.3).

The calculation of $f(0)$ has been shown to depend on the coefficients A_{em} which must either be calculated numerically from the equations given or from the Ray-leigh approximations discussed earlier. The Rayleigh approximation, $ka \ll 1$, is the statement that the scatterer is small compared with the wavelength of the incident radiation, but the Rayleigh condition, as RAYLEIGH used it, was applied to a single scatterer, and not to an assembly of scatterers. The density or separa-tion of scatterers from one another is not a part of the Rayleigh scattering ap-proximation, and it is for this reason that the scattering density expression $\left(\dfrac{n_0 f(\vartheta)}{k_1^2}\right)$ is necessary in connection with many scatterers and multiple scattering problems. One may have multiple scattering and at the same time the Rayleigh approximation may be appropriate; this means that the wavelength is appreciably larger than the scatterer size and at the same time the scatterers may be densely packed. Since

$$\alpha = \mathrm{Im}\,\beta \quad \text{and} \quad \frac{\omega}{V'} = \mathrm{Re}\,\beta$$

one finds on looking for the real and imaginary parts of β that in the Rayleigh approximation the first three coefficients A_{e0}^*, A_{e1}^*, and A_{e2}^* do not, in many cases of interest, provide an imaginary part for β. Such a situation arises only because the Rayleigh approximation has been used, i.e., $ka \ll 1$ or $a \ll \lambda$. In general, however, all the A_{em}^* have both real and imaginary parts.

In any case, with or without the Rayleigh approximation, there is always a real part of β which means that the phase velocity V' in the medium with scat-terers present is different from the phase velocity V of the matrix material with no scatterers present.

It is evident that the multiple scattering analysis yields a velocity difference caused by the presence of scatterers while the analysis involving independent scatterers, with individual scattering cross section γ, did not provide a way of recognizing such a velocity difference. The reason for this difference in the results appears to be connected with the fact that the adjustment of phases or the boundary conditions at large distances from a group of scatterers has been handled differently than in the case of the single scatterer. The question is not one that has been solved or even clearly stated thus far.

The application of the scattering results to experimental situations[1] has been carried out in relatively few cases in ultrasonic work with solids. At lower megacycle frequencies scattering by precipitates in alloys has provided some information regarding the magnitude and type of changes to be expected, although not thus far of a type that provides a really satisfactory check on the theory. The velocity change encountered before and after irradiation [35], [36] of a material by fast neutrons has shown good agreement with that predicted by scattering theory despite the fact that such damage regions are not spherical in shape. Using attenuation and velocity measurements for neutron damage in silicon, and with frequencies to 1000 Mc/sec, a value of 100 Å was determined for the scatterer size, i.e. the size of the damage region.

As higher and higher frequencies come into use, the influence of scattering by defects becomes more important. The use of scattering of high frequency stress waves by defects in solids as a means of studying the presence and behavior of such defects is certain to increase.

13. Evaluation of coefficients in scattering cross section for compressional waves. The coefficients A_m and B_m for an elastic sphere using compressional waves [8] are found, for $m > 0$, from the four equations:

$$
\begin{aligned}
A_{em} k_1 a\, h_{m+1}(k_1 a) &+ B_{em} m \varkappa_1 a\, h_{m+1}(\varkappa_1 a) - C_{em} k_2 a\, j_{m+1}(k_2 a) - \\
&- D_{em} m \varkappa_2 a\, j_{m+1}(\varkappa_2 a) = (-i)^{m-1}(2m+1)\frac{1}{k_1}[k_1 a\, j_{m+1}(k_1 a)],
\end{aligned}
\tag{13.1}
$$

$$
\begin{aligned}
A_{em} h_m(k_1 a) &- B_{em}[(m+1) h_m(\varkappa_1 a) - \varkappa_1 a\, h_{m+1}(\varkappa_1 a)] - C_{em} j_m(k_2 a) + \\
&+ D_{em}[(m+1) j_m(\varkappa_2 a) - \varkappa_2 a\, j_{m+1}(\varkappa_2 a)] \\
&= (-i)^{m-1}(2m+1)\frac{1}{k_1} j_m(k_1 a),
\end{aligned}
\tag{13.2}
$$

$$
\begin{aligned}
A_{em}[(\varkappa_1 a)^2 h_m(k_1 a) &- 2(m+2) k_1 a\, h_{m+1}(k_1 a)] + B_{em} m[(\varkappa_1 a)^2 h_m(\varkappa_1 a) - \\
&- 2(m+2)\varkappa_1 a\, h_{m+1}(\varkappa_1 a)] - C_{em} p[(\varkappa_2 a)^2 j_m(k_2 a) - 2(m+2)\times \\
&\times k_2 a\, j_{m+1}(k_2 a)] - D_{em} p\, m[(\varkappa_2 a)^2 j_m(\varkappa_2 a) - 2(m+2)\varkappa_2 a\, j_{m+1}(\varkappa_2 a)] \\
&= (-i)^{m-1}(2m+1)\frac{1}{k_1}[(\varkappa_1 a)^2 j_m(k_1 a) - 2(m+2) k_1 a\, j_{m+1}(k_1 a)],
\end{aligned}
\tag{13.3}
$$

$$
\begin{aligned}
A_{em}[(m-1) h_m(k_1 a) &- k_1 a\, h_{m+1}(k_1 a)] - \\
&- B_{em}\left[\left(m^2 - 1 - \frac{\varkappa_1^2 a^2}{2}\right) h_m(\varkappa_1 a) + \varkappa_1 a\, h_{m+1}(\varkappa_1 a)\right] - \\
&- C_{em} p[(m-1) j_m(k_2 a) - k_2 a\, j_{m+1}(k_2 a)] - \\
&- D_{em} p\left[\left(m^2 - 1 - \frac{\varkappa_2^2 a^2}{2}\right) j_m(\varkappa_2 a) + \varkappa_2 a\, j_{m+1}(\varkappa_2 a)\right] \\
&= (-i)^{m-1}(2m+1)\frac{1}{k_1}[(m-1) j_m(k_1 a) - k_1 a\, j_{m+1}(k_1 a)],
\end{aligned}
\tag{13.4}
$$

where μ_2, μ_1 are the shear moduli, and $p = \mu_2/\mu_1$.

---

[1] Some of the earliest ultrasonic work [25] to [29] in the megacycle range of frequencies was done on scattering by grain boundaries in polycrystalline aluminum and steels [34]. The results have yielded only empirical relations, and the principal value of these results is that of showing the difficulties and complexities of the problems of interpretation of such data.

For $m=0$, A_{e0} and C_{e0} are given by

$$A_{e0}\,h_1(k_1\,a) - C_{e0}\,\frac{k_2}{k_1}\,j_1(k_2\,a) = \frac{i}{k_1}\,j_1(k_1\,a), \qquad (13.5)$$

$$\begin{aligned}
A_{e0}\,[(\varkappa_1\,a)^2\,h_0(k_1\,a) - 4k_1\,a\,h_1(k_1\,a)] - C_{e0}\,p\,[(\varkappa_2\,a)^2\,j_0(k_2\,a) - 4k_2\,a\,j_1(k_2\,a)] \\
= \frac{i}{k_1}\,[(\varkappa_1\,a)^2\,j_0(k_1\,a) - 4k_1\,a\,j_1(k_1\,a)].
\end{aligned} \right\} \qquad (13.6)$$

B_{e0} and D_{e0} may be obtained from Eqs. (13.2) and (13.4) with $m=0$. A_{e0} and C_{e0} can be explicity evaluated.

For the case of Rayleigh scattering with $a \ll 1$,

$$A_{e0} \approx \frac{1}{3}\left[1 - \frac{3\,(\varkappa_1^2/k_1^2)}{3\,p\,(\varkappa_2^2/k_2^2) - 4\,(p-1)}\right]\frac{1}{k_1}\,(k_1\,a)^3, \qquad (13.7)$$

$$A_{e1} \approx i\,\frac{\varrho_2 - \varrho_1}{3\,\varrho_1}\,\frac{1}{k_1}\,(k_1\,a)^3, \qquad (13.8)$$

$$B_{e1} \approx -\frac{\varkappa_1^2}{k_1^2}\,A_{e1}, \qquad (13.9)$$

and for $m>1$,

$$\begin{aligned}
A_{em} = (-\,i)^m\,2^{2m}\,(4m^2 - 1)\,\frac{(m!)^2}{(2m!)^2}\,\times \\
\times\,\frac{p-1}{(p-1) + \dfrac{1}{m}\left[(m+1)\,p + \dfrac{2m^2+1}{2m-2}\right]\dfrac{\varkappa_1^2}{k_1^2}}\,\frac{1}{k_1}\,(k_1\,a)^{2m-1},
\end{aligned} \right\} \qquad (13.10)$$

$$B_{em} = -\frac{1}{m}\left(\frac{\varkappa_1}{k_1}\right)^{m+1}A_{em}. \qquad (13.11)$$

14. Evaluation of coefficients in scattering cross section for transverse waves.
The calculation of the coefficients [31] involved in the normalized scattering cross section γ_N [Eq. (11.8)] for an elastic sphere of constants $\varrho_2,\lambda_2,\mu_2$ and radius a embedded in a matrix defined by constants $\varrho_1,\lambda_1,\mu_1$, proceeds as follows:

$$a_n = \frac{\mu_2\,F_1(a)\,\dfrac{d}{da}\left[\dfrac{S_1(a)}{a}\right] - \mu_1\,S_1(a)\,\dfrac{d}{da}\left[\dfrac{F_1(a)}{a}\right]}{\mu_1\,S_1(a)\,\dfrac{d}{da}\left[\dfrac{B_1(a)}{a}\right] - \mu_2\,B_1(a)\,\dfrac{d}{da}\left[\dfrac{S_1(a)}{a}\right]}, \qquad (14.1)$$

$$e_n = \frac{\mu_1\,F_1(a)\,\dfrac{d}{da}\left[\dfrac{B_1(a)}{a}\right] - \mu_1\,B_1(a)\,\dfrac{d}{da}\left[\dfrac{F_1(a)}{a}\right]}{\mu_1\,S_1(a)\,\dfrac{d}{da}\left[\dfrac{B_1(a)}{a}\right] - \mu_2\,B_1(a)\,\dfrac{d}{da}\left[\dfrac{S_1(a)}{a}\right]}. \qquad (14.2)$$

b_n,d_n,f_n, and g_n are obtained from the remaining four equations given by:

$$\begin{pmatrix} E_{11} & E_{12} & E_{13} & E_{14} \\ E_{21} & E_{22} & E_{23} & E_{24} \\ E_{31} & E_{32} & E_{33} & E_{34} \\ E_{41} & E_{42} & E_{43} & E_{44} \end{pmatrix}\begin{pmatrix} b_n \\ d_n \\ f_n \\ g_n \end{pmatrix} = \begin{pmatrix} \delta_1 \\ \delta_2 \\ \delta_3 \\ \delta_4 \end{pmatrix}, \qquad (14.3)$$

where

$$\begin{aligned}
E_{11} &= D_2(a), & E_{21} &= D_2(a), \\
E_{12} &= D_3(a), & E_{22} &= D_3(a), \\
E_{13} &= -R_1(a), & E_{23} &= -T_2(a), \\
E_{14} &= -R_2(a), & E_{24} &= -T_3(a),
\end{aligned}} \right\} \tag{14.4}$$

$$E_{31} = (\lambda_1 + 2\mu_1) \frac{2n(n+1)}{2n+1} \frac{d}{da} A_1(a) + \lambda_1 \frac{4n(n+1)}{2n+1} \frac{A_1(a)}{a} +$$
$$+ \lambda_1(1 - 2n^2) \frac{B_2(a)}{a} + \lambda_1 n(n+1) \frac{D_2(a)}{a},$$

$$E_{32} = (\lambda_1 + 2\mu_1) \frac{2n(n+1)}{2n+1} \frac{d}{da} A_2(a) + \lambda_1 \frac{4n(n+1)}{2n+1} \frac{A_2(a)}{a} +$$
$$+ \lambda_1(1 - 2n^2) \frac{B_3(a)}{a} + \lambda_1 n(n+1) \frac{D_3(a)}{a},$$

$$E_{33} = -\left[(\lambda_2 + 2\mu_2) \frac{2n(n+1)}{2n+1} \frac{d}{da} R_1(a) + \lambda_2 \frac{4n(n+1)}{2n+1} \frac{R_1(a)}{a} +\right.$$
$$\left. + \lambda_2(1 - 2n^2) \frac{S_2(a)}{a} + \lambda_2 n(n+1) \frac{T_2(a)}{a}\right],$$

$$E_{34} = -\left[(\lambda_2 + 2\mu_2) \frac{2n(n+1)}{2n+1} \frac{d}{da} R_2(a) + \lambda_2 \frac{4n(n+1)}{2n+1} \frac{R_2(a)}{a} +\right.$$
$$\left. + \lambda_2(1 - 2n^2) \frac{S_3(a)}{a} + \lambda_2 n(n+1) \frac{T_3(a)}{a}\right],$$

$$E_{41} = \mu_1 \left\{ a \frac{d}{da} \left[\frac{D_2(a)}{a}\right] - \frac{A_1(a)}{a} \right\},$$

$$E_{42} = \mu_1 \left\{ a \frac{d}{da} \left[\frac{D_3(a)}{a}\right] - \frac{A_2(a)}{a} \right\},$$

$$E_{43} = -\mu_2 \left\{ a \frac{d}{da} \left[\frac{T_2(a)}{a}\right] - \frac{R_1(a)}{a} \right\},$$

$$E_{44} = -\mu_2 \left\{ a \frac{d}{da} \left[\frac{T_3(a)}{a}\right] - \frac{R_2(a)}{a} \right\};$$

$$\left. \begin{aligned} & \end{aligned} \right\} \tag{14.5}$$

$$\delta_1 = -E_1(a), \qquad \delta_2 = -G_1(a),$$

$$\delta_3 = -\left[(\lambda_1 + 2\mu_1) \frac{2n(n+1)}{2n+1} \frac{d}{da} E_1(a) + \lambda_1 \frac{4n(n+1)}{2n+1} \frac{E_1(a)}{a} +\right.$$
$$\left. + \lambda_1(1 - 2n^2) \frac{F_2(a)}{a} + \lambda_1 n(n+1) \frac{G_1(a)}{a}\right],$$

$$\delta_4 = -\mu_1 \left\{ a \frac{d}{da} \left[\frac{G_1(a)}{a}\right] - \frac{E_1(a)}{a} \right\}.$$

$$\left. \begin{aligned} & \end{aligned} \right\} \tag{14.6}$$

$$\begin{aligned}
A_1(r) &= -(i)^{n+1}(2n+1) \frac{1}{\varkappa_1 r} h_n(\varkappa_1 r), \\
A_2(r) &= \frac{(i)^n(2n+1)}{n(n+1)} \frac{1}{k_1} \frac{d}{dr} h_n(k_1 r),
\end{aligned} \right\} \tag{14.7}$$

$$\begin{aligned}
B_1(r) &= \frac{(i)^n(2n+1)}{n(n+1)} h_n(\varkappa_1 r), \\
B_2(r) &= -(i)^{n+1} \frac{1}{\varkappa_1 r} \frac{d}{dr} [r\, h_n(\varkappa_1 r)], \\
B_3(r) &= (i)^n \frac{1}{k_1 r} h_n(k_1 r),
\end{aligned} \right\} \tag{14.8}$$

$$D_1(r) = -(i)^n h_n(\varkappa_1 r),$$

$$D_2(r) = (i)^{n+1} \frac{(2n+1)}{n(n+1)} \frac{1}{\varkappa_1 r} \frac{d}{dr} [r\, h_n(\varkappa_1 r)],$$

$$D_3(r) = -(i)^n \frac{(2n+1)}{n(n+1)} \frac{1}{k_1 r} h_n(k_1 r),$$

(14.9)

$$E_1(r) = -(i)^{n+1}(2n+1) \frac{1}{\varkappa_1 r} j_n(\varkappa_1 r),$$

(14.10)

$$F_1(r) = (i)^n \frac{(2n+1)}{n(n+1)} j_n(\varkappa_1 r),$$

$$F_2(r) = -(i)^{n+1} \frac{1}{\varkappa_1 r} \frac{d}{dr} [r j_n(\varkappa_1 r)],$$

(14.11)

$$G_1(r) = (i)^{n+1} \frac{(2n+1)}{n(n+1)} \frac{1}{\varkappa_1 r} \frac{d}{dr} [r j_n(\varkappa_1 r)],$$

$$G_2(r) = -(i)^n j_n(\varkappa_1 r),$$

(14.12)

$$R_1(r) = -(i)^{n+1}(2n+1) \frac{1}{\varkappa_2 r} j_n(\varkappa_2 r),$$

$$R_2(r) = (i)^n \frac{(2n+1)}{n(n+1)} \frac{1}{k_2} \frac{d}{dr} j_n(k_2 r),$$

(14.13)

$$S_1(r) = (i)^n \frac{(2n+1)}{n(n+1)} j_n(\varkappa_2 r),$$

$$S_2(r) = -(i)^{n+1} \frac{1}{(\varkappa_2 r)} \frac{d}{dr} [r j_n(\varkappa_2 r)],$$

$$S_3(r) = (i)^n j_n(k_2 r),$$

(14.14)

$$T_1(r) = -(i)^n j_n(\varkappa_2 r),$$

$$T_2(r) = (i)^{n+1} \frac{(2n+1)}{n(n+1)} \frac{1}{\varkappa_2(r)} \frac{d}{dr} [r j_n(\varkappa_2 r)],$$

$$T_3(r) = -(i)^n \frac{(2n+1)}{n(n+1)} \frac{1}{k_2 r} j_n(k_2 r).$$

(14.15)

III. Thermoelastic effects.

15. Physical description of the effect. One cause of ultrasonic attenuation present in nearly all crystalline solids is the thermoelastic effect [1], [37]. The thermoelastic effect is usually small and frequently overshadowed by other mechanisms. In particular the thermoelastic effect is usually smaller than the dislocation effect; the two effects can however be separated for special orientations and conditions of study. The quantitative discussion here is that given by Lücke for the thermoelastic effect at high frequencies.

The thermoelastic effect arises from the fact that when a material is subjected to a stress, the resulting strain is in general accompanied by a change in temperature. If the strain is homogeneous throughout the specimen, the temperature change will be the same everywhere in the specimen. If the strain is not homogenous, as, for example, when a longitudinal ultrasonic wave is propagated through a solid[1], temperature gradients will be set up between regions of compression and of rarefaction. This will lead to a flow of heat, accompanied by a production of entropy and a dissipation of energy, which will result in an attenuation of the

---

[1] A pure shear wave does not give rise to any thermoelastic effect.

wave amplitude. This type of attenuation is frequency dependent. At very high frequencies the time per cycle is not sufficient for appreciable heat flow to occur between adjacent regions of compression and rarefaction; the process is thus essentially adiabatic and no attenuation occurs. At very low frequencies thermal equilibrium between regions of compression and rarefaction is approached; the process is thus essentially isothermal and reversible and again no attenuation occurs. When the period of the applied stress becomes comparable with the relaxation time of the heat transfer process, attenuation is observed and reaches a maximum when $\omega\tau = 1$, where ω is the angular frequency of the stress and τ is the relaxation time, i.e. the time of relaxation or equalization of temperature fluctuations.

16. Phenomenological analysis. A relatively simple treatment [38] of the thermoelastic effect is based on the relation between stress, strain and strain rate, that describes the behavior of a standard linear (viscoelastic) solid. Such a solid shows an elastic after effect and an attenuation. This relation is of the form:

$$\left(\frac{1}{\tau}\right)\sigma + \dot{\sigma} = \frac{M_1}{\tau}\,\varepsilon + M_0\,\dot{\varepsilon}. \tag{16.1}$$

Here σ is the stress, ε the strain, M_0 the unrelaxed ("true") modulus, M_1 the time dependent or relaxed modulus and τ the stress relaxation time at constant strain.

In the following it is assumed that

$$\frac{\Delta M}{M_0} = \frac{M_0 - M_1}{M_0} \ll 1. \tag{16.2}$$

The strain ε will be described as a function of displacement s and since only compressional waves cause thermoelastic effects, the compressional strains are given by Eq. (2.2),

$$\varepsilon_{ii} = \frac{\partial s_i}{\partial x_i}.$$

The equation of motion in the viscoelastic solid is Eq. (3.1),

$$\sigma_{ij,j} = \varrho\,\ddot{s}_i \quad (i, = 1, 2, 3)$$

and, since there are only diagonal non-vanishing elements of the tensor components σ_{ii} in the case of compressional waves, the equations of motion become

$$\varrho\,\frac{\partial^2 s_i}{\partial t^2} = \frac{\partial \sigma_{ii}}{\partial x_i}.$$

Then on combining Eqs. (3.1) and (2.2) with (16.1), differentiated twice with respect to time, one obtains the equation

$$\frac{1}{\tau}\frac{\partial^2\sigma_{ii}}{\partial t^2} + \frac{\partial^3\sigma_{ii}}{\partial t^3} = \frac{M_1}{\tau\varrho}\cdot\frac{\partial^2\sigma_{ii}}{\partial x_i^2} + \frac{M_0}{\varrho}\cdot\frac{\partial^3\sigma_{ii}}{\partial x_i^2\,\partial t}. \tag{16.3}$$

This equation admits solutions of the form

$$\sigma_{ii} = A\,e^{-\alpha x_i}\cdot\cos(k_i\,x_i - \omega t) \tag{16.4}$$

or

$$\sigma_{ii} = A\,e^{-\alpha' t}\cdot\cos k_i\,x_i\cos\omega t \tag{16.5}$$

where

$$\alpha v = \alpha'$$

which define an attenuated travelling wave and an attenuated standing wave.

When a body shows an elastic aftereffect under constant stress or strain, a wave propagating through the solid will be attenuated, as shown by Eqs. (16.4) or (16.5). Here α is the attenuation, ω the angular frequency, and k_i the propagation factor for the x_i direction. The relation for α and α' is

$$\alpha v = \alpha' \approx \frac{1}{2\tau} \frac{\Delta M}{M_0} \frac{\omega^2 \tau^2}{1 + \omega^2 \tau^2},$$ (16.6)

$$v_g = \left[\frac{M_0}{\varrho} \left\{ 1 - \frac{\Delta M}{M_0} \frac{(1 - \omega^2 \tau^2)}{(1 + \omega^2 \tau^2)^2} \right\} \right]^{\frac{1}{2}}$$ (16.7)

where v is the phase velocity ω/k.

It is seen from the above that the maximum change in attenuation or velocity will occur only when $\omega \tau \cong 1$.

An adiabatic elastic deformation, with a longitudinal component, leads to a change in temperature (thermoelastic effect). Because of the thermal expansion due to this temperature change, the elastic modulus M_{ad}, appears to be different from that for isothermal deformation, M_1. Some time is required for a flow of heat to occur; hence when the deformation is sufficiently fast it can be considered adiabatic even when the solid is not insulated from its surroundings. If L is the length of the path through which heat must flow (in the case of a periodic stress L is essentially the wave length) the thermoelastic temperature differences become negligible after a time

$$\tau \equiv \frac{L^2}{D} = \frac{L^2 \varrho c_p}{\gamma} = \frac{D}{v^2} = \frac{\gamma}{\varrho c_p v^2}$$ (16.8)

where D is the thermal diffusivity, c_p the specific heat at constant pressure and γ the heat conductivity.

After the relaxation time τ, the strain is determined by the modulus M_1, so that $\dfrac{\Delta M}{M} = \dfrac{M_{\mathrm{ad}} - M_1}{M_1}$.

Consider a stress wave σ_{xx} propagating in the x direction of a body whose dimensions in the y and z directions are small compared to the wavelength (low frequency approximation). In this case the lateral strains $\varepsilon_{y'y'}$ and ε_{zz} will be fully developed over the whole cross section of the specimen. For isotropic materials the increments of the strain components due to increments of stress and temperature will be given by:

$$d\varepsilon_{xx} = \frac{d\sigma_{xx}}{E_T} + \beta\, dT,$$ (16.9)

$$d\varepsilon_{yy} = d\varepsilon_{zz} = -\mu \frac{d\sigma_{xx}}{E_T} + \beta\, dT$$ (16.10)

where E_T is Young's modulus for isothermal deformation, μ is Poisson's ratio, and β is the linear thermal expansion coefficient. The work connected with this change in strain is

$$dW = \sigma_{xx}\, d\varepsilon_{xx}.$$ (16.11)

The components ε_{yy} and ε_{zz} do not contribute anything to the work because the stress is zero in the y and z directions.

The expression for the elastic work, in conjunction with the first and second law of thermodynamics yields

$$\frac{\varrho c_\sigma}{T} dT = \left(\frac{\partial \varepsilon_{xx}}{\partial T} \right)_\sigma d\sigma_{xx} = \beta\, d\sigma_{xx}$$ (16.12)

where c_σ is the specific heat at constant stress, which may be taken equal to c_p for small stresses. Combining Eq. (16.12) with Eq. (16.9) gives:

$$d\varepsilon_{xx} = \frac{d\sigma_{xx}}{E_{\mathrm{ad}}} = \frac{1}{E_T} - \frac{\beta^2 T}{c_p} d\sigma_{xx} \tag{16.13}$$

where E_{ad} is YOUNG's modulus for adiabatic deformation. Finally one obtains

$$\frac{\Delta M}{M} = \frac{E_{\mathrm{ad}} - E_T}{E_T} = \frac{\beta^2 T}{\varrho\, c_p} E_{\mathrm{ad}} \approx \frac{\beta^2 T}{\varrho\, c_p} E_T. \tag{16.14}$$

These equations are also valid for anisotropic media if one chooses for β and E the values in the direction of the propagating wave.

As a second limiting case consider a stress wave propagating, in the x direction, in a solid whose dimensions in the y and z direction are large compared to the wavelength (high frequency approximation). Under these circumstances the displacements and accelerations in the y and z directions would be large (even for small ε_{yy} and ε_{zz}). It is assumed therefore that there is no net lateral contraction and the strain components ε_{yy} and ε_{zz} vanish everywhere. The changes in elastic strain, indicated by a prime, must compensate the changes in thermal expansion. In the isotropic case the above leads to the expressions:

$$\left.\begin{aligned} d\varepsilon'_{yy} &= d\varepsilon'_{zz} = -\beta\, T, \\ d\varepsilon_{xx} &= d\varepsilon'_{xx} + \beta\, dT. \end{aligned}\right\} \tag{16.15}$$

Then, using the equations:

$$\left.\begin{aligned}
\sigma_{xx} &= \frac{E_T(1-\mu)}{(1+\mu)(1-2\mu)}\,\varepsilon'_{xx} + \frac{E_T\,\mu}{(1+\mu)(1-2\mu)}\,(\varepsilon'_{yy} + \varepsilon'_{zz}), \\
\sigma_{yy} &= \frac{E_T(1-\mu)}{(1+\mu)(1-2\mu)}\,\varepsilon'_{yy} + \frac{E_T\,\mu}{(1+\mu)(1-2\mu)}\,(\varepsilon'_{zz} - \varepsilon'_{xx}), \\
\sigma_{zz} &= \frac{E_T(1-\mu)}{(1+\mu)(1-z\mu)}\,\varepsilon'_{zz} + \frac{E_T\,\mu}{(1+\mu)(1-z\mu)}\,(\varepsilon'_{yy} + \varepsilon'_{xx})
\end{aligned}\right\} \tag{16.16}$$

one obtains

$$d\varepsilon_{xx} = \frac{(1+\mu)(1-2\mu)}{E_T(1-\mu)}\,d\sigma_{xx} + \left(\frac{1+\mu}{1-\mu}\right)\beta\,dt = \frac{1}{E_T'}\,d\sigma_{xx} + \beta'\,dT. \tag{16.17}$$

This equation is of the same form as Eq. (16.9), except that E_T and β are replaced by the effective quantities E_T' and β'. Since the work is given by Eq. (16.11) (here the strain components ε_{yy} and ε_{zz} are zero), Eqs. (16.12) to (16.14) also hold in this case, provided that E_T and β are replaced by E' and β'. One finally obtains

$$\frac{\Delta M}{M} = \frac{(1+\mu)}{(1-2\mu)(1-\mu)}\,\frac{\beta^2\, T\, E_T}{\varrho\, c_p}. \tag{16.18}$$

17. Calculations for cubic and hexagonal crystals. A similar treatment is also possible for anisotropic materials. Here the effective quantities E' and β' must be found for different directions. In this case the general equations for all crystal systems are complicated because of the anisotropy in the thermal expansions as well as the stress-strain relations. Explicit expressions in the case of cubic and hexagonal crystal systems have been derived for longitudinal waves propagating in principal directions. These will now be discussed in some detail.

$\alpha)$ *Cubic crystals.* In the cubic system β is independent of direction. For propagation in a $\langle 100 \rangle$ direction one obtains

$$\sigma_{xx} = c_{11}\,\varepsilon'_{xx} + c_{12}\,(\varepsilon'_{yy} + \varepsilon'_{zz}) \tag{17.1}$$

and, by combining with Eq. (4.3), one finds

$$d\varepsilon_{xx} = \frac{1}{c_{11}} d\sigma_{xx} + \frac{c_{11} + 2c_{12}}{c_{11}} \beta\, dT.$$ (17.2)

For propagation in the $\langle 110 \rangle$ direction with z parallel to a $\langle 100 \rangle$ and y parallel to another $\langle 110 \rangle$ direction one finds:

$$\sigma_{xx} = \tfrac{1}{2}(c_{11} + c_{12} + 2c_{44})\,\varepsilon'_{xx} + \tfrac{1}{2}(c_{11} + c_{12} - 2c_{44})\,\varepsilon'_{yy} + c_{12}\,\varepsilon'_{zz}$$ (17.3)

and

$$d\varepsilon_{xx} = \frac{2}{c_{11} + c_{12} + 2c_{44}} d\sigma_{xx} + \frac{2(c_{11} + 2c_{12})}{c_{11} + c_{12} + 2c_{44}} \beta\, dT.$$ (17.4)

For propagation along $\langle 111 \rangle$ (y and z axes arbitrary) one obtains:

$$\sigma_{xx} = \tfrac{1}{3}(c_{11} + 2c_{12} + 4c_{44})\,\varepsilon'_{xx} + \tfrac{1}{3}(c_{11} + 2c_{12} - 2c_{44})\,(\varepsilon'_{yy} + \varepsilon'_{zz})$$ (17.5)

and

$$d\varepsilon_{xx} = \frac{3}{c_{11} + 2c_{12} + 4c_{44}} d\sigma_{xx} + \frac{3(c_{11} + 2c_{12})}{c_{11} + 2c_{12} + 4c_{44}} \beta\, dT.$$ (17.6)

β) *Hexagonal crystals.* For hexagonal crystals the dependence of β on direction must be taken into account. Let $\beta_\parallel$ and $\beta_\perp$ be respectively the expansion coefficient parallel and perpendicular to the hexagonal axis. For propagation in the basal plane (with z direction coinciding with the hexagonal axis) one obtains:

$$d\varepsilon'_{yy} = -\beta_\perp\, dT; \qquad d\varepsilon'_{zz} = -\beta_\parallel\, dT,$$ (17.7)

$$d\varepsilon_{xx} = d\varepsilon'_{xx} + \beta_\perp\, dT,$$ (17.8)

$$\sigma_{xx} = c_{11}\,\varepsilon'_{xx} + c_{12}\,\varepsilon'_{yy} + c_{13}\,\varepsilon'_{zz},$$ (17.9)

$$d\varepsilon_{xx} = \frac{1}{c_{11}} d\sigma_{xx} + \left(\frac{c_{12}\beta_\perp + c_{13}\beta_\parallel}{c_{11}} \right) dT$$ (17.10)

and similarly, for propagation along the hexagonal axis (x direction along hexagonal axis and y and z directions arbitrary):

$$\sigma_{xx} = c_{33}\,\varepsilon'_{xx} + c_{13}(\varepsilon'_{yy} + \varepsilon'_{zz}),$$ (17.11)

$$d\varepsilon_{xx} = \frac{1}{c_{33}} d\sigma_{xx} + \left(\frac{2c_{13}\beta_\perp + c_{33}\beta_\parallel}{c_{33}} \right) dT.$$ (17.12)

The coefficients of $d\sigma_{xx}$ and dT in Eqs. (16.17), (17.2), (17.4), (17.6), (17.10) and (17.12) will be designated by $1/E'$ and β' respectively. These coefficients must be introduced into Eq. (16.18) instead of $1/E_T$ and β in order to obtain numerical values of $\Delta M/M$. Values of $\Delta M/M$ calculated in this way are listed in the table.

Even for metals with the highest thermal conductivity, the relaxation time τ [Eq. (16.8)] is of the order of 10^{-11} sec, hence $\omega\tau \ll 1$ for the frequencies used in megacycle measurements. Consequently, Eq. (16.6) for α' can be simplified to

$$\alpha = \frac{\Delta M}{M}\, \tau\, \frac{\omega^2}{2}.$$ (17.13)

Then using Eqs. (16.8) and (16.14) in Eq. (17.13) one finds

$$\alpha' = \frac{\omega^2}{2}\, \frac{(\beta)^2\, T\gamma}{c_p^2\, \varrho}.$$ (17.14)

Values of α', thus computed, are listed in the table.

Table. *Thermoelastic attenuation in some cubic crystals. The attenuation* α' *is calculated for a frequency of* 100 Mc/sec.

Ma-terial	⟨100⟩ - Direction			⟨110⟩ - Direction			⟨111⟩ - Direction		
	$1/\tau$ (sec^{-1})	$\Delta M/M$	(db/μsec)	$1/\tau$ (sec^{-1})	$\Delta M/M$	(db/μsec)	$1/\tau$ (sec^{-1})	$\Delta M/M$	(db/μsec)
Al	$0.42 \cdot 10^{12}$	$3.30 \cdot 10^{-2}$	0.135	$0.44 \cdot 10^{12}$	$3.37 \cdot 10^{-2}$	0.122	$0.45 \cdot 10^{12}$	$3.53 \cdot 10^{-2}$	0.064
Cu	0.165	2.43	0.251	0.19	3.17	0.147	0.23	3.42	0.063
Ag	0.067	3.18	0.817	0.083	3.93	0.531	0.088	1.31	0.231
Pb	0.18	5.62	0.546	0.22	6.87	0.365	0.23	2.29	0.153
α-Fe	1.22	1.27	0.018	1.57	1.63	0.011	3.69	0.54	0.0049
W	0.40	0.38	0.016	0.40	0.38	0.016	0.40	0.127	0.0097
Ge	0.68	0.21	0.0054	0.82	0.258	0.0037	0.67	0.27	0.0021
Si	1.35	0.25	0.0032	1.58	0.29	0.0023	1.65	0.31	0.0013
NaCl	5.0	2.89	0.0099	4.4	2.56	0.0127	4.2	0.85	0.0096
KCl	3.2	2.22	0.0118	2.4	1.63	0.0219	2.1	0.54	0.0221

If, in addition, the relation

$$c_p - c_v = \frac{9\beta^2 \, T}{\varrho \, K} \tag{17.15}$$

is used (K is compressibility), Eq. (17.14) may be written

$$\alpha' = \frac{2\pi^2}{9} \, \gamma \, K \left(\frac{c_p - c_v}{c_p^2} \right) \frac{(\beta^1)^2}{\beta^2} \, v^2 . \tag{17.16}$$

Of particular interest in this connection are some measurements made on zinc by P.C. WATERMAN [38] (Fig. 1). These measurements show for the propagation direction along the hexagonal axis:

1. calculated thermoelastic attenu-ation,

2. measured ultrasonic attenuation with longitudinal or compressional waves,

3. measured ultrasonic attenuation using transverse or shear waves.

The transverse wave has, in case 3 above, all of the shear component in the slip plane which is the basal plane normal to the hexagonal axis. The compressional wave, on the other hand, has no component of shear stress in the

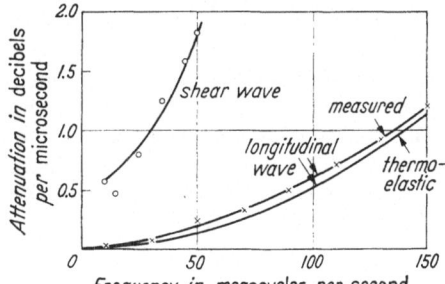

Fig. 1. Ultrasonic attenuation in zinc single crystals along the hexagonal axis as a function of frequency. After LUCKE [38]

slip plane and as a consequence has no component of dislocation damping. The agreement of compressional wave attenuation with the thermoelastic attenuation in this case is to be expected. This example is one of very few showing a clear cut separation of dislocation attenuation from thermoelastic attenuation and at the same time an absolute measurement of attenuation.

IV. Dislocation damping.

18. Description of the model for dislocation damping. The interaction between dislocations and high frequency alternating stress waves has received a rapidly increasing amount of attention both with regard to experiment and with regard to the construction of a dislocation model capable of accounting for the observed changes in the propagation behavior of ultrasonic stress waves in the presence of dislocations. The qualitative features of such interaction have been well demonstrated in many experiments, and one of the main problems that still exist in this area is that of determining the extent to which the present theory is in quantitative agreement with experiments. Part of the problem is that of designing better experiments, and, at the same time, determining better values of the constants involved.

That the damping loss and modulus changes observed in metals might arise from dislocations was suggested originally by T.A. Read [39]. J.S. Koehler [40] developed the idea that a dislocation line segment might vibrate under the influence of an alternating stress field and behave like a driven, damped, vibrating string. A. Granato and K. Lücke [41] made significant improvements in physical and analytical details of the model by Koehler [40]. Additional ideas for different damping models have been proposed by Nowick [42], Weertman [43] and Salkovitz [44] and others. The effect of crystal anisotropy on the interaction of dislocations with applied stress waves as well as the effect on elastic constants has been discussed by De Witt and Koehler [45].

The dislocation damping model discussed by A. Granato and K. Lücke [41] remains the best available thus far. There may be details of this model which will be modified in the future, but the main features of the model seem to be established. The model in question will be discussed in what follows.

It is assumed that an undeformed single crystal of high purity contains a dislocation network involving edge and screw type dislocations. The crystal may or may not be considered to have impurity atoms, but it is considered to have point defects such as vacancies and interstitials.

The dislocation loop lengths are assumed to be determined by the network character or dislocation intersections. The intersection spacings determine the longer loop lengths L_N. Intersections of dislocations form pinning of the dislocations which is strong relative to the weaker pinning of point defects or impurities which also pin at points lying between the intersections. The point defect pinning determines loop lengths L_c shorter than the intersection loop lengths L_N. This model is also considered to have a distribution of looplengths, and several possible distributions of looplengths have been considered.

An external stress is applied to the crystal containing dislocations, and there results an elastic strain ε_{el} plus a dislocation strain ε_{dis} where the dislocation strain is that additional strain caused by dislocation displacements. The dislocation stress strain law is in general a function of frequency and is non-linear.

The action of a dislocation under increasing stress is considered to consist of a "bowing out" of the dislocation loops with lengths L_c to the point where breakaway occurs and the loop length increases rapidly, and completely from L_c to L_N until the network loop length L_N is reached. At this stress field where the network loop length L_N is reached following breakaway it is assumed that dislocation multiplication may begin. It is assumed that the network pinning is sufficiently strong so that no breakaway of network loop lengths occurs. The breakaway of the L_c loop lengths may occur either as a result of the alternating stress fields, if the amplitude is sufficient, or as the result of a static stress applied to deform the material.

The dislocation damping behavior may be of two types. The first is that at low stresses where the shorter loop length L_c is present and the loss, which is frequency dependent and of a resonance type, occurs by some damping mechanism which produces a phase difference between the stress and strain. The second type of loss involves a larger stress amplitude so that the loops of length L_c may break away from the weaker point defect pinning and become loops of length L_N.

Following breakaway and further "bowing out" of the L_N loop the unloading part of the cycle occurs allowing the L_N loops to collapse and be repinned with length L_c. For small enough stresses the unpinning does not occur, and this is the condition that is considered to prevail at high frequencies. The question of the actual amplitude of displacement of the dislocation loops remains unsettled although the experimental evidence for breakaway is rather conclusive.

The model under discussion thus far is a zero temperature model, and for the dislocation to break away at zero temperature a sufficiently high stress must be applied to make the energy barrier disappear. In other words, there exists a threshold stress for mechanical breakaway. See Sect. 23 for discussion and for definition of energy barrier. When temperature is introduced into the analysis of breakaway the dislocation can overcome the energy barrier by thermal activation at much lower mechanical stresses as long as there is a second potential minimum into which the dislocation can jump. For low enough stresses, however, this second minimum does not appear to exist so that even for thermal breakaway a stress threshold does seem to exist. See Sect. 23 for further discussion of the thermal breakaway of a dislocation.

The following analysis is that given by GRANATO and LÜCKE for the zero temperature model.

19. Equations of motion and solutions. The attenuation and modulus change caused by dislocations when a stress wave propagates through a solid can be calculated starting with the equations of motion (3.1) which, with the use of (2.2) leads to the form:

$$\frac{\partial^2 \sigma_{ij}}{\partial x_j^2} = \varrho \frac{\partial^2}{\partial t^2} \varepsilon_{ij}. \tag{19.1}$$

The strain ε consists of the elastic strain ε_{el} and the dislocation strain ε_{dis} arising from the motion of the dislocations under the alternating stress.

$$\varepsilon = \varepsilon_{el} + \varepsilon_{dis}. \tag{19.2}$$

The elastic strain is then [cf. Eq. (2.5)]

$$\sigma_{ij} = c_{ijkl} \varepsilon_{kl},$$

or for simplicity in this discussion

$$\sigma = G \varepsilon_{el}. \tag{19.3}$$

The dislocation strain produced by a loop of length l in a cube of unit dimensions may be represented by $\bar{\xi} l b$ where b is the magnitude of the Burgers vector and $\bar{\xi}$ is the average displacement of a dislocation of length l and this is given by:

$$\bar{\xi} = \frac{1}{l} \int_0^l \xi(y)\, dy, \tag{19.4}$$

where y is the coordinate along the dislocation line. Consequently, if $\varLambda$ is the total length of movable dislocation line in the unit cube, the dislocation strain will be

$$\varepsilon_{dis} = \frac{\varLambda b}{l} \int_0^l \xi(y)\, dy. \tag{19.5}$$

The displacement behavior of the dislocation under the alternating stress is determined by the model chosen, and the model adopted is that of a driven,

12*

damped, vibrating dislocation line as used by Koehler [40]. The differential equation is that for a vibrating string

$$A \frac{\partial^2 \xi}{\partial t^2} + B \frac{\partial \xi}{\partial t} - C \frac{\partial^2 \xi}{\partial y^2} = b\sigma, \tag{19.6}$$

where $\xi = \xi(x, y, t)$ is the displacement of an element of the dislocation from its equilibrium position and the boundary conditions are

$$\xi(x, 0, t) = \xi(x, l, t) = 0.$$

The displacement ξ is assumed to lie in the slip plane. A is the effective mass per unit length, B is the damping force per unit length, and C is an effective tension [63] in a bowed out dislocation. $b\sigma$ is the driving force per unit length of the dislocation exerted by the applied shear stress.

The constants in the differential equation are given by

$$A = \pi \varrho b^2, \quad C = \frac{2G b^2}{\pi (1 - \nu)},$$

where ϱ is the density of the material and b, as before, is the magnitude of the Burgers vector, G is the shear modulus, and ν is the Poisson's ratio. B, the damping constant, has been estimated in at least two ways, one from heat developed in the neighborhood of an oscillating dislocation [46], the other from scattering of thermal phonons by dislocations [47]. There seem to be no fully satisfactory derivations of B. Combining the equations above leads to two equations, one a partial differential equation, the other an integral differential equation:

$$\frac{\partial^2 \sigma_{xx}}{\partial x^2} - \frac{\varrho}{G} \frac{\partial^2 \sigma_{xx}}{\partial t^2} = \frac{A \varrho b}{l} \frac{\partial^2}{\partial t^2} \int_0^l \xi(y)\, dy, \tag{19.7}$$

$$A \frac{\partial^2 \xi}{\partial t^2} + B \frac{\partial \xi}{\partial t} - C \frac{\partial^2 \xi}{\partial y^2} = b\sigma(x, y, t), \tag{19.8}$$

with the boundary conditions $\xi(x, 0, t) = \xi(x, l, t) = 0$. Of the possible solutions those for which σ is periodic in time and independent of y are of most interest. The dislocations are considered to be normal to the propagation direction.

A trial solution of the form

$$\sigma = \sigma_0\, e^{-\alpha x}\, e^{i[\omega(t - x/v)]} \tag{19.9}$$

leads to

$$\xi = \frac{4 b \sigma_0}{A} \sum_{n=0}^{\infty} \frac{1}{2n+1} \sin\left(\frac{(2n+1)\pi y}{l}\right) \frac{e^{i(\omega t - \delta n)}}{((\omega_n^2 - \omega^2)^2 + (\omega d)^2)^{\frac{1}{2}}} \tag{19.10}$$

where

$$d = \frac{B}{A}, \quad \omega_n = (2n+1)\frac{\pi}{l}\left(\frac{C}{A}\right)^{\frac{1}{2}}, \quad \text{and} \quad \delta_n = \arctan \frac{\omega d}{(\omega_n^2 - \omega^2)}.$$

It turns out that this form for $\xi = \xi(x, y, t)$ has advantages over a closed form which can also be obtained. Use of the first term of this series yields good quantitative results for attenuation and good qualitative results for velocity. Additional terms are needed for quantitative high frequency velocity calculations.

20. Attenuation and velocity. The above expressions for ξ and σ lead to the expressions for attenuation and velocity as follows:

$$\alpha(\omega) = \frac{1}{v}\left(\frac{4G\,b^2}{\pi^4\,C}\right)\omega_0^2\,\Lambda L^2\,\frac{\omega^2\,d}{(\omega_0^2-\omega^2)^2+(\omega\,d)^2},\tag{20.1}$$

$$v(\omega) = v_0\left[1-\left(\frac{4G\,b^2}{\pi^4\,C}\right)\omega_0^2\,\Lambda L^2\,\frac{\omega_0^2-\omega^2}{(\omega_0^2-\omega^2)^2+(\omega\,d)^2}\right]$$

where $\left.\rule{0pt}{2.5em}\right\}\tag{20.2}$

$$v_0 = \sqrt{\frac{G}{\varrho}} \quad\text{and}\quad \omega_0 = \frac{\pi}{L}\left(\frac{C}{A}\right)^{\frac{1}{2}}.$$

It is worth noting from the expressions for $\alpha(\omega)$, $v(\omega)$, ξ, and σ that only the dislocation displacement component in phase with the stress contributes to the velocity change $\Delta v = v_0 - v(\omega)$. The displacement component out of phase with the applied stress causes the attenuation $\alpha(\omega)$. The relation of attenuation to other loss expressions was given in Eqs. (8.6), (8.9), (8.10), together with various conversions in units, in relations (8.5), (8.7), and (8.8). The decrement, for example, is connected with the attenuation by the relation (8.6):

$$\delta\text{ (nepers)} = \alpha\text{ (nepers/cm)} \cdot \lambda\text{ (cm)}$$

and with the Q of the system by the relations (8.9) and (8.10). Expressions for decrement and modulus are of somewhat more use where one is concerned with kilocycle measurements and standing wave techniques. Since the concern in this article is with high frequency work, primarily with pulse echo methods, the discussion is carried on in terms of attenuation and velocity.

The fractional change in velocity is frequently of more interest than the velocity alone, and from the previous expression this is

$$\frac{\Delta v}{v_0} \equiv \frac{v_0 - v(\omega)}{v_0} = \frac{4G\,b^2}{\pi^4\,C}\,\omega_0^2\,\Lambda L^2\,\frac{(\omega_0^2-\omega^2)}{(\omega_0^2-\omega^2)^2+(\omega\,d)^2}\tag{20.3}$$

and since

$$\frac{\Delta v}{v} = \frac{1}{2}\frac{\Delta G}{G}$$

the fractional modulus change arising from the dislocation damping effect is given directly by $2\Delta v/v$.

It is for some purposes convenient to put $\alpha(\omega)$ and $v(\omega)$ in a slightly different form as follows:

$$\alpha(\omega) = \frac{1}{v}\left(\frac{4G\,b^2}{\pi^4\,C}\right)\Lambda L^2\,d\,\frac{\left(\frac{\omega}{\omega_0}\right)^2}{\left(1-\frac{\omega^2}{\omega_0^2}\right)^2+\left(\frac{\omega}{\omega_0}\right)^2\left(\frac{d}{\omega_0}\right)^2},\tag{20.4}$$

$$\frac{\Delta v}{v_0} = \left(\frac{4G\,b^2}{\pi^4\,C}\right)\Lambda L^2\,\frac{\left(1-\frac{\omega^2}{\omega_0^2}\right)}{\left(1-\frac{\omega^2}{\omega_0^2}\right)^2+\left(\frac{\omega}{\omega_0}\right)^2\left(\frac{d}{\omega_0}\right)^2},\tag{20.5}$$

where ω/ω_0 is called the normalized frequency. It is also convenient, in high frequency work, to express the attenuation in units of decibels per centimeter (db/cm) or decibel per microsecond (db/μsec) rather than in nepers/cm. The choice of (db/μsec) has advantages, and these units will be used in what follows. The conversion is given in Eq. (8.7),

$$\alpha\text{ (db/}\mu\text{sec)} = 8.68\times10^{-6}\,v\text{ (cm/sec)}\,\alpha\text{ (nepers/cm)},$$

so that the attenuation expression will be written

$$\alpha(\omega) = 8.68 \times 10^{-6} \left(\frac{4G\,b^2}{\pi^4\,C}\right) \Lambda L^2 d \left\{ \frac{\left(\frac{\omega}{\omega_0}\right)^2}{\left(1 - \frac{\omega^2}{\omega_0^2}\right)^2 + \left(\frac{\omega}{\omega_0}\right)^2 \left(\frac{d}{\omega_0}\right)^2} \right\}. \tag{20.6}$$

It is found that the experimental results can be accounted for in terms of the theory if values of B range from about 5×10^{-5} to about 8×10^{-4} depending on the material, values of L from about 10^{-3} to 10^{-5}. When the dislocation density becomes too large the loop lengths become too small to observe damping effects. In other words, the dislocations are then effectively completely pinned. The experimental evidence in connection with higher dislocation densities is very meager.

The frequency response of attenuation and fractional velocity change as a function of the normalized frequency depends upon the damping value $d = B/A$ relative to the resonant frequency ω_0. If B is sufficiently large relative to A $(d \gg \omega_0)$ the case is one of large damping while if $d \ll \omega_0$ the case is one of small damping.

If, from Eq. (20.4), the expression

$$\frac{\alpha(\omega)}{\frac{1}{v}\left(\frac{4G\,b^2}{\pi^4\,C}\right)\Lambda L^2 \omega_0}$$

is plotted as a function of ω/ω_0 using

$$\frac{d}{\omega_0} \frac{\left(\frac{\omega}{\omega_0}\right)^2}{\left(1 - \frac{\omega^2}{\omega_0^2}\right)^2 + \left(\frac{\omega}{\omega_0}\right)^2 \left(\frac{d}{\omega_0}\right)^2} = F(\omega) \tag{20.4a}$$

for various cases of large $(d/\omega_0 \gg 1)$ and small $(d/\omega_0 \ll 1)$ damping, the result is as shown in Fig. 2. A similar plot may be obtained from Eq. (20.5) by displaying

$$\frac{\frac{\Delta v}{v_0}}{\left(\frac{4G\,b^2}{\pi^4\,C}\right)\Lambda L^2}$$

as a function of ω/ω_0 using

$$\frac{1 - \frac{\omega^2}{\omega_0^2}}{\left(1 - \frac{\omega^2}{\omega_0^2}\right)^2 + \left(\frac{\omega}{\omega_0}\right)^2 \left(\frac{d}{\omega_0}\right)^2} = G(\omega) \tag{20.5a}$$

as shown in Fig. 3. As seen from the attenuation-frequency plot the attenuation has a resonance behavior and the resonance frequency depends on C, A, and loop length L. The resonance shape depends on the type of damping which in turn depends on B and A and ω_0. The maximum $F(\omega)$ occurs at $\omega = \omega_0$.

The attenuation frequency dependence $F(\omega)$, Eq. (20.4a), can be put in slightly different and sometimes more useful form for the case of large damping, i.e. $(d/\omega_0 \gg 1)$

$$F(\omega) = \frac{d}{\omega_0} \frac{\left(\frac{\omega}{\omega_0}\right)^2}{1 + \left(\frac{\omega}{\omega_0}\right)^2 \left[\left(\frac{d}{\omega_0}\right)^2 - 2\right] + \left(\frac{\omega}{\omega_0}\right)^4}.$$

Then, if in addition to the large damping condition, we inquire about the frequency region such that $\omega/d \ll 1$, it is seen that

$$F(\omega) \approx \frac{d}{\omega_0} \frac{\left(\frac{\omega}{\omega_0}\right)^2}{1 + \left(\frac{d}{\omega_0}\right)^2 \left(\frac{\omega}{\omega_0}\right)^2} \approx \left(\frac{\omega_0}{d}\right) \left\{ \frac{\left(\frac{\omega}{\omega_m}\right)^2}{1 + \left(\frac{\omega}{\omega_m}\right)^2} \right\} \qquad (20.4\,b)$$

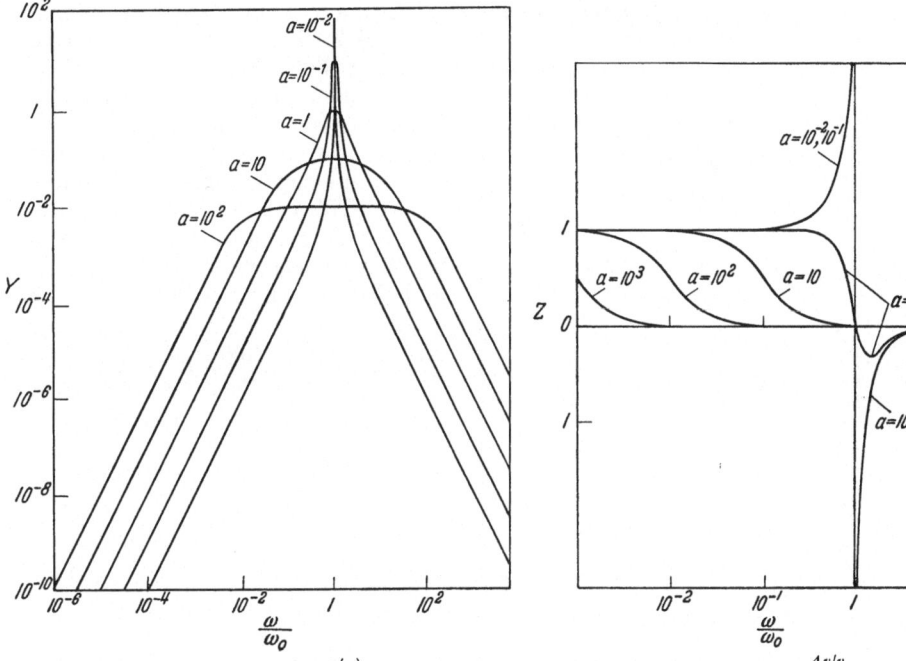

Fig. 2. Dependence of $Y = \dfrac{\alpha(\omega)}{\frac{1}{v}\left(\frac{4Gb^2}{\pi^4 C}\right) A L^2 \omega_0}$ on ω/ω_0 for various values of $a = d/\omega_0$.

Fig. 3. Dependence of $Z = \dfrac{\Delta v/v_0}{\left(\frac{4Gb^2}{\pi^4 C}\right) A L^2}$ on ω/ω_0 for various values of $a = d/\omega_0$.

where

$$\omega_m = \frac{\omega_0^2}{d}$$

is the frequency at which the attenuation is half of maximum for the case of large damping. Since $\omega_0^2 = \dfrac{\pi^2 C}{L^2 A}$, this means that

$$\omega_m = \frac{\pi^2 C}{L^2 B}.$$

Returning to the attenuation relation Eq. (20.6) with the last form of $F(\omega)$ in terms of ω_m, one finds

$$\alpha = \left[8.68 \times 10^{-6} \left(\frac{4Gb^2}{\pi^4 C}\right) A L^2 \omega_m \right] \left[\frac{\left(\frac{\omega}{\omega_m}\right)^2}{1 + \left(\frac{\omega}{\omega_m}\right)^2} \right] \quad \text{(db/μsec)} . \qquad (20.7)$$

The factor in the first set of brackets will be called α_m; it is the maximum attenuation value which occurs when $\omega \to \infty$ for the large damping case. At

$\omega = \omega_m$, $\alpha = \tfrac{1}{2}\alpha_m$, hence Eq. (20.7) may be written

$$\alpha = \alpha_m \frac{\left(\dfrac{\omega}{\omega_m}\right)^2}{1 + \left(\dfrac{\omega}{\omega_m}\right)^2}.$$
(20.8)

A plot of log (α/α_m) as a function of log (ω/ω_m) results in a curve shown in Fig. 4. The normalized attenuation has an initial slope of 2. The corresponding curve for decrement is also shown and the initial slope in this case is 1. It is evident at this point that for frequencies well below ω_m the attenuation α depends on dislocation density to the first power and loop length L to the fourth power. This can be seen by considering $(\omega/\omega_m)^2 \ll 1$ in the frequency term which then becomes simply $(\omega/\omega_m)^2$ and this combined with α_m leads to

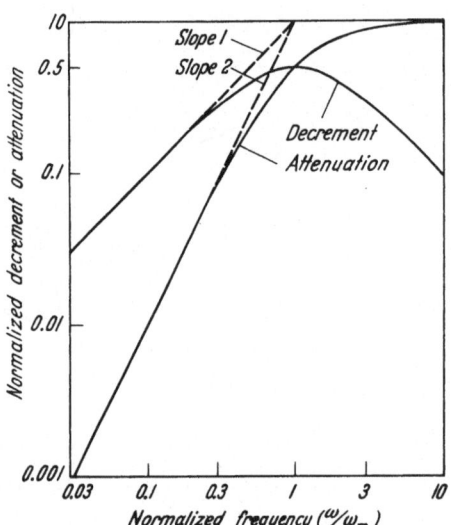

Fig. 4. Normalized attenuation and decrement as a function of ω/ω_m for the case of large damping (Granato and Truell).

$$\left.\begin{aligned} \alpha = 8.68 \times 10^{-6} &\left(\frac{4G\,b^2}{\pi^4\,C}\right) \times \\ \times &\left(\frac{B}{\pi^2\,C}\right) \Lambda L^4\,\omega^2 \quad (\text{db}/\mu\text{sec}) \end{aligned}\right\}$$
(20.9)

or

$$\left.\begin{aligned} \alpha = \frac{1}{v} &\left(\frac{4G\,b^2}{\pi^4\,C}\right) \times \\ \times &\left(\frac{B}{\pi^2\,C}\right) \Lambda L^4\,\omega^2 \;(\text{nepers/cm}). \end{aligned}\right\}$$
(20.10)

Using the same approximations regarding frequency, namely $(\omega/\omega_m)^2 \ll 1$, one sees from Eq. (20.5) that the frequency part of the equation approaches unity under these conditions, hence that

$$\frac{\Delta v}{v_0} = \left(\frac{4G\,b^2}{\pi^4\,C}\right) \Lambda L^2.$$
(20.11)

It is seen that the loop length L and the dislocation density Λ of a material can be determined from measurements of ultrasonic velocity and attenuation if the attenuation is caused by dislocation damping[1]. Since $\Delta v = v_0 - v(\omega)$ it is possible to measure Δv if the purely elastic (without dislocation effects) velocity v_0 can be measured. v_0 can, in principle at least, be obtained by measuring the velocity at a frequency sufficiently high that the dislocations cannot follow the stress wave driving force and such measurements have been made at frequencies even as low as hundreds of megacycles. $v(\omega)$ is, of course, the velocity at some much lower frequency, well below resonance, where dislocation damping mechanisms operate. From data on $\Delta v/v_0$ and α the two Eqs. (20.9) and (20.11)

---

[1] The expressions for α and for $\Delta v/v_0$ contain the shear wave modulus G which is the purely elastic modulus for a perfect single crystal, and G is not affected by dislocation changes.

allow calculation of L and Λ; the results for low frequencies are:

$$L = \frac{1}{2\nu} \left[\frac{10^6}{8.68} \frac{C}{B} \frac{\alpha \left(\dfrac{db}{\mu sec} \right)}{\dfrac{\Delta v}{v}} \right]^{\frac{1}{2}}, \tag{20.12}$$

$$\Lambda = 8.68 \times 10^{-6} \left(\frac{\pi^4 B}{G b^2} \right) \nu^2 \left(\frac{\left(\dfrac{\Delta v}{v_0} \right)^2}{\alpha \left(\dfrac{db}{\mu sec} \right)} \right). \tag{20.13}$$

The procedure just described will yield values of L and Λ for a given state of the material under proper conditions. When, however, the material is undergoing changes produced, for example, by deformation or irradiation, the measurement of α and $v(\omega)$ as a function of the deformation or irradiation, as it proceeds, will yield changes in L and Λ as a function of the process. It is, of course, assumed that the processes consist of changes in dislocation damping and that the dislocation density is not so large that the dislocations interfere with each other except for the intersection pinning. In any case the loop length L and the dislocation density Λ are values for those dislocations whose vibration can be "seen" by the ultrasonic stress wave.

21. Distribution of dislocation loop lengths. Thus far no consideration has been given to the fact that not all dislocations can be expected to have the same loop length. In view of the fact that the attenuation depends on the fourth power of the loop length, it might be expected that the results might be sensitive to the distribution of loop lengths. Various distributions have been studied in detail, and it has been found that qualitative results are changed very little except for the fact that the large increase in damping near resonance does not occur for a distribution of loop lengths other than a delta distribution function. Detailed comparison of the results shows that for a distribution of loop lengths one can quite well use a single "effective" loop length. The effective loop length is larger than the average because of the fourth power dependence of attenuation on loop length, which means that the longer loop lengths are weighted more heavily and thereby influence the attenuation more than shorter loop lengths do. It is found, for example, that the effective loop length L_e for an exponential distribution is 3.3 times the average value.

The distribution function used by KOEHLER for this type of calculation was

$$N(l)\, dl = \frac{\Lambda}{L^2} e^{-\frac{l}{L}}\, dl, \tag{21.1}$$

where $N(l)\, dl$ is the number of loops with length lying between l and $l+dl$ and where L is the average loop length.

The attenuation resulting from this distribution is $\alpha(l)$ averaged over the distribution $N(l)\, dl$

$$\alpha = \int_0^\infty \alpha(l)\, N(l)\, dl = \frac{\Lambda}{L^2} \int_0^\infty \alpha(l)\, e^{-\frac{l}{L}}\, dl. \tag{21.2}$$

Similarly the fractional velocity difference (or modulus) for the exponential distribution is obtained from

$$\frac{\Delta v}{v} = \frac{\Lambda}{L^2} \int_0^\infty \left(\frac{\Delta v}{v}\, (l) \right) e^{-\frac{l}{L}}\, dl, \tag{21.3}$$

where $\alpha(l)$ from Eq. (20.4) and $\dfrac{\Delta v}{v}\, (l)$ from Eq. (20.5) are used.

The result of this analysis can be obtained in terms of complex exponential integral functions, and the low frequency behavior of the attenuation and fractional velocity difference may be found by expansions of the function.

The expressions to be evaluated are

$$\alpha = \frac{1}{v} \frac{4G\,b^2}{A^3} C \frac{A\,B}{\omega^4\,L^2} \int_0^\infty \frac{x^5\,e^{-D\beta x}\,dx}{[(1-x^2)^2 + \beta^2\,x^4]} \tag{21.4}$$

and

$$\frac{\Delta v}{v_0} = \frac{4G\,b^2}{A^2} C \frac{A}{\omega^4\,L^2} \int_0^\infty \frac{x^3\,(1-x^2)\,e^{-D\beta x}}{[(1-x^2)^2 + \beta^2\,x^4]}\,dx,$$

where

$$x = \frac{\omega}{\pi}\left(\frac{A}{C}\right)^{\frac12} l, \quad \beta = \frac{d}{\omega}, \quad D = \frac{\pi}{L\,d}\left(\frac{C}{A}\right)^{\frac12}.$$

The evaluation of these integrals is given in Ref. [41]. The result of evaluating (21.4) is

$$\alpha = \frac{1}{v} \frac{4G\,b^2}{\pi^4\,C} A L^2 \left\{ \frac{\frac{\omega_0^2}{\omega^2}\frac{B}{A}}{1+\left(\frac{B}{\omega A}\right)^2} + \frac{\frac{\omega_0^2}{\omega^4}\frac{B}{A}}{\left[1+\left(\frac{B}{\omega A}\right)^2\right]^2} \sum_{n=1}^{4} A_n\,e^{-\frac{\omega_0}{\omega}x_n}\left(-\mathrm{Ei}\left[\frac{\omega_0}{\omega}x_n\right]\right) \right\}. \tag{21.5}$$

In Eq. (21.5) $\omega_0^2 = \frac{\pi^2}{L^2}\frac{C}{A}$ has the same meaning as before in terms of the other quantities but it is no longer to be considered as a resonance frequency in the same way it was with a single loop length. As an alternative the quantity $\frac{\omega_0}{\omega}x_n$ can be written $D\beta x_n$. The x_n are as follows:

$$x_1 = \frac{1}{\sqrt{2(1+\beta^2)}}\left\{\sqrt{\sqrt{1+\beta^2}+1} + i\sqrt{\sqrt{1+\beta^2}-1}\right\}, \tag{21.6}$$
$$x_2 = -x_1, \quad x_3 = x_1^*, \quad x_4 = -x_1^*.$$

A similar expression has been obtained for $\Delta v/v_0$ from Eq. (21.5):

$$\frac{\Delta v}{v_0} = \frac{4G\,b^2}{A^2} C A \frac{1}{\omega^4\,L^2}\left\{ \frac{1}{1+\left(\frac{B}{\omega A}\right)^2} \sum_{n=1}^{4} C_n\,e^{-\frac{\omega_0}{\omega}}\left[-\mathrm{Ei}\left(\frac{\omega_0}{\omega}x_n\right)\right] - \right.$$
$$\left. - \frac{1}{1+\left(\frac{B}{\omega A}\right)^2}\left(\frac{\omega^2}{\omega_0^2}\right) - \frac{1}{\left[1+\frac{B}{\omega A}\right]^2} \sum_{n=1}^{4} A_n\,e^{-\frac{\omega_0}{\omega}}\left[-\mathrm{Ei}\left(\frac{\omega_0}{\omega}x_n\right)\right] \right\}. \tag{21.7}$$

The evaluation of $[-\mathrm{Ei}(-z)]$ for particular cases in the attenuation expression (21.5) was handled as follows:

If $\beta = d/\omega \to \infty$, the case is one of low frequency or large damping or both. In this case successive integration by parts yields the semiconvergent series

$$[-\mathrm{Ei}(-z)] = \int_z^\infty \frac{e^{-y}}{y}\,dy \approx \left[\sum_{K=0}^\infty (-1)^K \frac{K!}{z^{K+1}}\right] e^{-z}.$$

The terms $K=0, 1, 3, 4, 5, 7, 9$ yield no contribution to this sum. The term $K=2$ yields a contribution which, however, is cancelled exactly by one of the other terms preceding this sum. The term $K=6$ gives the first low frequency term and the term $K=8$ gives the second term which is independent of B. The

$K=10$ term gives a term depending on B which is of the same size but opposite in sign to the $K=8$ term. The $K=10$ term has the same frequency dependence as the $K=8$ term. This calculation leads to the following result: For the case in which $d/\omega \gg 1$, i.e., $d/\omega \to \infty$, it is found that for the low frequency case

$$\alpha = \frac{1}{v} \frac{4G b^2}{\pi^6 C^2} (5!) \Lambda L^4 B \omega^2 \left[1 + \frac{84 L^2 A \omega^2}{\pi^2 C} - \frac{3042 L^4 B^2 \omega^2}{\pi^4 C^2}\right], \qquad (21.8)$$

$$\frac{\Delta v}{v_0} = \frac{4G b^2}{\pi^4 C} (3!) \Lambda L^2. \qquad (21.9)$$

Except for the factors 3! and 5!, Eq. (21.9) and the first term of (21.8) are the same as Eqs. (20.10) and (20.11).

22. Strain amplitude effects.
Thus far nothing has been incorporated into dislocation damping theory about the effect of strain amplitude on the value of α and $\Delta v/v$. As long as one is concerned with attenuation well into the megacycle region it is unlikely that the strain amplitude will become large enough to arouse detectable strain amplitude effects. In order to have strain-amplitude effects present it is necessary that some breakaway occur—breakaway of some of the dislocations from their weaker (point defect) pinning so that L_c, the loop length determined by the weaker pinning, should increase until the network loop length L_N is reached. Breakaway occurs when the dislocation strain reaches some particular value. This breakaway effect is responsible for a hysteresis type of loss usually referred to as the frequency independent, strain amplitude dependent loss. The loss is caused by unpinning during one part of the stress cycle and the elastic collapse and repinning of the loops during another part of the stress cycle and the same for the other half of the alternating stress cycle. The expression (decrement) for this loss derived by GRANATO and LÜCKE is as follows:

$$\delta_H = \frac{c_1}{\varepsilon_0} e^{-\frac{c_2}{\varepsilon_0}}, \qquad (22.1)$$

where

$$c_1 = \Omega K \left(\frac{8G b^2}{\pi^3 C}\right) \frac{\Lambda L_N^3}{\pi L_c^2} \varepsilon' a,$$

$$c_2 = \frac{K \varepsilon' b}{L_c},$$

and ε' is the fractional difference in size between the pinning defects and the atoms of the material. K is a parameter concerned with breakaway; it is a function of the orientation of the propagation direction with respect to the slip planes. a is the lattice parameter, and the strain amplitude is ε_0. Ω is an orientation factor concerned with the resolved shear stress on the slip plane, i.e. the stress effective in moving dislocations.

A corresponding relation for the fractional modulus difference was also derived [41] and is equal to the decrement

$$\left(\frac{\Delta G}{G}\right)_H = \delta_H = \frac{c_1}{\varepsilon_0} e^{-\frac{c_2}{\varepsilon_0}} \qquad (22.2)$$

or

$$\gamma = \frac{\delta_H}{\left(\frac{\Delta G}{G}\right)_H} = 1.$$

23. Thermal effects in dislocation damping.
The effect of temperature on dislocation damping at megacycle frequencies is contained in the specific damping

constant B as long as one is concerned only with high frequencies and strains sufficiently low so that breakaway of dislocations from pinning agents is not induced by the strain. At megacycle frequencies, especially high megacycle frequencies, this strain condition is usually satisfied.

At lower frequencies, in the kilocycle range, the effect of temperature enters in another way having to do with the combination of stress and thermal breakaway of dislocations from pinning agents such as impurities or vacancies. The result is a dislocation damping which is strain-amplitude dependent and in which thermal activation will aid the applied stress in the unpinning process.

Consequently, there are two problems: one is the theoretical basis of the damping constant B and in particular its dependence on temperature; the other is the thermal activation in connection with the breakaway of dislocations. The first problem has been mentioned at the beginning of the discussion of dislocation damping where it was pointed out that the damping constant B had been estimated [47] from calculations concerned with the scattering by dislocations of the lattice vibrations or phonons.

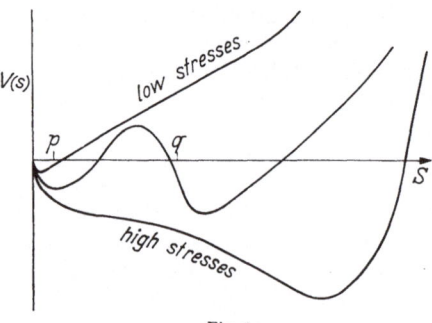

Fig. 5.

A derivation of B by Leibfried [47] yielded an expression

$$B = \left(\frac{3kz}{10v_t a^2}\right) T, \qquad (23.1)$$

where a is the lattice spacing, v_t is the transverse wave velocity, z is the number of atoms in an elementary cell, and k is Boltzmann's constant. B is thus expected to depend linearly on temperature. Evidence for such a temperature dependence between helium and room temperatures is given by the work of Alers and Thompson [57], who show that for copper B and B/A are linearly related to the temperature. The basis of Leibfried's result is the interaction or scattering of thermal vibration waves by dislocations. A somewhat different loss mechanism has been proposed by Mason [46], which leads to a slightly different dependence. The predicted magnitude from Mason's results is about 5×10^{-5} which is about a factor of ten too small. There are difficulties with all of the attempts to evaluate B from fundamentals.

The second problem concerned with the thermal breakaway of dislocations has been discussed by Teutonico, Granato and Lücke[1]. Since this problem arises only at relatively low (kilocycle) frequencies, it will be outlined only briefly here.

In order to extend the present dislocation damping theory to elevated temperatures, it is necessary to examine the possible static equilibrium positions or energy states of a pinned dislocation line as a function of the applied stress.

In the case of a dislocation with a single pinning point—a dislocation with its end fixed and a pinning agent at the center point—the pinning is the result of the interaction of the stress fields of the dislocation and the point defect. The interaction energy has been calculated by Cottrell from classical elasticity and an analytical expression is available.

---

[1] Thesis L. J. Teutonico, Brown University 1958 and as yet unpublished work of Teutonico, Granato, and Lücke; private communication. See also Bull. Amer. Phys. Soc., Ser. II **7**, No. 3, 223 Abstracts U_1, U_2 (1962).

This interaction term between an impurity or other point defect and a dislocation introduces into the analysis a potential barrier. Into the symmetric double loop problem an approximate representation of the Cottrell interaction force is introduced and to the associated potential energy is added the line energy of the dislocation and the strain energy of the medium. The expression for the total potential energy yields three equilibrium configurations depending on the stresses involved. Fig. 5 shows the potential $V(s)$ where s is the dislocation displacement at the loop midpoint. At intermediate stresses a second energy minimum enters with an energy barrier between the two minima. At sufficiently high stresses this energy barrier disappears and only the second minimum exists at larger values of s.

The physical picture is that for breakaway to occur without thermal assistance (i.e. at zero temperature) a stress sufficiently high must be applied so that the energy barrier disappears. When, however, a temperature effect is introduced it is possible for the dislocation to surmount the energy barrier by thermal activation at lower stresses as long as there is a second stable potential minimum into which it may go. For sufficiently low stresses, however, the second minimum does not exist; consequently there is the important feature that for thermal breakaway a stress is necessary and a stress threshold exists. An energy diagram depicting the situation when two stable levels are present, that is when a sufficient stress is applied, is shown in Fig. 6. U_1 is the activation energy for the dislocation element to jump in one direction and U_2 is that for the reverse direction.

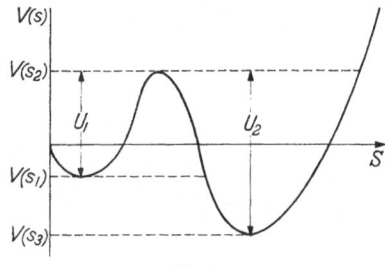

Fig. 6.

In addition to the stable state or configurational questions for the dislocation, particularly the double loop model used here as an example, one must have a description of the kinetics of the dislocation breakaway, and for this one sets up a rate equation. Consider a solid in which there are double loops all of length $2l$ and each with an energy diagram as in Fig. 6, such that thermal breakaway is possible. Let f be a function representing the fraction of dislocations that has broken away and that is occupying the second minimum. The rate at which dislocations enter the second state from the first is

$$(1 - f)\, v_1\, e^{-U_1/kT},$$

where U_1 is the barrier energy as in Fig. 6 and v_1 is an effective incidence frequency associated with the dislocation in the first state. The rate at which dislocations go from the second state to the first state is given by

$$- f v_2\, e^{-U_2/kT}$$

where U_2 is as indicated in Fig. 6 and v_2 is an effective frequency of incidence on the barrier for the transition from state 2 to state 1. The rate equation is then

$$\frac{df}{dt} = (1 - f)\, v_1\, e^{-U_1/kT} - f v_2\, e^{-U_2/kT}$$

or

$$\frac{df}{dt} + Pf = Q, \tag{23.2}$$

where

$$P = v_1 e^{-U_1/kT} + v_2 e^{-U_2/kT},$$

$$Q = v_1 e^{-U_1/kT}.$$

A solution of this differential equation is

$$f(t) = e^{-\int P dt} \left[\int Q e^{\int P dt} dt + K \right]. \tag{23.3}$$

P and Q are time dependent if the stress is time dependent. One can examine the situation for constant stress; the result is for the region of interest

$$f = \frac{Q}{P} (1 - e^{-Pt}), \qquad Q \approx P, \tag{23.4}$$

and the constant stress case gives one some idea about the solutions for more complicated cases.

Here the relaxation time associated with f is

$$\tau = \frac{1}{P}. \tag{23.5}$$

In order for breakaway to occur to an extent $e^{-1} f$ an interval of time Δt is necessary such that

$$P \Delta t \approx 1. \tag{23.6}$$

Now if Δt is less than the period of a cycle with an oscillating stress of angular frequency ω, the time interval Δt is proportional to $1/\omega$ or $\Delta t = c/\omega$ where c is a numerical factor of order of unity. Then

$$P \Delta t = v_1 e^{-U_1/kT} (c/\omega) = 1 \tag{23.7}$$

and

$$U_1 = kT \log_e \frac{v_1 c}{\omega}; \tag{23.8}$$

but U_1 can also be expressed in terms of the Cottrell pinning energy U_0 (≈ 0.1 ev) and of the mechanical breakaway stress σ_M:

$$U_1 \approx U_0 \left(1 - \frac{\sigma}{\sigma_M} \right)^2. \tag{23.9}$$

Hence

$$U_0 \left(1 - \frac{\sigma}{\sigma_M} \right)^2 = kT \log_e \frac{v_1 c}{\omega} \tag{23.10}$$

and

$$\sigma = \sigma_M \left(1 - \sqrt{\frac{kT}{U_0} \log_e \frac{v_1 c}{\omega}} \right), \tag{23.11}$$

where σ is the new breakaway stress expressed in terms of the mechanical breakaway stress.

In the zero temperature theory the decrement δ is proportional to $e^{-\sigma_M/\sigma_0}$ and with temperature and thermal breakaway effects introduced into the problem the factor becomes $e^{-\sigma_B/\sigma_0}$ where σ_B is the value of σ given above in terms of $\sigma_M, U_0, v_1, \omega$, and the temperature T. σ_0 is the maximum stress amplitude during a cycle. This discussion is intended only to show what factors are involved and how they enter the problem. The details of the calculation are given in the references.

In view of the fact that, in the high frequency work, strain amplitude effects are of relatively less importance, the matter will not be discussed further. The experimental work appears to agree well with the theory and the reader is referred to Refs. [48] and [49], for examples. This work has been carried out in the kilocycle range of frequencies.

Before leaving this part of the discussion of the theory for this dislocation loss mechanism, it should be noted that σ and ε have the meaning of shear stress and shear strain. Many experiments are carried out with longitudinal stresses, and even when shear stresses are used the component resolved on the slip plane is not that applied directly. In any case, there are geometry and orientation factors which must be considered in order to take account of the orientation relations between the direction of propagation of the wave, shear or longitudinal, and the slip plane and the slip direction in a crystal. There are usually many slip planes and the effect of all slip systems must be considered for a given orientation of the ultrasonic wave. Such calculations are simple in principle but involved in detail. The net effect is taken into account by the factor Ω which has been mentioned in c_1 of Eq. (22.1). The factor Ω should also appear as a factor in all of the preceding equations for α and $\Delta v/v_0$, e.g. Eqs. (20.1) through (20.11) and (21.4) through (21.9). Detailed discussion of such calculations have been omitted here. It was pointed out that for many cases a value of about $\frac{1}{25}$ is appropriate.

24. Some selected experimental results. The experimental results and their connection with the theory of dislocation damping at megacycle frequencies are available in essentially two areas of work. One of these is concerned with the study of attenuation and velocity changes while the sample is undergoing deformation. The second is concerned with attenuation and velocity measurements as they are affected by irradiation of the specimen. The irradiation case is itself composed of separate cases depending on the type of material and the irradiation used. Two examples of the effect of radiation induced defects will be discussed in what follows; these examples show the changes in attenuation and velocity measured during irradiation. One of these examples is that of the reactor irradiation of copper, the other is one of the irradiation of sodium chloride by Cobalt 60 gamma rays. The second example mentioned will be discussed next.

$\alpha)$ *Irradiation effects in NaCl.* It has been found in many experiments that when single crystals of sodium chloride are irradiated with Cobalt 60 gamma rays the ultrasonic attenuation decreases and the velocity increases as a function of the irradiation time or total irradiation flux. The magnitude of attenuation decrease or velocity increase depends on the deformation of the sample or, in other words, on the dislocation density and the loop lengths. The range of ultrasonic frequencies used in this work extends from low kilocycle frequencies [51] to [54] to about 300 megacycles/sec. Controlling the temperature during the experiment is very important because a few degrees change can cause appreciable changes in sodium chloride and the velocity changes cannot be reliably measured if the temperature is not held constant within a degree or two. Figs. 7 and 8 show typical attenuation and velocity data for a sample of sodium chloride which has been deformed approximately 1%. The velocity change was about 0.9% corresponding to an attenuation change of 1.3 db/μsec. Another example showed a velocity change of 0.3% corresponding to an attenuation change of 0.75 db/μsec. A very few crystals have been found where essentially no attenuation change occurred on irradiation. Presumably these crystals had a very low dislocation density. Using the data of Figs. 7 and 8 in relations (20.12) and (20.13) yields a dislocation density of 1.2×10^7 and an initial loop length of 1.3×10^{-4} cm where values of $R = 10^{-4}$, $G - 1.25 \times 10^{11}$ dynes/cm² and $b - 3 \times 10^{-8}$ cm have been used. The frequency used was 15 Mc/sec. During the course of the irradiation loop length decreased from the initial value of 1.3×10^{-4} cm to 4.7×10^{-5} cm after 22 min of irradiation. The irradiation rate for the cases discussed here was 3500 roentgens/hour.

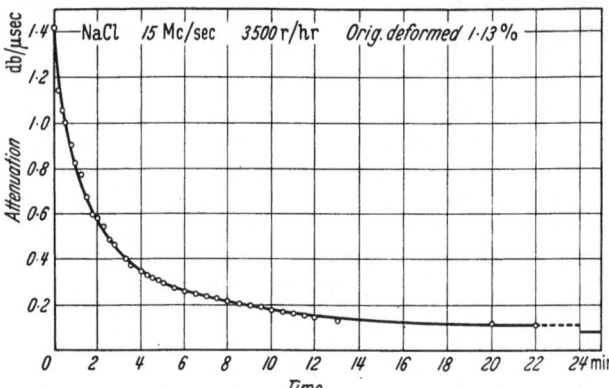

Fig. 7. Attenuation as a function of irradiation time for deformed NaCl single crystal. After Truell [53].

It can be seen, from relations (20.12) and (20.13), that for a given frequency the ratio $R = \dfrac{\Delta v/v_0}{\sqrt{\alpha}}$ should be constant as long as the dislocation density Λ does not change. It is assumed that Λ does not change during gamma irradiation; the gamma irradiation is assumed to provide point defects capable of pinning the dislocations during the irradiation.

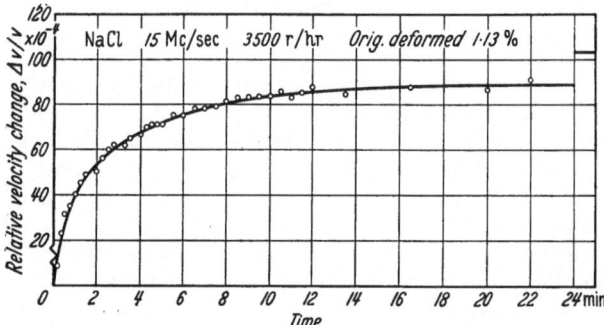

Fig. 8. Fractional change in velocity as a function of irradiation time for deformed NaCl single crystal. After Truell [53].

In experiments of this type the ratio R as a function of irradiation appears as in Fig. 9 where it is seen that R is constant after about a minute or two of irradiation. The decrease in R, in this case, from 90×10^{-4} to about 72×10^{-4} in the first minute of irradiation may be the consequence of having the resonant frequency ω_0 shift from a lower to a higher value as the loop lengths shorten during the pinning process so that initially the condition $\omega/\omega_m \ll 1$ may not have been satisfied sufficiently well to make the simplified frequency relations valid. There is evidence from dispersion measurements on sodium chloride that the resonant frequency, after deformation but before irradiation, lies near 35 Mc/sec. Deformation has the effect of lowering the value of the resonant frequency; and in the example of the dispersion work the effect was to shift the resonant frequency from about

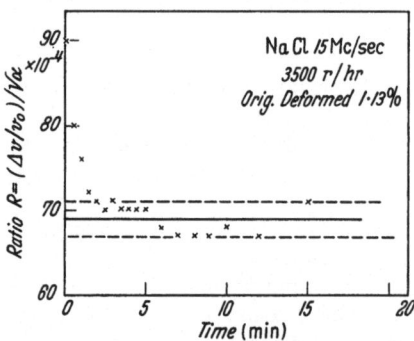

Fig. 9. The ratio R as a function of irradiation time for deformed NaCl single crystal. After Truell [53].

75 Mc/sec to about 35 Mc/sec as the result of a deformation of roughly 0.1 %. Since the measurements during irradiation were made at a frequency of 15 Mc/sec, it is seen that if the resonant frequency was near 35 Mc/sec the condition $\omega/\omega_m \ll 1$

definitely was not satisfied; it is further evident that, as pinning proceeds, if the resonant frequency went to 75 Mc/sec or higher, the condition would have been much more nearly met.

The assumption that it is the decrease in loop lengths, without any change in dislocation density, that causes the attenuation decrease and the velocity increase leads to the question of how the pinning process occurs and by what form of defects. Information about the concentration of these point defects can be obtained from the attenuation and velocity measurements.

The decrease in loop length should depend on the concentration or the number of effective pinning agents $c(t)$ added during irradiation to one (initial) loop length (L_0). Then the loop length at any time t after irradiation begins will be

$$L = \frac{L_0}{1 + c(t)}, \qquad (24.1)$$

where L_0 is the initial loop length at the time $(t=0)$ irradiation begins and $c(t)$ is the number of pinning points added to one loop length as a function of time.

Rewriting Eq. (20.9) in the form

$$\alpha = K_1 \omega^2 \Lambda L^4 \qquad (24.2)$$

where

$$K_1 = 8.68 \times 10^{-6} \left(\frac{4 G\, b^2}{\pi^4\, C} \right) \left(\frac{B}{\pi^2\, C} \right),$$

as seen by comparing Eqs. (20.9) and (24.2) and using (24.2) with (24.1), the result is

$$\alpha = \alpha_0 \left\{ \frac{1}{1 + c(t)} \right\}^4 \qquad (24.3)$$

where $\alpha_0 = K_1 \omega^2 \Lambda L_0^4$ is the initial attenuation before irradiation begins.

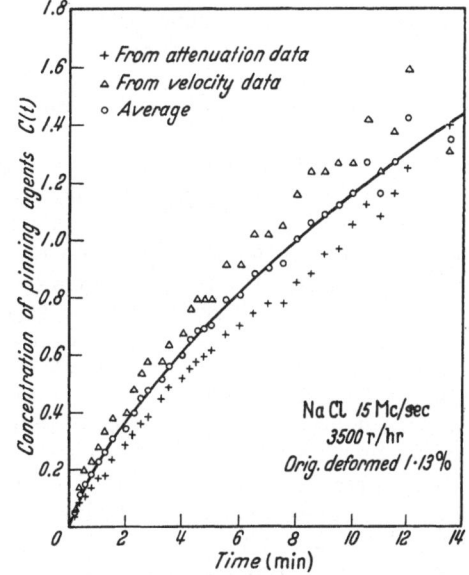

Fig. 10. Concentration of effective pinning agents as a function of irradiation time for deformed NaCl single crystal. After Truell [53].

At any time t during irradiation the concentration $c(t)$ is from (24.3) given by

$$c(t) = \left(\frac{\alpha}{\alpha_0} \right)^{-\frac{1}{4}} - 1. \qquad (24.4)$$

Thus $c(t)$ is given in terms of the initial attenuation at $t=0$ and the attenuation at any later time t.

In a similar way $c(t)$ can be calculated from velocity measurements. From Eq. (20.11),

$$\frac{\Delta v}{v_0} = K_2 \Lambda L^2, \qquad \text{where} \qquad K_2 = \left(\frac{4 G\, b^2}{\pi^4\, C} \right), \qquad (24.5)$$

and this relation together with (24.1) yields

$$c(t) = \left\{ \frac{\left(\frac{\Delta v}{v_0} \right)}{\left(\frac{\Delta v}{v_0} \right)_0} \right\}^{-\frac{1}{2}} - 1, \qquad \text{where} \qquad \left(\frac{\Delta v}{v_0} \right)_0 = K_2 \Lambda L_0^2. \qquad (24.6)$$

From the data presented in Figs. 7 and 8, $c(t)$ calculated from both (24.4) and (24.6) is shown in Fig. 10 as well as an average value of $c(t)$ shown by the solid line.

$c(t)$ as a function of t is not linear either in this case or in a number of other cases like this one. It might have been expected that as a first approximation $c(t)$ would have the form

$$c(t) = \frac{n\,r\,L_0}{\Lambda}\,t, \qquad (24.7)$$

where n is the number of pinning points per cm³ produced by one roentgen of irradiation per unit time and r is the irradiation rate in roentgens per unit time, Λ/L_0 is the number of loop lengths per cm³, and $n\,r$ the number of pinning points per cm³ per unit time with r roentgens of irradiation. Consequently, $(n\,r\,L_0/\Lambda)$ is the number of additional pinning points added on the average to each loop length L_0 per unit time.

If a rough linear approximation is used on the curve of Fig. 10, i.e. $c(t) = \beta\,t$, β has a value of about 0.2 on the early part of the curve which, with $\Lambda = 1.2 \times 10^7$, $L_0 = 1.3 \times 10^{-4}$, $r = 60$ roentgens per minute leads to $n = 3 \times 10^8$ defects/min/roentgen effective in pinning.

It is found from such curves as that of Fig. 10 that the concentration of pinning points added during an irradiation period of 20 minutes does not exceed 10^{11} per cm³. In such a period of 20 minutes the dislocations are nearly completely pinned as far as attenuation measurements can see the matter.

An equal density of color centers formed would be well below a value that can at present be studied by optical absorption measurements. It is an experimental fact that the attenuation decrease and velocity increase are essentially complete long before any measurable coloring takes place in sodium chloride. Consequently color centers cannot be ruled out as the pinning agents in the dislocation damping changes despite the fact that no measurable coloring occurs in the time that the dislocations are completely or nearly completely pinned.

The preceding work indicates the number of point defects needed to account for the observed attenuation and velocity changes. The study of attenuation and velocity as a function of gamma irradiation under different states of deformation and temperature [54], [55] yields information about the point defects themselves. For example, attenuation measured as a function of gamma ray irradiation shows a strong sensitivity to deformation where between 0 and 1.5% deformation the larger the deformation the more rapidly the attenuation decreases with gamma ray irradiation. Fig. 11 shows the deformation effect just mentioned.

The results of attenuation-irradiation measurements at room temperature, dry ice temperature and liquid nitrogen temperature show that there is little if any effect of temperature on the rate at which the attenuation decreases with gamma ray irradiation. There appears in this case to be almost no thermal activation involved in the process of pinning the dislocations. Allowing for greater uncertainty than seems to be present in the measurements, it is difficult to see how there can be a thermal activation energy exceeding 0.01 to 0.02 electron volts.

The deformation and temperature effects just mentioned have the effect of narrowing down somewhat the possible pinning process.

The following processes seem to be possible:

1. The production of point defects (by the ionizing radiation) at or very near the dislocations in sufficient numbers so that little or no migration is necessary in order to have the point defects interact with and pin dislocations.

2. The dislocations themselves move far enough to "pick up" the point defects which pin them.

3. The point defects, after being created, move to the dislocations and pin them, but with very little or no thermal activation—perhaps an electrical interaction process. Electrical interactions between dislocations and point defects have been discussed in connection with the motion of dislocations during the cyclic stressing of sodium chloride [81].

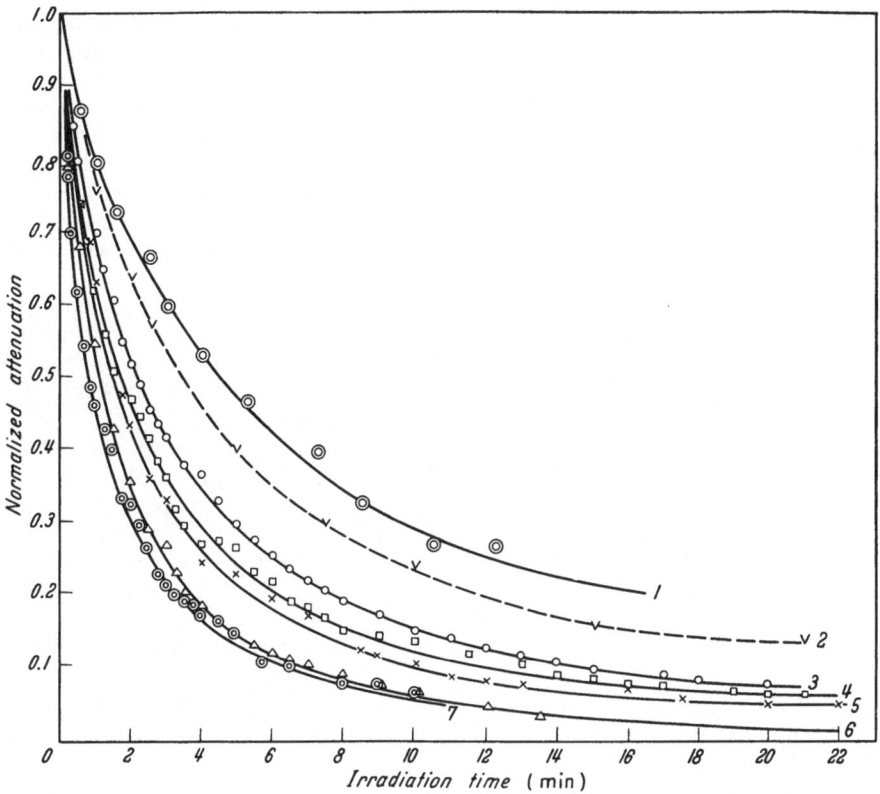

Fig. 11. Normalized attenuation as a function of irradiation time for deformed single crystals of NaCl. Curves numbered *1* to *7* in order of increasing plastic deformation from zero to 1.33%. After TRUELL [54].

There may, of course, be some combination of the above possibilities. It does not seem possible at present to select a single model from the several possible models.

β) Irradiation effects in copper. Another type of experiment making use of radiation induced defects to study dislocation damping changes is that of THOMPSON and HOLMES [56], who measured decrement and modulus changes in neutron irradiated high purity copper. These measurements, made at about 28° C (in reactor), showed that the L^4 and L^2 pinning relations are valid for this case for the attenuation and modulus changes. Fig. 12 shows the decrement decrease and modulus increase observed in the 10 to 30 kilocycle range of frequencies. The modulus change is about 3.7% for this particular copper crystal. The values of dislocation density and loop length as well as B, C, and A from dislocation theory are consistent with the data from these experiments. There is, however, some difficulty between the frequency dependence observed as compared with that predicted by the theory at low (kilocycle) frequencies.

13*

In the megacycle range of frequencies the work of Alers and Thompson [57] has been concerned with the effect of neutron irradiation and temperature (4.2 to 250° K) on the dislocation damping characteristics of pure copper single crystals. Again the effect of reactor irradiation is that of removing dislocation damping effects by immobilizing or pinning the dislocation. The attenuation decreases and the velocity increases as a function of irradiation. The velocity change is, however, found to be considerably smaller at megacycle frequencies than at kilocycle frequencies.

Fig. 13 shows the attenuation-frequency behavior for annealed copper before and after neutron irradiation. The measurements were made at 195° K in the [110] direction with compressional waves. The fact that the damping is inversely proportional to frequency shows that a maximum must exist at a frequency well below the lowest frequency shown in Fig. 13. It has not only been shown that there is a peak in the dislocation damping values but that the peak moves to higher frequencies when the temperature is lowered.

This work of Alers and Thompson on copper has produced experimental evidence that B, the specific damping, does vary linearly with temperature as predicted by Leibfried [47]. By making etch pit counts on the crystal (4×10^6 pits/cm²) and using this as an estimate of the dislocation density they found B to have a value of between

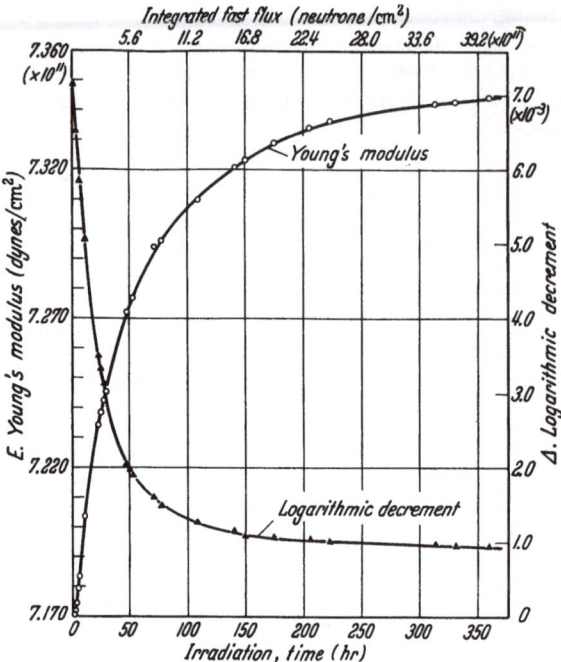

Fig. 12. Young's modulus and logarithmic decrement as a function of irradiation time with flux of 3.1×10^6 neutrons per cm²/sec. After Thompson and Holmes [56].

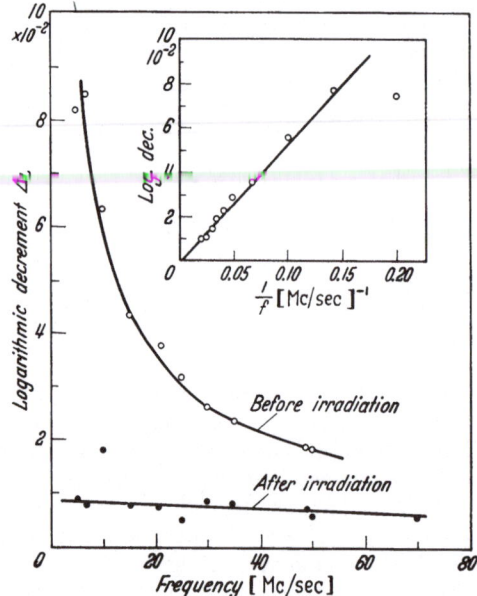

Fig. 13. Frequency dependence of attenuation of compressional wave propagating in $\langle 110 \rangle$ direction at 195° K in a well annealed pure copper single crystal before and after neutron irradiation. The inset shows the contribution to the attenuation removed by the irradiation plotted against the reciprocal of the frequency. After Alers and Thompson [57].

1×10^{-4} and 8×10^{-4} dynes sec/cm² at room temperature depending on the crystals. The range of average loop lengths was from 1×10^{-4} to 3×10^{-4} cm, and they obtained a relation for copper

$$\frac{L^2}{C} = \frac{1.3 \times 10^3}{\Lambda} \; cm^2/dyne$$

as well as a numerical value

$$\frac{4 b^2 \Lambda L^2}{\pi^4 C} = 3.5 \times 10^{14} \; cm^2/dyne$$

for the coefficient of Eq. (20.5).

ALERS and THOMPSON also estimate, by use of the previous numerical value together with $\Lambda = 4 \times 10^6$ and a measured strain amplitude [58] of the ultrasonic waves of 1×10^{-7}, that the mean displacement of the dislocation in these experiments was about 4×10^{-8} cm or roughly the lattice spacing.

GRANATO and STERN [59] have studied the attenuation as a function of both loop length and frequency in pure copper by measuring the change in attenuation as a function of gamma irradiation at various frequencies from 5 to 45 Mc/sec. The results are in some ways similar and in some ways different from those of ALERS and THOMPSON [57].

GRANATO and STERN found that a maximum existed in the dislocation damping in their copper crystal and that it originally appeared at about 3 Mc/sec. The maximum moved continuously to higher frequencies as the gamma irradiation progressed. The amplitude of the maximum decreased with irradiation. They found that the orientation and temperature dependence were consistent with the dislocation damping model in that only shear stresses on the slip systems[1] are effective in producing damping. The fact that the maximum in the damping curve shifts to higher frequencies for lower temperatures is an effect consistent with resonance under conditions of large damping. Because this maximum in the data of GRANATO and STERN is much broader than the present theory predicts, they assumed that there are two components to the damping, one for each of two types of dislocations—edge and screw. They found that the results can be described with the assumption of two components each of which behaves as predicted by the theory for random distributions of pinning points. The data could be represented for all times and frequencies by the sum of two maxima each of which moved to higher frequencies at different rates. The two components are called "fast" and "slow" corresponding to the relative rates at which the central frequencies of the maximum shift with irradiation.

The characteristics of the two dislocation components, as they influence the damping, is specified mainly by L^2/C [e.g. Eq. (4.3)] where L is average loop length and C the effective dislocation line tension. The two components for edge and screw dislocations may or may not have different values of L, but C must differ if there are to be two types (edge and screw) of dislocations. The temperature dependence of L^2/C is found to be different for the two types of dislocations.

There is some evidence that the edge dislocation is the "fast" component and the screw dislocation the "slow" component. The fast component has a larger effective cross section for point defects and a larger temperature dependence for L^2/C. The temperature dependence of L^2/C observed in the megacycle range seems to be consistent with that found from velocity or modulus measurements made in the kilocycle range of frequencies [60].

---

[1] A slip system consists of a slip plane and slip direction.

γ) *Plastic deformation effects in aluminum.* As an example of a dislocation damping effect involving deformation and orientation, a series of experiments will be described in which ultrasonic attenuation and velocity have been measured in aluminum single crystals as a function of deformation and orientation [61]. The aluminum single crystals[1] were oriented for single slip and for polyslip, and they were deformed in tension at room temperature up to a total strain of about 1%. The attenuation and velocity measurements were made continuously during the deformation. Compressional and transverse waves were used at frequencies around 10 Mc/sec. On crystals oriented for single slip, these experiments show that, in the easy glide range of strain hardening, dislocation multiplication is confined to the primary slip system. The fact that the end of easy glide is associated with dislocations multiplying in other slip systems is also clearly shown by the ultrasonic measurements. An initial sharp increase in attenuation appears to be associated with dislocation breakaway preceding dislocation multiplication. Stress-strain measurements made concurrently with the attenuation-strain and velocity-strain measurements show from their orientation dependence how the operation of four, six, and eight glide systems affects the dislocation damping, hence the attenuation and velocity.

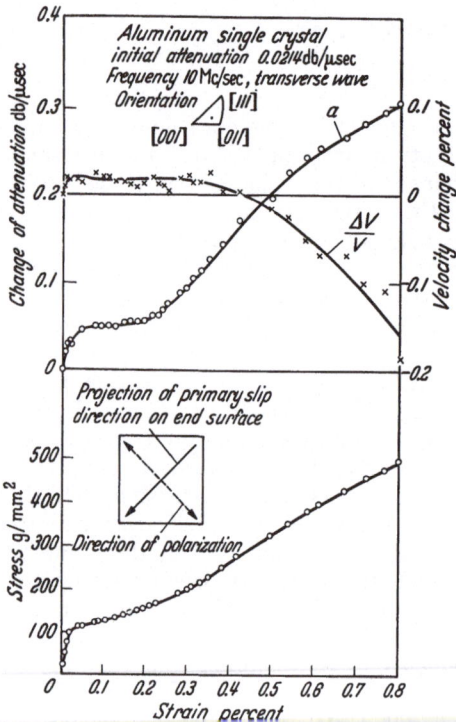

Fig. 14. Stress, shear wave attenuation, and velocity change as a function of strain for "⟨0.5⟩" orientation. The polarization direction of the shear wave is perpendicular to the projection of primary direction on end surface of sample. After Hikata, Chick, Elbaum and Truell [61].

The aluminum used in the single crystals had an approximate purity of 99.995%. Ultrasonic attenuation and velocity measurements were made along the crystallographic directions ⟨100⟩, ⟨110⟩, ⟨111⟩ and "⟨0.5⟩". The "⟨0.5⟩" orientation means that one of the twelve possible {111} ⟨110⟩ glide systems of the face-centered cubic crystal is inclined at 45° to the long axis of the sample so that the resolved shear stress on the one favored slip plane is 0.5 of the stress along the axis of the sample.

The results of the tensile deformation of a sample with ⟨0.5⟩ orientation are shown in Fig. 14. The plane of polarization or particle displacement of a 10 Mc/sec transverse wave was made perpendicular to the primary glide direction and to its projection on the end face as shown in Fig. 14. The particle displacement of the ultrasonic wave does not, in this case, have any component in the primary glide direction. As a consequence, when dislocation multiplication occurs and is confined only to this particular easy glide plane and direction, the ultrasonic wave will not see or detect the changes and so, in this case, the attenuation will not increase as long as this easy glide system is the only one in which appreciable

---

[1] The crystals had the shape of bars $5'' \times \frac{3}{8}'' \times \frac{3}{8}''$.

dislocation multiplication takes place. This is what occurs in the attenuation-strain pattern of Fig. 14 in the strain range between 0.05 % and 0.25 %. The initial sharp rise below 0.05 % strain is attributed to breakaway of dislocations from weak pinning, hence an increase in loop length which occurs before appreciable increase in dislocation density. The level section shows the easy glide dislocation behavior to which the shear or transverse wave is not coupled and the subsequent attenuation increase corresponds to a third stage in which additional glide systems come into

action. Substantial dislocation multiplication begins to occur in these additional slip systems and further, dislocations on one slip system interfere with or cut into dislocations of other slip systems. During this deformation the velocity increased a small amount and remained nearly constant during easy glide and then decreased rapidly in later stages of deformation.

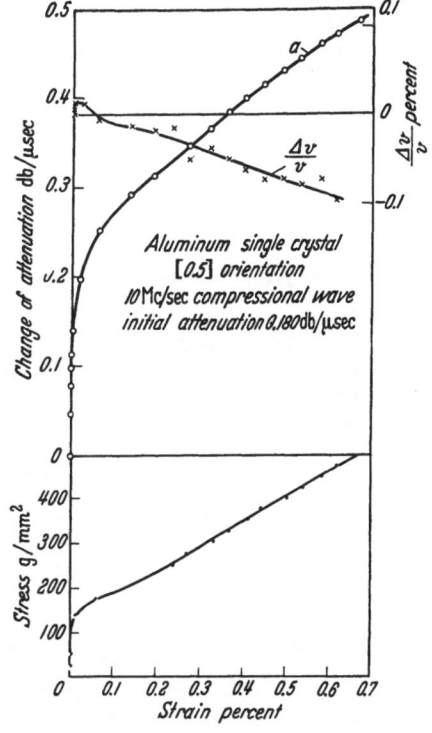

With the $\langle 0.5 \rangle$ orientation and a compressional wave the results are quite different. Fig. 15 shows that the attenuation increases rapidly with increasing strain and no indications of easy glide are present as they were in Fig. 14. The corresponding velocity change is also different in that it decreases before easy glide begins.

The sharp difference in the attenuation-strain behavior between transverse and compressional waves is explained by the fact that with the $\langle 0.5 \rangle$ orientation compressional waves have shear components along all of the glide systems, and particularly in the primary one, while with this orientation the transverse waves, properly polarized, have no shear component in the primary slip system as explained above. The separation of

Fig. 15. Stress, compressional wave attenuation, and velocity change as a function of strain for "$\langle 0.5 \rangle$" orientation. After HIKATA, CHICK, ELBAUM and TRUELL [61].

the effects of primary and secondary glide systems during plastic deformation of face centered cubic metal single crystals seems to be accomplished by these methods.

Experimental results of attenuation-strain and velocity-strain behavior for compressional waves in $\langle 100 \rangle$, $\langle 110 \rangle$, and $\langle 111 \rangle$ orientations were also obtained, and an example for the $\langle 100 \rangle$ direction is shown in Fig. 16. Similar results for $\langle 110 \rangle$ and $\langle 100 \rangle$ orientations were also obtained with transverse waves. In all cases the attenuation increases rapidly from the beginning of deformation, and in all cases the velocity shows a maximum from 0.02 to 0.01 % strain.

The location and amplitude of the maxima vary for different orientations, but their presence can be explained in terms of dislocation damping theory.

In order to examine these velocity changes and in particular to account for an initial velocity increase, it is convenient to rearrange slightly Eqs. (20.5) and (20.6). The change consists of expressing the frequency dependent part $G(\omega)$

(20.5a) in terms of ω_0/ω instead of ω/ω_0. The result is

$$\frac{\Delta v}{v_0} = \Omega \left(\frac{4G\,b^2\,A}{\pi^2\,A\,\omega^2}\right)\left(\frac{(y^2-1)}{(y^2-1)^2+\left(\frac{d}{\omega}\right)^2}\right), \tag{24.8}$$

where $y=\omega_0/\omega$, and use has been made of

$$\omega_0^2\,L^2 = \pi^2\,C/A$$

in obtaining the result.

The frequency dependent part of the attenuation $F(\omega)$, Eq. (20.4), when thus expressed leads to

$$\alpha = 8.68\times10^{-6}\,\Omega\left(\frac{4G\,b^2}{\pi^2}\,\frac{A}{A}\right)\frac{d}{\omega^2}\left(\frac{1}{(y^2-1)^2+\left(\frac{d}{\omega}\right)^2}\right). \tag{24.9}$$

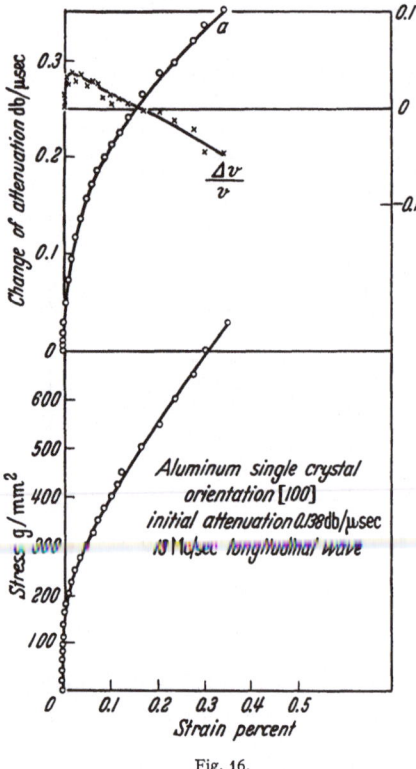

The values of A, B, and C are regarded as constants for a particular material and, for sufficiently small deformation, Ω and Λ can be regarded as constant, i.e. in the strain range in which the velocity is increasing. It is therefore assumed that in this low strain region the changes in attenuation and velocity are caused by an

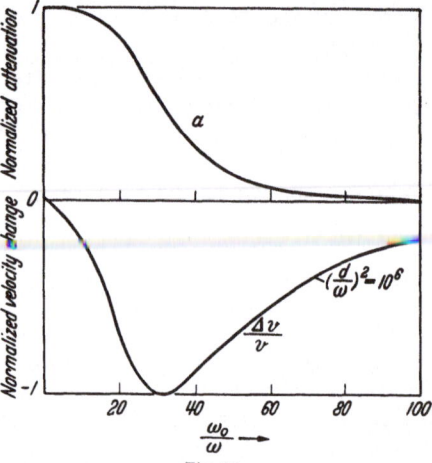

Fig. 16.

Fig. 17.

Fig. 16. Stress, compressional wave attenuation, and velocity change as a function of strain for ⟨100⟩ orientation. After Hikata, Chick, Elbaum and Truell [61].

Fig. 17. Expressions (24.8) and (24.9) as a function of $y=\omega_0/\omega$ for $(d/\omega)^2 = 10^6$. Attenuation is normalized to its maximum value and fractional velocity change $\Delta v/v$ is normalized to its minimum value. After Hikata, Chick, Elbaum and Truell [61].

increase in loop length L. The attenuation and velocity, normalized to their maximum values, are plotted in Fig. 17 from Eqs. (24.8) and (24.9) for a value of $(d/\omega) = 10^3$ where the frequency is 10 Mc/sec. $\Delta v/v_0$ has a minimum value at

$$\left(\frac{\omega_0}{\omega}\right)^2 = 1 + \frac{d}{\omega}, \tag{24.10}$$

but since d/ω is in this case large ($\approx 10^3$) in comparison with unity, Eq. (24.10) can be written approximately as

$$\omega_0 \approx \sqrt{d\,\omega}.\tag{24.10a}$$

As the deformation increases there is a region, shown in Fig. 17, in which ω_0 can vary and at the same time allow the attenuation and velocity to increase simultaneously. As the deformation increases at low deformation the loop length must increase, which means that ω_0 must decrease according to Eq. (20.2),

$$\omega_0 = \frac{\pi}{L}\sqrt{\frac{C}{A}}.$$

Consequently, in Fig. 17 it can be seen that the velocity and attenuation can both increase as ω_0/ω decreases only if one starts at a value of ω_0/ω at or below the minimum in $\Delta v/v_0$. In other words, the resonant frequency ω_0 cannot lie above this minimum if the velocity is to increase during the initial stages of deformation. If the resonant frequency ω_0 does lie above the minimum, the velocity will then decrease while the attenuation increases in the earliest stages of deformation.

Materials of high purity would normally have longer loop lengths, before deformation, than would the same materials with lower purity. Consequently it is to be expected that the increasing velocity effect would not be observed in low purity materials or materials where for any reason the loop length is too short.

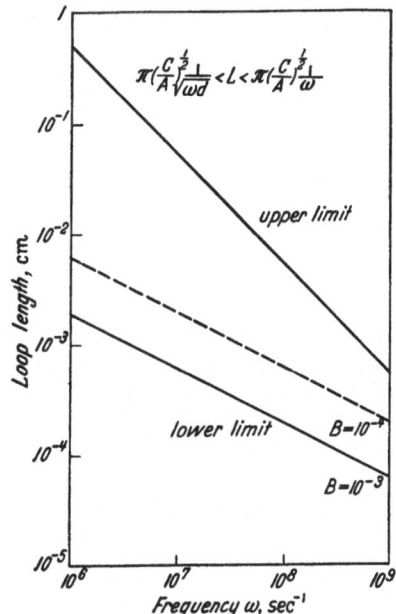

Fig. 18. Upper and lower limits on loop length as a function of ultrasonic frequency ω for which an increase in both attenuation and velocity, with deformation, can be obtained. The lower limit depends on the damping constant B. After Hikata, Chick, Elbaum and Truell [61].

Upper and lower limits can be determined for L in order that ω_0/ω lie between unity and the value at which the maximum occurs in the $\Delta v/v_0$ curve of Fig. 17. Using the relation (20.2) in the form

$$L = \frac{\pi}{\omega_0}\sqrt{\frac{C}{A}}$$

together with the fact that $\omega \leqq \omega_0 \leqq \sqrt{\omega d}$ as shown above in Eq. (24.10a), it follows that

$$\pi\left(\frac{C}{A}\right)^{\frac12}\frac{1}{\sqrt{\omega d}} \leqq L \leqq \pi\left(\frac{C}{A}\right)^{\frac12}\frac{1}{\omega}.\tag{24.11}$$

Fig. 18 shows the limits of L as a function of frequency. The upper limit on L is a plot of the right-hand side of (24.11) as a function of ω. The lower limit on L depends on $d = B/A$, hence on the values of B and A. Since B seems to lie between 10^{-4} and 10^{-3}, the left-hand side of (24.11) or the lower limit on L is given for these two values of B. The limits of L for this material and a frequency of 13 Mc/sec are 3×10^{-4} cm and 8×10^{-3} cm.

On the basis of the preceding discussion one can explain the concurrent increase of attenuation and velocity in the early stages of deformation. This analysis also provides the explanation for those experiments where no velocity increase has been observed. In the latter experiments less pure aluminum was used and usually in polycrystalline form. Under these circumstances the initial loop length would be smaller than the lower limit derived above and would lie too high in frequency.

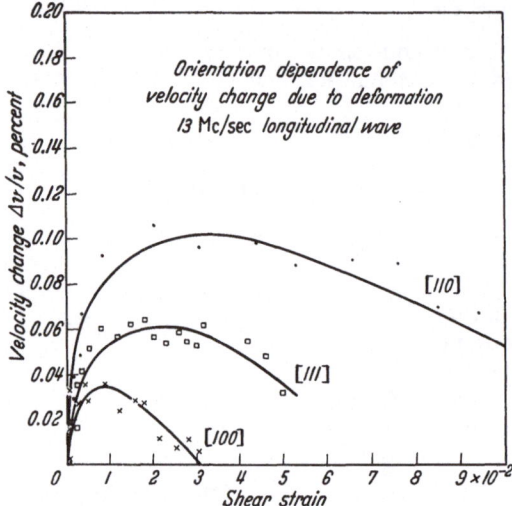

Fig. 19. Compressional wave velocity change $\Delta v/v$ for $\langle 100 \rangle$, $\langle 111 \rangle$, and $\langle 110 \rangle$ orientations as a function of shear strain. After Hikata, Chick, Elbaum and Truell [61].

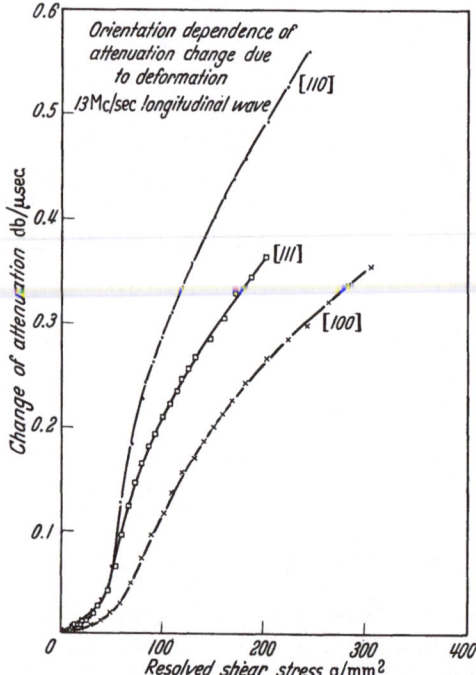

Fig. 20. Compressional wave attenuation change for $\langle 100 \rangle$, $\langle 111 \rangle$ and $\langle 110 \rangle$ orientation as a function of resolved shear stress. After Hikata, Chick, Elbaum and Truell [61].

The reason why the velocity eventually decreases as deformation continues is simply that the loop lengths cannot continue to increase. As soon as sufficient dislocation multiplication occurs the dislocations in different glide systems must interact and interfere in a way to cause a shortening of the loop length. The effect of this loop length shortening is to cause the velocity to decrease. Although the velocity decreases at this stage of deformation, the attenuation continues to increase because of the increase in dislocation density. Eventually, however, the attenuation also decreases as the deformation becomes larger.

The position and magnitude of each maximum of the $\Delta v/v_0$ versus deformation curve should depend on how many slip systems are operating and to what extent, hence on orientation. Among the three crystal directions $\langle 100 \rangle$, $\langle 110 \rangle$, and $\langle 111 \rangle$ one would assume that interaction or interference of slip systems would occur first with the $\langle 100 \rangle$ orientation, then with $\langle 111 \rangle$, and finally with the $\langle 110 \rangle$ orientation. The $\langle 100 \rangle$ orientation has eight equally favored slip systems while the $\langle 111 \rangle$ orientation has six equally favored slip systems and the $\langle 110 \rangle$ orientation has four. In Fig. 19 the fractional velocity changes are plotted as a function of the resolved shear strain, and it is apparent that the position and magnitude of the maxima stand in the expected relation to one another based on the orientation considerations mentioned.

As the deformation increases the assumption that Ω and Λ are constant is obviously not valid. As dislocation multiplication becomes appreciable the distribution of dislocations on each slip system is altered and the orientation factor Ω then depends on deformation as well as on crystallographic orientation and the type of ultrasonic wave used.

Fig. 20 shows the relative attenuation behavior for the three orientations $\langle 100 \rangle$, $\langle 110 \rangle$, and $\langle 111 \rangle$. Again the attenuation differences for the three orientations is in the proper relation to agree with the highest attenuation for the orientation involving the fewest slip systems.

25. Bordoni peaks. Metals with the face centered cubic structure exhibit a characteristic "peak" in the damping of stress waves, at temperatures in the vicinity of one third the Debye temperature. These peaks were originally observed by BORDONI [62], [63], in lead, aluminum, silver and copper.

Extensive studies of the Bordoni peaks have subsequently been carried out by numerous investigators, for example, NIBLETT and WILKS [64], CASWELL [68], PARÉ [65], EINSPRUCH and TRUELL [66], THOMSON and HOLMES [67], BORDONI et al. [69], and BRUNER [70].

The main features of these peaks can be summarized as follows:

1. The peak occurs in both single crystals and polycrystalline specimens, after slight plastic deformation.

2. The height of the peak increases and the peak maximum shifts slightly to higher temperatures with increasing amount of plastic deformation; these effects saturate at about 2 to 3 percent plastic strain.

3. The peak disappears upon annealing at elevated temperature.

4. The height of the peak and the temperature of the peak maximum are independent of the wave amplitude.

5. Impurities reduce the height of the peak and shift the peak maximum slightly toward lower temperatures; similar effects are observed as a result of neutron irradiation.

6. In addition to the main peak, a much smaller, subsidiary peak is frequently observed at lower temperatures.

7. The temperature at which the peak maximum occurs depends on the frequency of the wave; this temperature decreases with decreasing frequency.

8. The peaks originate from thermally activated processes and exhibit, in a general way, an exponential temperature dependence of the relaxation time according to an Arrhenius type equation. However, they cannot be described in terms of a single relaxation process; a range of relaxation times is involved in each case.

The above features indicate that the Bordoni peaks are due to a relaxation process which involves dislocations. The fact that the peak position is independent of prestrain suggests that an "intrinsic" dislocation mechanism is involved, which does not depend on a particular length of dislocation lines. The relaxation character of the effect is further supported by the observation that both the height and the position of the peak are essentially independent of the wave amplitude. This is in contrast with the dislocation damping mechanisms discussed previously, where hysteresis and resonance phenomena involving dislocation loop lengths were considered.

As was mentioned before, the temperature at which the peak occurs can be related to the driving frequency ω, by an expression of the form:

$$\omega = \omega_0 \, e^{-\frac{E}{kT}} . \tag{25.1}$$

Hence the activation energy E and the attempt frequency ω_0 can, in principle, be found from the changes of the temperature corresponding to the peak maximum, as a function of driving frequency. If the peak arose from a simple relaxation process with single values of activation energy and attempt frequency, the dissipation at a frequency ω would be given by:

$$\frac{1}{Q} = 2\,\frac{1}{Q_m}\cdot\frac{(\omega/\omega_m)}{1+(\omega/\omega_m)^2} \tag{25.2}$$

where $\frac{1}{Q_m}$ corresponds to the maximum dissipation and $\omega_m = A\,\exp\left(-\dfrac{E}{kT}\right)$.

The half-width of the peak given by expression (25.2) is then determined by the activation energy E and attempt frequency A. In fact, the observed peaks are frequently twice as wide as the values deduced on the basis of the activation energy obtained from the shift in the temperature of the peak maximum, as a function of frequency. This indicates that the relaxation process must involve either a range of activation energies, or a range of attempt frequencies, or both. It is thus clear that only some average values of the activation energy are determined experimentally. CASWELL [68] has shown that a Gaussian distribution of activation energies, with a standard deviation of 16 degrees, will fit the shape of the observed peaks. On the other hand, THOMPSON and HOLMES [67] have suggested on the basis of their experiments that a spectrum of discrete activation energies is involved, whereas BORDONI et al. [69] suggest that there is a spectrum of attempt frequencies.

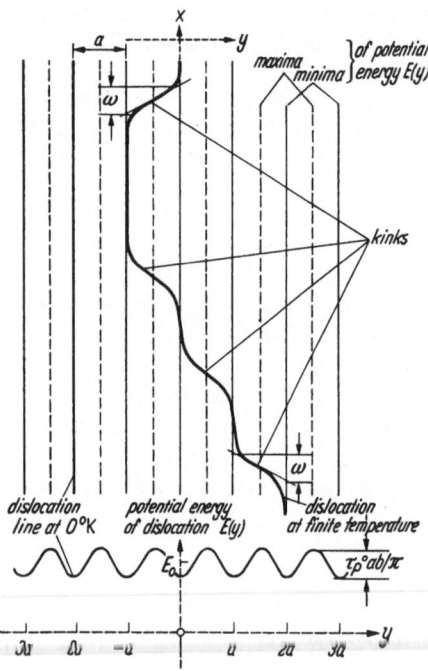

Fig. 21. The potential energy of a dislocation line due to the Peierls stress τ_p. Figure not to scale. After SEEGER, DONTH, PFAFF [72].

The values of E and ω_0 found experimentally from expression (25.1) are in the range of 0.05 to 0.2 ev and 10^8 sec^{-1} to 10^{12} sec^{-1} respectively.

The earliest theoretical interpretation of Bordoni peaks was suggested by MASON [71], who proposed that the relaxation process arises from the jumping of dislocations parallel to close packed directions between two adjacent positions of energy minimum (two Peierls wells [75]).

SEEGER et al. [72] have extended this idea on the basis of SEEGER's earlier suggestion [73], [74] that the relaxation process involves the displacement of small bulges or kinks in the dislocations rather than long segments extending between two strong pinning points, as originally suggested by MASON. The theories of SEEGER and SEEGER et al. will now be summarized briefly.

Consider first a dislocation line lying along a close packed direction of its glide plane (see Fig. 21) that is in a position of minimum energy. In the absence of thermal fluctuations, a shear stress equal to the Peierls stress [76] is required in order to displace it by one interatomic distance, that is to its adjacent position of energy minimum. At finite temperatures, however, the dislocation will not lie along a single potential energy minimum. Instead it will contain kinks (see

Fig. 21) and bulges consisting of pairs of kinks of opposite sign. Stresses much smaller than the Peierls stress will be sufficient to move such kinks sideways, that is in a direction parallel to the dislocation line, where the potential barriers opposing motion are very small. Under the influence of such stress, new pairs of kinks of opposite sign can also form with the aid of thermal energy, hence with a temperature dependent frequency. If the frequency v of the applied stress is large compared with the frequency v_f at which pairs of kinks are generated, kink pair formation will not contribute to the strain and no energy dissipation will take place. If v is small compared with the frequency of thermal formation of kinks, again no energy dissipation will occur. Only when the values of v and v_f are comparable will there be an appreciable contribution to energy loss and a maximum in dissipation will occur when the two frequencies are equal. To calculate the magnitude of the dissipation one must find the rate of formation v_f of the kink pairs.

In his original treatment, SEEGER assumed that this rate is given by an Arrhenius type equation

$$v_f = A \, e^{-\frac{E}{kT}}, \tag{25.3}$$

where E is the energy of formation of a kink pair and A is the attempt frequency. A was taken to be equal to the frequency of oscillation of the dislocation in the Peierls potential well:

$$A \approx \frac{1}{2\pi a} \sqrt{\frac{2\tau_p}{\varrho}} \tag{25.4}$$

where τ_p is the Peierls stress, ϱ the bulk density, and a the lattice spacing. SEEGER further assumed that $E = 2W_k$, where W_k is the additional energy associated with a single kink in a dislocation which is otherwise parallel to a close packed direction in the crystal. On this basis the following expression for E was obtained:

$$E = 2\,W_k = \frac{4a}{\pi} \sqrt{\frac{2E_0\,a\,b\,\tau_p}{\pi}} \tag{25.5}$$

where b is the Burgers vector, E_0 is the energy per unit length of dislocation line, and the other symbols were defined before.

This treatment yielded the right orders of magnitudes for A and E_0, but it was subsequently realized that the derivation was not correct, because the Arrhenius equation does not apply to the formation of kink pairs in dislocations. This is due to the fact that kink pair formation is a collective process involving tens or hundreds of atoms. A more complete treatment was subsequently given by SEEGER et al. [72], based on an expression for the rate of formation of kinks derived by DONTH [77] from the theory of stochastic processes.

The main conclusions of this treatment are as follows. The shape of a dislocation line $y(x, t)$, where t is the time and the x and y coordinates are defined in Fig. 21, is determined approximately by the differential equation:

$$E_0 \frac{\partial^2 y}{\partial x^2} - m \frac{\partial^2 y}{\partial t^2} = b\,\tau_p^0 \, \mathrm{Sin} \frac{2\pi y}{a} - b\,\tau, \tag{25.6}$$

where m is the mass per unit length of dislocation, τ the resolved shear stress and τ_p^0 the Peierls stress in the absence of thermal or quantum mechanical activation. On the basis of Eq. (25.6) the energy of kink formation W_k is given by Eq. (25.5); the kink width w and the critical (unstable) equilibrium separation d_{cr} of the kinks under a stress τ (where $\tau \ll \tau_p^0$) are:

$$w = \left(\frac{\pi\, a\, E_0}{2b\, \tau_p^0}\right)^{\frac{1}{2}}, \tag{25.7}$$

$$d_{cr} = \frac{w}{\pi} \log \frac{16\, \tau_p^0}{\pi\, \tau}. \tag{25.8}$$

A study of the solutions of Eq. (25.6) for $\tau = 0$ shows that there are "normal modes" of dislocation oscillations characterized by an energy W. These modes are ordinary harmonic waves if $W \ll 2W_k$, but they are related to the formation of pairs of kinks of opposite sign when $W \approx W_k$. The interaction between these modes is calculated subject to the assumption that kinks of opposite sign separated by a distance larger than d_{cr} are independent of each other, since they are torn apart by the applied stress τ. From this treatment the mean frequency ν_f with which dislocation bulges are formed, and the activation energy E of the process, were found to be given by:

$$\log \frac{\nu_f}{B} = F_1(r, \alpha); \tag{25.9}$$

$$B = \frac{\pi^2}{32} \frac{G\, b^2}{a^2\, v^2} \frac{kT}{m^{\frac{3}{2}}\, E_0^{\frac{1}{2}}}; \tag{25.10}$$

$$E \equiv -\frac{d \log \nu_f}{d(1/kT)} = kT\, F_2(r, \alpha). \tag{25.11}$$

Here v is the appropriate wave velocity (for screw dislocations in an isotropic medium this is the velocity of transverse waves), G is the shear modulus, $F_1(r, \alpha)$ and $F_2(r, \alpha)$ are functions of the parameters

$$r = \frac{2W_k}{kT}; \tag{25.12}$$

$$\alpha = 1 - \frac{\pi\, \tau}{8\, \tau_p^0}. \tag{25.13}$$

Using rate theory in a way similar to Mason [71], the maximum value of dissipation associated with this relaxation process was shown by Seeger et al. [72] to be given by

$$\left(\frac{1}{Q}\right)_{max} = \frac{p}{2\sqrt{1+p}}, \tag{25.14}$$

where

$$p = \frac{2N_0\, a\, b^2\, A\, G\, w}{kT}\left(\frac{1}{3}\frac{d_{cr}}{w} + 0.3\right). \tag{25.15}$$

Here N_0 is the number per unit volume of dislocation loops (of average length L) contributing to the relaxation process and A is the area swept out by one dislocation during the process. A lower limit of $(1/Q)_{max}$ is obtained when $A = L\, a$. An upper limit of $(1/Q)_{max}$ can be obtained from the assumption that each dislocation sweeps out the largest possible area which is compatible with the applied stress, the line energy E_0 and the loop length L; in this case

$$\left(\frac{1}{Q}\right)_{max} = \frac{b^2\, L^3\, N_0\, G}{48\, E_0} \approx \frac{L^3\, N_0}{24}. \tag{25.16}$$

The theory of Seeger et al., outlined above, accounts qualitatively for most of the main features of the Bordoni peak. It leads to an activation energy which is an intrinsic property of the dislocations and is thus independent of the separation of pinning points, so that the temperature at which the peak occurs is approximately independent of both the degree of prestrain and the presence of impurities. The activation energy is also independent of the strain amplitude. The theory does not predict any dependence of the peak height on frequency, which is in agreement with the experimental evidence available to date.

Several of the Bordoni peak features, however, are not accounted for explicitly by this theory. In particular, the theory leads to a peak only about half as wide as that observed. Qualitatively, this difficulty can be overcome on the basis that the Peierls stress should be different for screw, edge and mixed type dislocations, with a corresponding range of activation energies for the relaxation process. (SEEGER had proposed specifically that the presence of the main and subsidiary peaks can be attributed to two distinct dislocation directions with respect to their Burgers vectors.) The concept of a range of activation energies due to different dislocation types is also consistent with the slight shift of the peak to higher temperature with increasing amounts of plastic deformation; the latter may be expected to affect differently dislocations of different types. The same applies to the slight shift of the peak to lower temperature with increasing impurity content; impurity atoms will interact differently with different types of dislocations.

The absence of the Bordoni peak in fully annealed specimens constitutes another difficulty. One is led to assume that this state corresponds to an insufficient dislocation density. The initial increase of dissipation with plastic deformation is associated with an increase in the number of dislocations. The dislocation density continues to increase for larger prestrain, but the average length of loops will decrease and the dissipation will reach a steady value. The decrease of the Bordoni peak with increasing impurity content, or with neutron irradiation may be attributed to pinning of dislocations with a resulting reduction of the effective loop length.

Some modifications of the SEEGER et al. [72] theory were proposed by LOTHE [78] and by MASON [79].

BRUNER [70], in an attempt to explain the absence of Bordoni peaks in metals with the body centered cubic structure, proposed a different mechanism for the relaxation process. He criticized the theory of SEEGER et al. on the basis that it should be equally applicable to both face centered and body centered cubic structure. BRUNER [70] suggested that since in face centered cubic structures a dislocation is dissociated into two partials, each partial could assume a position of equilibrium with respect to a point defect located in a plane adjacent to the slip plane of the dissociated dislocation. The relaxation process would then arise from the jumping, under the applied stress, of the pair of partial dislocations between the two positions of equilibrium. BRUNER's mechanism does not account successfully for many of the experimentally observed features of Bordoni peaks, such as, for example, the width of the peak and the influence of impurities and of neutron irradiation. In addition, the quantitative estimates of the activation energy and the value of energy dissipation involve assumptions the validity of which is not established. Nevertheless, the absence of Bordoni peaks in metals with the body centered cubic structure remains unexplained; perhaps it may be tentatively attributed to an insufficient purity of the metals with this structure investigated so far.

BRAILSFORD [80] has proposed a different dislocation model to account for the Bordoni peak. In contrast with the theory of SEEGER et al., BRAILSFORD predicts that the relaxation process originates from inherently kinked dislocations whose average direction does not coincide with a close packed direction of the crystal. In this model the behavior of a dislocation under an applied stress is described in terms of a redistribution of abrupt kinks along its length. Kink

diffusion is considered to be a thermally activated process of which the characteristic relaxation time depends on dislocation line length as well as on attempt
frequency and activation energy for kink diffusion. This model accounts successfully for the decrease in the peak height and slight lowering of the peak temperature with increasing impurity content or with neutron irradiation. Furthermore,
assuming an exponential distribution of dislocation line length, Brailsford
calculated activation energies and peak widths which are in good agreement with
the observed values. The change of dissipation with increasing plastic deformation
is less convincingly explained. More experimental evidence is undoubtedly needed
before the validity of this theory can be fully assessed.

V. Stress wave interaction with conduction electrons in metals.

26. Conditions for interaction. Ultrasonic attenuation by the interaction
between stress waves and conduction electrons in metals becomes appreciable
at temperatures below about 10° K. At higher temperatures other mechanisms,
discussed in different parts of this paper, are usually dominant.

Interaction between electrons and lattice waves of thermal origin have been
a part of the theory of metals for a long time, but only relatively recently has it
been recognized that the same approach is applicable in the case of ultrasonic
waves. The main differences between the two cases arise from the different
frequency ranges involved and from the fact that thermal vibrations have a distribution of frequencies, while the ultrasonic case is essentially a "monochromatic"
one (a single frequency is involved).

It must be pointed out, however, that recently [83] ultrasonic frequencies
of the order of 10^{10} cycles/sec have been produced, and it may thus become
possible to approach thermal frequencies ($\sim 10^{13}$ cycles/sec) in ultrasonic experiments. The importance of this aspect becomes more apparent when one considers
wavelengths, rather than frequencies. This is so because interactions between
ultrasonic waves and electrons become of particular importance when the ultrasonic wavelength is comparable with the electron mean free path. In this connection it is worth considering briefly the orders of magnitude of these variables.
Below the Debye temperature, important thermal waves have frequencies ν
given by $h\nu \approx kT$ (where h is Planck's constant, k is Boltzmann's constant and
T is the absolute temperature), which correspond to wavelengths of about 10^{-6} cm
at 10° K. This is to be compared with ultrasonic wavelengths, at a frequency of
100 Mc/sec, of about 10^{-3} cm, in most solids. At these temperatures the mean
free path of an electron, in metals carefully purified by the best techniques available
at the present time, is of the order of 10^{-2} cm. Thus, in this case the electron mean free
path is longer than either the thermal wavelengths or the ultrasonic wavelengths in
the frequency range easily accessible in experiments. On the other hand, in very
impure metals the thermal wavelength can be longer than the electron mean free
path.

27. Early observations. The first experimental observations of an electronic
effect in the attenuation of ultrasonic waves in metals are attributed to Bömmel [84] and to MacKinnon [85], [86], who were investigating ultrasonic attenuation in superconductors. In the course of their studies they found that the
ultrasonic attenuation increased markedly at low temperatures (several degrees
absolute) and then decreased almost discontinuously upon cooling through the
superconducting transition temperature. Shortly after Bömmel's and MacKinnon's experiments, Morse [87], Mason [88], and Kittel [89], using semi-
classical treatments, showed that the large increase of the attenuation in a normal

metal could be attributed to the interaction between electrons and the ultrasonic wave. Subsequent application of the BARDEEN-COOPER-SCHRIEFFER [90] theory[1] of superconductivity by MORSE and BOHM [91] provided a successful explanation for the temperature dependence of the ultrasonic attenuation in superconductors. The BCS theory predicts, among other features, an energy gap in the electron distribution, at the superconducting transition temperatures. The magnitude of this gap was determined experimentally in tin and indium by MORSE and BOHM [91], using the ultrasonic attenuation method. More recently it has been predicted by COOPER [92] that the energy gap could be an anisotropic function of crystalline direction. The existence of this anisotropy was demonstrated experimentally for tin by MORSE, OLSEN and GAVENDA [93].

28. Early qualitative interpretation. Specific treatments of the ultrasonic attenuation by conduction electrons will now be considered. Both semi-classical and quantum mechanical calculations, with various degrees of refinement, are now available. The common feature of the semi-classical treatments is that they are all based on the free electron model of a metal.

MASON [88] considered that the viscosity of the electron gas produced attenuation in the same way as in the case of acoustic propagation in an ordinary gas or liquid. MASON then applied the standard expressions for sound attenuation due to viscosity, which give for the amplitude attenuation of transverse waves:

$$\alpha_t = \frac{2\pi^2 v^2 \eta}{\varrho\, v_t^3}, \tag{28.1}$$

where v is the frequency, η is the viscosity, ϱ the density and v_t the transverse wave velocity. For longitudinal waves a similar expression is obtained:

$$\alpha_l = \frac{8\pi^2 v^2 \eta}{3\varrho\, v_l^3}, \tag{28.2}$$

where v_l is the longitudinal wave velocity. The viscosity η can be expressed, from kinetic theory, as $\eta = \dfrac{l\, N\, m\, \bar{v}}{3}$, where N is the number of particles per unit volume, m their mass, l their mean free path and $\bar{v}$ their average velocity. In the case of electrons, $\bar{v}$ is interpreted as the mean velocity $\bar{v}_F$, derived from the Fermi distribution, and expressions (28.1) and (28.2) become:

$$\alpha_t = \frac{\pi^2 N\, m\, \bar{v}_F}{5\varrho\, v_t^3}\, v^2\, l, \tag{28.3}$$

$$\alpha_l = \frac{4\pi^2 N\, m\, \bar{v}_F}{15\varrho\, v_l^3}\, v^2\, l. \tag{28.4}$$

MORSE'S [87] calculation was based on a relaxation mechanism. He assumed that in the absence of any elastic strain the Fermi surface was spherical (as required by the free electron model of a metal) and determined the effect of lattice strains on the electron distribution and the way the electron distribution adjusts to periodic changes of strain. If a uniform longitudinal strain ε_x (in the x direction) is imposed on an isolated element of metal containing N electrons per unit volume, two extreme situations are possible: (1) The deformation takes place slowly and the Fermi surface remains spherical, because scattering processes (due to impurities, imperfections and thermal vibrations) have time to restore the electron equilibrium distribution (relaxation is rapid compared to the rate of straining). (2) The deformation occurs rapidly compared to the rate of restoring the equi-

---

[1] In the following we use the abbreviation: BCS theory.

librium distribution by the scattering processes, and only the momentum components in the direction of ε_x react rapidly at first (an "adiabatic" shift of quantum states is considered). In this case an elongated, or ellipsoidal, Fermi surface initially results. Eventually scattering processes will tend to restore the equilibrium spherical distribution at a rate that is characterized by a relaxation time τ. Morse showed, by a simple argument, that the increment of electron momentum in the x direction, due to ε_x, was $\frac{8}{15} N E_F \varepsilon_x$ wheres E_F is the Fermi energy. For a harmonic time variation of the strain, the process can be described by a relaxational longitudinal elastic constant $c_l(\omega, \tau)$ of the form:

$$c_l(\omega, \tau) = c_l^0 + \frac{i \omega \tau}{1 + i \omega \tau} \frac{8}{15} N E_F$$

where c_l^0 is the usual equilibrium constant. When $\omega \tau \ll 1$, this expression reduces to:

$$c_l(\omega, \tau) \approx c_l^0 \left[1 + \frac{i \omega \tau}{\varrho v_l^2} \left(\frac{8}{15} N E_F \right) \right]$$

where v_l is the longitudinal wave velocity. Since $c_l^0 = \varrho_0 v_l^2$, the propagation constant is $q = \omega (\varrho_0/c_l^0)^{\frac{1}{2}}$, and the coefficient of amplitude attenuation is given by:

$$\alpha = \frac{2}{15} \frac{N m v_F}{\varrho_0 v_l} q l_e \qquad (q l_e < 1) \tag{28.5}$$

where v_F is the Fermi velocity. Morse points out [94] that in this derivation, the condition $q l_e < 1$ is far more important physically than $\omega \tau < 1$. This is so because the treatment is based on the assumption of homogeneous strain, which would not be satisfied unless the electron mean free path were much less than the ultrasonic wave length.

Morse's result is equivalent to Mason's; both are subject to the same restriction $q l_e < 1$, and both predict an attenuation proportional to the square of the frequency.

Pippard [95], again using the free electron model, extended these treatments to arbitrary values of $q l_e$. He further clarified the problem by pointing out that the displacements of the lattice ions by a stress wave introduce an internal electric field. This field is electrostatic for longitudinal waves and electromagnetic for transverse waves.

Pippard's argument may be summarized as follows. A lattice wave travelling through a solid gives rise to variations of electric forces on the electrons. The lattice ions will undergo a periodic displacement with velocity v, but the electron density N may not remain constant, instead, it may show a periodic variation n. For example, in the compressed region of a longitudinal wave the excess of positive ion charge will attract electrons, the opposite being true for the expanded region. If the electron and ion densities do not follow each other exactly, space charges will be developed, which lead to a periodic electric field in the direction of wave propagation. In the case of transverse waves there are no density changes and hence no electric fields resulting from space charge. It is possible, however, that the lattice and electronic currents may not cancel each other, in which case magnetic fields are generated; these give rise, in turn, to electric fields by induction. There will also be relaxation effects, due to collisions of electrons with the lattice, which tend to restore equilibrium with the local surroundings. Pippard, in effect, calculates the net distortion of the Fermi surface due to the combined influence of the internal electric fields and collisions.

Any given electron at the Fermi surface (the only ones of interest here) travels with a velocity v_F, much greater than the ultrasonic velocity. The type of interaction between the electron and the ultrasonic wave will, therefore, depend on the electron mean free path l_e. When $q l_e < 1$, the situation remains essentially the same as in the simpler treatments discussed before. In this case, for longitudinal waves, PIPPARD's result for the attenuation is similar to Eq. (28.5) above. When $q l_e > 1$, the electron will be affected only if it moves at such an angle to the direction of the ultrasonic wave propagation, that it always remains in phase with the wave; in other words, when the drift velocity of the electron in the direction of the wave is equal to the ultrasonic velocity[1]. In this case ($q l_e > 1$) PIPPARD's calculations give an attenuation coefficient which varies linearly with frequency and is independent of $q l_e$:

$$\alpha' = \frac{\pi\, N\, m\, v_F}{12\, \varrho_0\, v_l^2}\, \omega \qquad (q\, l_e > 1).\tag{28.6}$$

PIPPARD's treatment shows that the overall dependence of the attenuation of longitudinal waves on $q l_e$ is given by

$$\frac{\alpha}{\alpha'} = \frac{6}{\pi}\left[\frac{q\, l_e\, A}{3(1-A)} - \frac{1}{q\, l_e}\right]\tag{28.7}$$

where $A = (\text{arc tan } q\, l_e)/(q\, l_e)$, and α' is the limiting attenuation when $q\, l_e \gg 1$.

29. More complete classical interpretation. A more complete and more general treatment of the attenuation was given by STEINBERG [96], who derived his results using BOLTZMANN's transport equation and the free electron model of a metal; he assumed, in addition, the knowledge of an effective relaxation time[2] τ.

STEINBERG's approach also allowed him to evaluate the change in velocity of the ultrasonic waves due to interaction with electrons. A summary of his treatment will now be presented.

A periodic distortion is considered in a conductor of density ϱ, having N conduction electrons and N lattice ions per unit volume. The motion of the lattice causes a departure of the electron distribution from thermal equilibrium; a kinetic stress, or negative momentum current density $\boldsymbol{\Theta}$ is thus set up in addition to the stress associated with the lattice strain. $\boldsymbol{\Theta}$ offers an impedance to lattice motion which leads to an attenuation of the stress wave and to a wave velocity v_s, different from the velocity v_0 determined in the absence of this impedance.

For either transverse or longitudinal waves the lattice displacements $\boldsymbol{s}$ are governed by the equation of motion:

$$\varrho\, \frac{\partial^2 \boldsymbol{s}}{\partial t^2} = \varrho\, v_0^2\, \nabla^2 \boldsymbol{s} + \nabla \cdot \boldsymbol{\Theta},\tag{29.1}$$

where $\boldsymbol{\Theta}$ is a stress tensor[3]. For sufficiently small lattice velocities $\boldsymbol{u} = \partial \boldsymbol{s}/\partial t$ the electron distribution is assumed to remain close to thermal equilibrium

---

[1] Since the Fermi velocity is between two and three orders of magnitude greater than the ultrasonic velocity, this means that the scattered electrons lie, for all practical purposes, in a ring around the Fermi surface, in a plane perpendicular to the direction of ultrasonic wave propagation.

[2] STEINBERG points out that a unique relaxation time does not exist, because at the lowest temperatures the rate at which interactions with thermal lattice vibrations restore thermal equilibrium to the electron distribution is not independent of the initial distribution. Only an effective relaxation time may, therefore, be used.

[3] This tensor $\boldsymbol{\Theta}$ as used here is in dyadic form

$$\begin{aligned}\boldsymbol{\Theta} = {}& \Theta_{11}\, \boldsymbol{i}\,\boldsymbol{i} + \Theta_{12}\, \boldsymbol{i}\,\boldsymbol{j} + \Theta_{13}\, \boldsymbol{i}\,\boldsymbol{k} + \\ & + \Theta_{21}\, \boldsymbol{j}\,\boldsymbol{i} + \Theta_{22}\, \boldsymbol{j}\,\boldsymbol{j} + \Theta_{23}\, \boldsymbol{j}\,\boldsymbol{k} + \\ & + \Theta_{31}\, \boldsymbol{k}\,\boldsymbol{i} + \Theta_{32}\, \boldsymbol{k}\,\boldsymbol{j} + \Theta_{33}\, \boldsymbol{k}\,\boldsymbol{k}.\end{aligned}$$

Alternatively Eq. (29.1) may be written in component form

$$\varrho\, \frac{\partial^2 s_i}{\partial t^2} = c_{ijkl}\, \frac{\partial^2 s_l}{\partial x_k\, \partial x_j} + \frac{\partial \Theta_{ij}}{\partial x_j}.$$

and Θ will be linearly related to the local lattice displacements. Thus in the case of a plane harmonic stress wave of angular frequency $\omega = v_s q$ Steinberg writes

$$\nabla \cdot \Theta = - i \varrho^2 v_0 q^2 Z(\omega) s. \qquad (29.2)$$

Here $Z(\omega)$ represents the dimensionless impedance associated with Θ; its value is to be determined from the transport equation.

If α is the attenuation coefficient of energy per unit length, then s varies as $\exp\left[i(\omega t - q' \cdot r)\right]$ where $q' = q - \frac{1}{2} i \alpha$, and the equation of motion requires that:

$$\frac{\omega^2}{v_0^2} - (q')^2 - i q^2 Z(\omega) = 0. \qquad (29.3)$$

The attenuation α and the modified velocity v_s can now be expressed in terms of a resistance $R(\omega)$ and a reactance $X(\omega)$ defined by $Z = R + iX$. The real and imaginary parts of Eq. (29.3) yield;

$$\alpha = q R(\omega) \qquad (29.4)$$

and

$$v_s = v_0 \left[1 - X(\omega) - \tfrac{1}{4} R^2(\omega)\right]^{\frac{1}{2}}. \qquad (29.5)$$

In the following it is assumed that all quantities associated with the stress wave may be considered to vary as the displacement s. Thus the internal electric field E, the electron current density j and the non-equilibrium electron density n can be related to the local lattice velocity u through Maxwell's equations by:

$$\frac{\partial}{\partial t}(j - N e u) = \frac{c^2}{4\pi} \nabla \times \nabla \times E, \qquad (29.6)$$

(c is the velocity of light and e is the electronic charge) and by the equation of continuity

$$\nabla \cdot j = \frac{\partial}{\partial t}(n e) = 0. \qquad (29.7)$$

At thermal equilibrium the electron velocity distribution $f(v)$ is the Fermi distribution:

$$f_0(v) = \{\exp\left[\tfrac{1}{2} m (v^2 - v_F^2)/kT\right] + 1\}^{-1}, \qquad (29.8)$$

where v_F is the velocity at the Fermi surface and k is Boltzmann's constant. When the lattice is in motion, the velocity distribution $f(v)$ is relaxed by collisions toward a displaced equilibrium referred to the moving lattice, and described by $f(|v - u|)$. Here it is assumed that the relaxation is exponential in time and is characterized by a relaxation time τ. For these conditions the Boltzmann transport equation, in the presence of a force per particle $F = eE$, can be written in the form:

$$\frac{e}{m}(E \cdot \nabla_v) f + (v \cdot \nabla) f + \frac{\partial f}{\partial t} = - \frac{f - f_0(|v - u'|)}{\tau}, \qquad (29.9)$$

where f is the unknown instantaneous distribution function and ∇_v is the gradient operator in terms of the velocity components. In adopting this form for the right-hand side of Eq. (29.9) one assumes that $(\partial f / \partial t)_c$ (the time rate of change of f due to collisions) may be substituted by $- \dfrac{\Delta f}{\tau}$.

Assuming small departure from thermal equilibrium, such that $(f - f_0) \ll f_0$, $u \ll v_F$, $n \ll N$, $(E \cdot \nabla_v)(f - f_0) \ll (E \cdot \nabla_v) f_0$, an approximate solution of the transport equation is found, for a given direction of stress wave propagation (z direction in this case):

$$f = f_0(|v - u|) - \psi(v) \frac{\partial f_0}{\partial v}, \qquad (29.10)$$

where

$$\psi = \frac{([e\,\boldsymbol{E}\,\tau/m] + \boldsymbol{u})\,(\boldsymbol{v}/v) + (n\,v_F^2/3\,N\,v)}{1 + i\,\omega\,\tau - i\,q'\,\tau\,v_z} - \left(\frac{\boldsymbol{v}\cdot\boldsymbol{u}}{v} + \frac{n\,v_F^2}{3\,N\,v}\right). \tag{29.11}$$

All the necessary quantities can now be obtained in terms of $\boldsymbol{u}$ by computing the current density:

$$\boldsymbol{j} = \left(\frac{2e\,m^3}{h^3}\right)\int \boldsymbol{v}\,f(\boldsymbol{v})\,d^3v \approx N\,e\,\boldsymbol{u} + \left(\frac{3N\,e}{4\pi\,v_F}\right)\int v_F\,\psi(\boldsymbol{v}_F)\,d\Omega. \tag{29.12}$$

From the above, the non-equilibrium momentum current density takes the form:

$$-\boldsymbol{\Theta} = \left(\frac{2m^4}{h^3}\right)\int \boldsymbol{v}\,\boldsymbol{v}\,[-f(\boldsymbol{v}) - f_0(|\boldsymbol{v} - \boldsymbol{u}|)]\,d^3v \approx \left(\frac{3N\,m}{4\pi\,v_F}\right)\int v_F\,v_F\,\psi(\boldsymbol{v}_F)\,d\Omega, \tag{29.13}$$

and using Eq. (29.2) one obtains the impedance:

$$Z(\omega) = \frac{i\,\boldsymbol{q}'\cdot\boldsymbol{\Theta}\cdot\boldsymbol{u}^*}{\varrho\,v_0^2\,v_s^{-1}\,q\,|u|^2}. \tag{29.14}$$

Eq. (29.14) may now be specialized for transverse and longitudinal waves. The case of transverse waves will be treated first.

For transverse displacements there will be no density change and hence $n = 0$. In the absence of an appreciable magnetic field, the electric field will take the direction of lattice displacements. Under these circumstances, STEINBERG finds

$$Z(\omega) = \frac{N\,m\,v_F}{\varrho\,v_0^2\,v_s^{-1}}\left\{\frac{1}{q\,l} - \left[\frac{(K_0^2 + q^2)\,(1 - G)}{G'\,K_0^2 + q'^2}\right]\right\} \tag{29.15}$$

here $G = \frac{2}{3}[1 - (1 + a^{-2})\,F]$, $F = 1 - a^{-1}\arctan a$, $G' = G_a/q_i$ where $a = q'\,l/(1 + i\omega\,\tau)$ and l is the mean free path of the electrons. $K_0 = [4\pi\,i\,\omega\,N\,e^2\,\tau/mc^2]^{\frac{1}{2}}$ where c is the velocity of light. It may be noted that the variable $|K_0|$ is essentially the reciprocal of the skin depth at the frequency under consideration.

For frequencies presently used in experimental studies, the approximations $\omega\,\tau \ll 1$, $|K_0^2| \gg q'^2$ and $|G'\,K_0^2| \gg |q'^2|$ are usually valid. In this case two frequency ranges may be distinguished: $q^2\,l^2 \ll 1$ and $q\,l > 1$.

When $q^2\,l^2 \ll 1$, the acoustic impedance is given approximately by:

$$Z(\omega) = \frac{N\,m\,v_F\,l\,\omega}{5\,\varrho\,v_0^2}\left[1 - \omega\,\tau\,\frac{\alpha}{q} - \frac{\alpha^2}{4q^2} - i\left(\frac{\alpha}{q} + \omega\,\tau\right)\right]. \tag{29.16}$$

On the basis of Eq. (29.4), the attenuation coefficient will be found from:

$$\frac{\alpha}{q} = \frac{N\,m\,v_F\,l\,\omega}{5\,\varrho\,v_0^2}\left(1 - \omega\,\tau\,\frac{\alpha}{q} - \frac{\alpha^2}{4q^2}\right). \tag{29.17}$$

The term $(N\,m\,v_F\,l\,\omega/5\,\varrho\,v_c^2)$ may be written in the form $\frac{1}{5}(m/M)\,(v_F^2/v_0^2)\,\omega\,\tau$, where M is the mass of a lattice ion. This term is generally of order $\omega\,\tau$, hence the attenuation per wavelength is small ($\frac{1}{2}\alpha \ll q$), and one obtains approximately

$$\alpha = \frac{N\,m\,v_F\,l\,\omega^2}{5\,\varrho\,v_0^2\,v_s}. \tag{29.18}$$

Using Eq. (29.5), an approximate value for the velocity v_s is then obtained:

$$v_s = v_0\left[1 + \frac{3\alpha^2}{4q^2}\left(1 + \frac{20\,M\,v_0^2}{3\,m\,v_F^2}\right)\right]^{\frac{1}{2}}. \tag{29.19}$$

The shift $(v_s - v_0)$ is positive and extremely small compared to v_0.

When $q\,l>1$ the acoustic impedance is approximately:

$$Z(\omega) = \frac{4N\,m\,v_F}{3\pi\,\varrho\,v_0^2\,v_s^{-1}}\left(1 - i\,\frac{\alpha}{2q}\right).$$

(29.20)

From Eq. (29.20) one obtains the attenuation coefficient and velocity:

$$\alpha = \frac{4N\,m\,v_F\,\omega}{3\pi\,\varrho\,v_0^2},$$

(29.21)

$$v_s = v_0\left(1 + \frac{\alpha^2}{4q^2}\right)^{\frac{1}{2}}.$$

(29.22)

In this case the attenuation varies linearly with frequency, unlike the case for $q^2 l^2 \ll 1$, where this variation is quadratic. It should be noted that here the attenuation depends only on equilibrium parameters of the conductor. The shift in velocity is still very small[1] and positive.

For still higher frequencies, where $|q'^2| \gg |K_0^2|$ and $\omega\,\tau \gg 1$, the impedance is given approximately by:

$$Z(\omega) = \frac{N\,m\,v_s^2}{\varrho\,\tau\,v_0^2}\left(1 - i\,\frac{3\pi\,v_s}{4v_F}\right).$$

(29.23)

The attenuation attains a limiting value:

$$\alpha = \frac{N\,m}{\varrho\,v_0^2\,v_s^{-1}\,\tau},$$

(29.24)

and the velocity is given by:

$$v_s = v_0\left(1 + \frac{3\pi\,v_s\,\alpha}{4v_F\,q}\right)^{\frac{1}{2}}.$$

(29.25)

The shift $(v_s - v_0)$ remains positive and small for further increases of frequency.

For longitudinal waves, the expression for the impedance is:

$$Z(\omega) = \frac{N\,m\,v_F}{\varrho\,v_0^2\,v_s^{-1}}\left\{\frac{q'}{q\,a}\left[\frac{a^2}{3}\left(\frac{1-F}{F}\right) - 1\right]\right\}.$$

(29.26)

When $\omega\,\tau \ll 1$, two frequency ranges can again be distinguished. For $q^2 l^2 \ll 1$ the acoustic impedance approaches the value:

$$Z(\omega) = \frac{4N\,m\,v_F\,l\,\omega}{15\varrho\,v_0^2}\left[\left(1 - \omega\,\tau\,\frac{\alpha}{q} - \frac{\alpha^2}{4q^2}\right) - i\left(\frac{\alpha}{q} + \omega\,\tau\right)\right].$$

(29.27)

Except for the numerical coefficient, $Z(\omega)$ is the same as for transverse waves. One obtains for the attenuation coefficient and for the velocity:

$$\alpha = \frac{4N\,m\,v_F\,l\,\omega^2}{15\varrho\,v_0^2\,v_s},$$

(29.28)

$$v_s = v_0\left[1 + \frac{3\alpha^2}{4q^2}\left(1 + \frac{5M\,v_0^2}{m\,v_F^2}\right)^{\frac{1}{4}}\right].$$

(29.29)

When $q\,l>1$, the acoustic impedance is approximately:

$$Z(\omega) = \frac{\pi\,N\,m\,v_F}{6\varrho\,v_0^2\,v_s^{-1}}\left(1 - i\,\frac{\alpha}{2q}\right).$$

(29.30)

In this case α and v_s are given by:

$$\alpha = \frac{\pi\,N\,m\,v_F\,\omega}{6\varrho\,v_0^2},$$

(29.31)

$$v_s = v_0\left(1 + \frac{\alpha^2}{4q^2}\right)^{\frac{1}{2}}.$$

(29.32)

---

[1] Usually less than one part in 10^5.

As can be seen from Eqs. (29.28), (29.29), (29.31) and (29.32), the qualitative behavior of α and v_s is the same as for transverse waves. A difference between transverse and longitudinal waves is noted, however, in the limit of very high frequencies. Unlike the case for transverse waves, where $\omega \tau \ll 1$, the impedance is still given by Eq. (29.30). Consequently α and v_s are also given by Eqs. (29.31) and (29.32).

It may be seen that the expressions for attenuation derived by STEINBERG are similar in form to those obtained by MORSE [87] and by MASON [88] for $q\,l < 1$, and by PIPPARD [96] for $q\,l > 1$. There are, however, differences in the numerical coefficients. The experimental results available to date agree well with the theoretical predictions, as far as the temperature dependence (through the electron mean free path) and the frequency dependence (for different frequency ranges) are concerned. However, the accuracy of these results does not appear to be sufficient to allow a clear distinction between the numerical coefficients found in various expressions for attenuation by conduction electrons. Two difficulties contribute prominently to this uncertainty. The first is due to the fact that so far many of the experiments were carried out on metals such as lead and tin, for which the free electron approximation is particularly inadequate. The second is due to the attenuation arising from causes other than interaction of the ultrasonic wave with conduction electrons. At very low temperatures this attenuation is usually temperature independent, but not independent of frequency; it may constitute a significant fraction of the total effect, particularly at high frequencies. Under such circumstances it may be difficult to separate the two effects and to obtain accurate values of the numerical coefficients for electron attenuation.

30. Quantum mechanical interpretation. It was mentioned earlier, that the interaction between electrons and ultrasonic waves can also be treated by quantum mechanical methods. An example of such calculations will now be presented, following an approximate treatment for the case of longitudinal waves, given by MORSE [94].

Any motion of a crystalline lattice can be represented as a superposition of the normal modes of the lattice. In particular, an ultrasonic wave of a given frequency can be approximated by one such mode, especially since in usual experiments one deals with ultrasonic radiation that consists essentially of a plane harmonic wave.

If the normal modes are expressed as standing waves, the displacement for a wave having a propagation vector $\boldsymbol{q}$, can be represented by

$$s = \frac{\varepsilon}{\sqrt{N}} \left[s \exp(i\,\boldsymbol{q}\cdot\boldsymbol{r}) + s^* \exp(-i\,\boldsymbol{q}\cdot\boldsymbol{r}) \right], \tag{30.1}$$

where ε is a unit vector in the direction of polarization and N is the number of atoms per unit volume. The displacement amplitude satisfies the differential equation

$$\ddot{s} + \omega^2 s = 0, \tag{30.2}$$

and the possible energies of the mode are thus given by

$$E_p = (n + \tfrac{1}{2})\hbar\omega, \tag{30.3}$$

where n is the quantum number and ω is the angular frequency.

Under the influence of a perturbation proportional to the amplitude, the only non-vanishing matrix elements of the problem are [97]:

$$(n-1\,|\,s\,|\,n) = \left(\frac{\hbar\,n}{2\omega\,M}\right)^{\frac{1}{2}};\qquad \varDelta E_p = -\hbar\,\omega,\tag{30.4}$$

$$(n+1\,|\,s^*\,|\,n) = \left(\frac{\hbar\,(n+1)}{2\omega\,M}\right)^{\frac{1}{2}};\qquad \varDelta E_p = \hbar\,\omega,\tag{30.5}$$

where M is the mass of the atoms under consideration. These relations describe the absorption and the emission of a phonon, respectively.

The displacements associated with a lattice mode introduce perturbations in the lattice potential within which the electrons move, and scattering of electrons results. The rate of energy transfer from the lattice mode (the ultrasonic wave, in this case) to the electron system can be obtained by calculating the net rate of phonon scattering, $\partial n/\partial t$ due to all the conduction electrons. This necessitates the evaluation of the matrix element

$$(\boldsymbol{k}',\,n\pm1\,|\,V\,|\,\boldsymbol{k},\,n),$$

which describes the scattering of an electron from state $\boldsymbol{k}$ to $\boldsymbol{k}'$, with the simultaneous emission or absorption of a phonon from the lattice mode.

A definition of the perturbed potential must now be found. Following Bardeen and Shockley [98], the energy of an electron in a deformed crystal can be expressed as

$$E(\boldsymbol{k},\,\boldsymbol{s}) = E_0 + \frac{\hbar^2\,k^2}{2m^*} + E_1\nabla\cdot\boldsymbol{s}\,,\tag{30.6}$$

where the first two terms are the same as in an undeformed lattice and the last term gives the effect of strain. E_1 has the same meaning as the lattice interaction constant used by Wilson [99]. (For simple metals $E_1 \approx \frac{2}{3}E_F$, where E_F is the Fermi energy.) The perturbing potential acting on the electrons is thus given by $V = E_1\nabla\cdot\boldsymbol{s}$. The matrix element $(\boldsymbol{k}'\,|\,V\,|\,\boldsymbol{k})$ can now be calculated using Bloch wave functions, giving:

$$(\boldsymbol{k}'\,|\,V\,|\,\boldsymbol{k}) = \frac{i\,\boldsymbol{q}\cdot\boldsymbol{\varepsilon}}{\sqrt{N}}\cdot\left\{\begin{matrix}s\\-s^*\end{matrix}\right\},\quad\text{where}\quad\left\{\begin{matrix}\boldsymbol{k}'&\boldsymbol{k}+\boldsymbol{q}\\\boldsymbol{k}'=\boldsymbol{k}-\boldsymbol{q}\end{matrix}\right\},\tag{30.7}$$

which correspond to absorption and emission of a phonon, respectively. Finally, combining Eqs. (30.4) and (30.5) with (30.7), one obtains the matrix element that describes scattering of an electron $\boldsymbol{k}$ to $\boldsymbol{k}'$ with simultaneous absorption or emission of a phonon:

$$(\boldsymbol{k}',\,n-1\,|\,V\,|\,\boldsymbol{k},\,n) = \frac{i\,q\,E_1}{\sqrt{N}}\left(\frac{\hbar\,n}{2\omega\,M}\right)^{\frac{1}{2}};\qquad \boldsymbol{k}' = \boldsymbol{k}+\boldsymbol{q},\tag{30.8}$$

$$(\boldsymbol{k}',\,n+1\,|\,V\,|\,\boldsymbol{k},\,n) = \frac{i\,q\,E_1}{\sqrt{N}}\left(\frac{\hbar\,(n+1)}{2\omega\,M}\right)^{\frac{1}{2}};\qquad \boldsymbol{k}' = \boldsymbol{k}-\boldsymbol{q}.\tag{30.9}$$

The corresponding transition probabilities per unit time can be written as:

$$P = \frac{2\pi}{\hbar}\,|(\boldsymbol{k}',\,n\pm1\,|\,V\,|\,\boldsymbol{k},\,n)|^2\,\delta(E'-E\pm\hbar\,\omega),\tag{30.10}$$

where $\delta(x)$ is the Dirac delta function and $E = E(\boldsymbol{k})$ is the electron energy. The argument of the δ-function in the two cases may be approximated by $\pm(q\cos\Theta\times dE/dk - \hbar\,\omega)$, where Θ is the angle between $\boldsymbol{k}$ and $\boldsymbol{q}$, the latter direction being taken as the polar direction in a spherical co-ordinate system. Since $\delta(-x)=\delta(x)$, one can write for the rate of energy loss per unit energy per unit volume:

$$-\frac{1}{n}\frac{\partial n}{\partial t}=\frac{\pi q^2 E_1^2}{\varrho_0 \omega}\sum_k \{f(\boldsymbol{k})\,[1-f(\boldsymbol{k}')]-f(\boldsymbol{k}')\,[1-f(\boldsymbol{k})]\}\times \\ \times \delta\left(q\cos\Theta\,\frac{dE}{dk}-\hbar\omega\right),$$
(30.11)

where $f(\boldsymbol{k})$ is the Fermi-Dirac distribution function and ϱ_0 is the density of the metal. The initial and final states have been weighted by their occupancy and their vacancy probabilities respectively. This sum can be expressed as an integral over the electron distribution, and one obtains:

$$-\frac{1}{n}\frac{\partial n}{\partial t}=\frac{q^2 E_1^2}{2\pi\varrho_0\omega}\int_0^\infty\int_0^\pi [f(\boldsymbol{k})-f(\boldsymbol{k}')]\,\delta\left(q\cos\Theta\,\frac{dE}{dK}-\hbar\omega\right)\operatorname{Sin}\Theta\,d\Theta\,k^2\,dk. \quad (30.12)$$

It should be emphasized that use of the δ-function implies significant scattering only of those electrons which satisfy the condition of $\cos\Theta\,(dE/dk)=\hbar\omega$. Since the group velocity of an electron is given by $v=\frac{1}{\hbar}\frac{dE}{dk}$, the scattered electrons have $v\cos\Theta=v_s$ where v_s is the ultrasonic wave velocity. Thus, appreciable scattering will only affect those electrons which have a group velocity, in the direction of the ultrasonic wave, equal to the ultrasonic velocity. This condition is identical with the one based on classical considerations and discussed previously in connection with PIPPARD's work.

The integral in Eq. (30.12) can be written in terms of the variable

$$y=\hbar\omega-q\cos\Theta\,\frac{dE}{dk}=\hbar\omega\left(1-\frac{v_x}{v_s}\right).$$

This substitution, followed by integration within the appropriate limits of y, leads to the expression

$$-\frac{1}{n}\frac{\partial n}{\partial t}=\frac{E_1^2}{2\pi\varrho_0 v_s}\int_0^\infty [f(E)-f(E')]\,k^2\left(\frac{dk}{dE}\right)^2 dE. \quad (30.13)$$

A further simplification can be introduced through

$$E'-E=\hbar\omega,\qquad f(E)-f(E')\approx\hbar\omega\,\frac{\partial f}{\partial E},$$

and Eq. (30.13) becomes

$$-\frac{1}{n}\frac{\partial n}{\partial t}=\frac{\hbar\omega E_1^2}{2\pi\varrho_0 v_s}\left[\frac{k^2}{(dE/dk)^2}\right]'. \quad (30.14)$$

The prime on the square bracket serves to emphasize that the quantity inside it is evaluated on the Fermi surface where $v\cos\Theta=v_s$.

The energy attenuation coefficient 2α, which is the fractional intensity loss per unit distance travelled by the wave, will be given by:

$$2\alpha=-\frac{1}{v_s n}\frac{\partial n}{\partial t}=\frac{E_1^2\hbar v}{\varrho_0 v_s^2}\left[\frac{k^2}{(dE/dk)^2}\right]'. \quad (30.15)$$

where v is the frequency of the ultrasonic wave. Using the substitution $\frac{dE}{dk}=\frac{\hbar^2 k}{m^*}$, where m^* is the effective mass, (30.15) becomes

$$2\alpha=\frac{(m^* E_1)^2 v}{\varrho_0 v_s^2 \hbar^3}. \quad (30.16)$$

In deriving this expression for the attenuation coefficient for an ultrasonic wave due to electron scattering, no other scattering processes are taken into account explicitly. It is clear, however, that when the equilibrium electron distribution is used, some additional (but unspecified) scattering is necessary in order to maintain a steady-state distribution which approximates the Fermi distribution.

It has been assumed implicitly, throughout this treatment, that the electron mean free path associated with these other scattering processes is long compared to the wavelength of the phonon. This is essentially a result of the uncertainty principle, as can be seen from the following. If l_e is the electron mean free path due to random scattering processes, and Δp is the uncertainty in the electron momentum, then according to the uncertainty principle $\Delta p \cdot l_e \geq \hbar$. In the scattering process, the change in electron momentum is $\hbar q$ which must be greater than the uncertainty Δp. Hence, $\hbar q \cdot l_e > \hbar$ or $q\, l_e > 1$. If this inequality were not satisfied, the δ-function approximation could not be used and appreciable scattering would not be confined only to those electrons for which $v \cos \Theta = v_s$.

Eq. (30.16) gives the attenuation coefficient for an ultrasonic wave due to electron scattering, derived by the usual quantum method. This may be compared with Eq. (28.6), derived by PIPPARD on the basis of classical reasoning, using the free electron model. It should be noted that Eq. (30.16) may be specialized to free electrons by substituting $E_1 = \dfrac{2}{3} E_F = \dfrac{1}{3} m\, v_F^2$, $N = \dfrac{8\pi}{3} \left(\dfrac{m\, v_F}{h}\right)^3$ and $m^* = m$. When these substitutions are made, Eq. (30.16) reduces to (28.6), i.e. the classical and quantum mechanical calculations agree exactly for the case of free electrons.

Quantum mechanical calculations of the electron scattering by ultrasonic waves for arbitrary electron mean free path (particularly for the case of $q\, l_e < 1$) are not available at the present time. It should also be emphasized that the treatment discussed above applies only to longitudinal waves.

31. Influence of magnetic field. Changes in ultrasonic attenuation, due to conduction electrons, as a function of magnetic field will be discussed next.

The magnetic field dependence of the attenuation arises from a change in the electron mean free path. Qualitatively, one may assume that a magnetic field, by causing the electrons to follow curved paths, reduces the distance between collisions in the direction of propagation. This results in reducing the attenuation, which depends upon the distance in the direction of propagation travelled by an electron between collisions, relative to a wavelength. This is supported by the following experimental observations. When $q\, l_e > 1$, the attenuation α varies approximately linearly with ω for the magnetic field $H = 0$; whereas the variation of α is proportional to ω^2 for sufficiently high magnetic fields. The latter occurs when the field is high enough to expect that the electron orbit diameters are smaller than the ultrasonic wavelength. Since, in the absence of a magnetic field, the attenuation varies as ω^2 when $q\, l_e < 1$, the application of a strong field appears to result in a shortening of the mean free path.

A more detailed treatment of this effect, based on the free-electron model, has been given by STEINBERG [100]. He showed that for transverse waves, with the magnetic field perpendicular to both the polarization and propagation directions, the field dependence of the attenuation for $q\, l_e < 1$ should be of the form:

$$\frac{\alpha(H)}{\alpha(0)} = \frac{1}{1 + (2\omega_c\, \tau)^2}, \qquad (31.1)$$

where $\omega_c = (e\, H)/(m\, c)$ is the cyclotron frequency. For a magnetic field parallel to the direction of polarization and perpendicular to the propagation direction, this relation becomes:

$$\frac{\alpha(H)}{\alpha(0)} = \frac{1}{1+(\omega_c \tau)^2}. \tag{31.2}$$

Eqs. (31.1) and (31.2) predict that for large fields the attenuation should vary as H^{-2}, which agrees quite well with experimental results when $q\, l_e < 1$, as shown

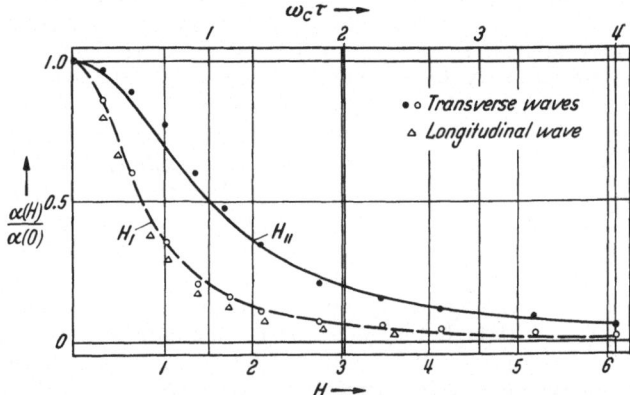

Fig. 22. Observed magnetic field dependence of the attenuation in copper at a frequency of 8.6 Mc/sec. Solid lines are calculated from Eqs. (31.1) and (31.2). After MORSE [94].

in Fig. 22. Moreover, although Eq. (31.1) was derived for the case of transverse waves, it is found experimentally (see Fig. 22) to hold also for longitudinal waves, when the field is perpendicular to the direction of propagation.

It is readily seen that the field for which $\frac{\alpha(H)}{\alpha(0)} = \frac{1}{2}$ corresponds to $\omega_c\,\tau = \frac{1}{2}$ in the case of Eq. (31.1) and to $\omega_c\,\tau = 1$ in the case of Eq. (31.2). The relaxation time τ of a sample can, therefore, be determined from the above relations, if the cyclotron mass is known from an independent measurement. It is likewise possible to determine directly the ratio of relaxation times, or mean free paths, of two different samples of the same metal by finding the value of magnetic fields for which $\frac{\alpha(H)}{\alpha(0)} = \frac{1}{2}$; the ratio of these fields is the inverse ratio of the mean free paths.

32. Application to Fermi surface study. Far more interesting, from the point of view of the electron theory of metals, is the attenuation in the range $q\, l_e > 1$, where an oscillatory variation of α with field strength is found (see Fig. 23).

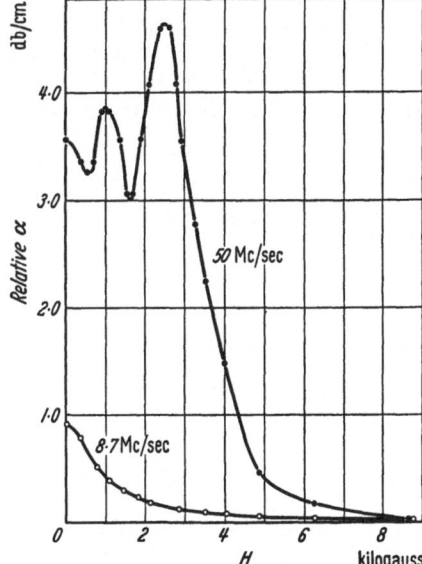

Fig. 23. Examples of extreme types of behavior of the magnetic field dependence of attenuation. These measurements were made in an indium polycrystal with magnetic field perpendicular to the direction of propagation of the longitudinal wave. After MORSE [94].

BÖMMEL [101] observed that for $q\, l_e > 1$ the attenuation was not a monotonically decreasing function of increasing field strength; a "bump" was observed in a plot of attenuation versus field strength. This "bump" was at first tentatively interpreted as a resonance-like effect arising when the electron mean free path was sufficiently long so that the electron orbit size matched the ultrasonic wavelength. Shortly after this observation was made, Pippard [102] proposed that this phenomenon could be used to investigate the shape of the Fermi surface. When the electron mean free path is greater than the ultrasonic wavelength, the electrons pass through many phases of the wave without being scattered, and in view of the large difference between the Fermi velocity and the ultrasonic velocity, the ultrasonic wave appears stationary to the electrons. It should be possible, therefore, to find the proper value of magnetic field such that the orbit diameter of the electron matches up with a multiple of the ultrasonic wavelength. As the magnetic field changes, the orbit diameter will change until it matches again a multiple of wavelength and one would expect periodic oscillations in the attenuation as a function of field strength H, with a period inversely proportional to H. Furthermore, the orbit diameter is proportional to the electron momentum in a direction perpendicular both to the magnetic field and to the ultrasonic wave. Thus, in principle, the periodicity of the attenuation for various field directions with respect to the axes of a single crystal could be used to determine the shape of the Fermi surface. Although Pippard's basic idea is now being successfully used, his prediction of the detailed conditions for maxima and minima in the attenuation did not turn out to be correct.

The first experiments on single crystals, utilizing the above principles, were carried out on copper by Morse and Gavenda [103] and on tin by Morse and Olsen [108]. More recently this effect was studied in bizmuth by Reneker [104] and in silver and gold by Morse, Myers and Walker [105].

A simple, semi-quantitative explanation of the oscillatory behavior of the attenuation as a function of magnetic field strength, when $q\,l>1$, can be given as follows. Suppose that the direction of wave propagation and the magnetic field are perpendicular to each other. If the field has a value H_1, such that the orbit diameter of the electrons just matches an integral number of wavelengths, then one can write

$$\frac{m}{r}\left(\frac{\hbar\,k}{m}\right)^2 = \frac{e}{c}\left(\frac{\hbar\,k}{m}\right)H_1 \tag{32.1}$$

and substituting $p=\hbar\,k$

$$r_1 = \frac{p\,c}{e\,H_1}. \tag{32.2}$$

Experimentally it is not important whether the above condition corresponds to a minimum or a maximum in attenuation, because if the field is now changed to a value H_2, such that the orbit diameter is one wavelength less, the same condition will apply, where

$$r_2 = \frac{p\,c}{e\,H_2}. \tag{32.3}$$

Hence:

$$r_2 - r_1 = \frac{p\,c}{e}\left(\frac{1}{H_2} - \frac{1}{H_1}\right) \tag{32.4}$$

or

$$\Delta r = \frac{p\,c}{e}\,\Delta\left(\frac{1}{H}\right), \tag{32.5}$$

where $\Delta(1/H)$ is the period in $1/H$. Since the diameter was assumed to have changed by one wavelength, $\Delta r=\lambda/2$, where λ is the wavelength, and by substitution in (32.5):

$$p = \frac{e}{2c\,\Delta\,(1/H\,\lambda)}. \tag{32.6}$$

Thus one can expect to obtain the Fermi momentum, for any chosen directions of wave propagation and magnetic field in a crystal, by measuring the attenuation periodicity of $1/(H \lambda)$. Experimentally it is observed that an attenuation maximum occurs when the orbit is an integral number of wavelengths.

More elaborate treatments of the subject have been given, for example, by KJELDAAS and HOLSTEIN [106] and by COHEN, HARRISON and HARRISON [107]. Within the restrictions of the free electron model these authors found more detailed expressions of the attenuation for longitudinal waves and for transverse waves with the plane of polarization both along and perpendicular to the magnetic field direction. All these treatments predict an oscillatory behavior of the attenuation as a function of magnetic field strength. In particular the attenuation is an oscillatory function of $\beta = 2q\, v_F/\omega_c$, where v_F is the Fermi velocity and ω_c is the cyclotron frequency. Using the relation $\omega_c = eH/(m^* c)$, where m^* is the cyclotron mass, one finds

$$\beta = 2q\, v_F\, m^*\, c/(e\, H) = 4\pi\, v_F\, m^*\, c/(e\, H\, \lambda).$$

Thus, the attenuation is again shown to be periodic in $1/(H \lambda)$, as predicted from the simpler model.

It was pointed out by COHEN, that this effect is not of the resonance type, but arises rather from the strength of the interaction between particular orbits and the electric fields due to the ultrasonic waves. He also indicates that although all the electrons on the Fermi surface contribute to the attenuation, electrons located near extremal sections are much more heavily weighted. Thus the oscillations arise from electron orbits which correspond to extremal sections of the Fermi surface. This is a most important consideration when attempts are made to determine the shape of the Fermi surface from the periodicity of the attenuation, for various combinations of field and propagation directions in a metal single crystal.

33. Application to superconductivity study. So far all considerations have been confined to metals in the normal state (as opposed to the superconducting state). In the following, attenuation in superconductors will be discussed.

Experiments by BÖMMEL on lead [84] and by MACKINNON [85], [86] on tin first established that the ultrasonic attenuation decreases rapidly, as a function of decreasing temperature, when a metal becomes superconducting. Unambiguous proof that conduction electrons in the normal state cause attenuation is provided in this connection; when a magnetic field of sufficient strength to destroy superconductivity is applied to the metal, the attenuation returns to the higher value observed above the transition temperature.

The above mentioned decrease in attenuation can be interpreted on the basis of the BCS theory of superconductivity [90]; it is also consistent with the observed increase of the lattice thermal conductivity in the superconducting state. According to the BCS theory, the normal electron energy distribution in the metal is changed in the superconducting state. This change is characterized by the appearance, at the Fermi energy, of a temperature dependent energy gap $2\mathscr{E}(T)$ in the density of states function $n(E)$. $n(E)$ assumes a modified form:

$$n_s(E) = \frac{n_n(0)\, E}{\sqrt{E^2 - \mathscr{E}^2}}, \tag{33.1}$$

where the subscripts s and n refer to the superconducting and normal states respectively and the energy E is measured from the Fermi energy (i.e. $E = 0$ corresponds to the Fermi level). In order to obtain an expression for the attenuation in the superconducting state, the modified form of the density function (33.1) is introduced into the expression for electron scattering [Eq. (30.13)] along with an additional term $(1 - \mathscr{E}^2/E^2)$ required by the BCS theory. When $\hbar\,\omega \ll E$, the modified form of Eq. (30.13) is quite simple and the ratio of attenuations in the superconducting and normal states turns out to be twice the Fermi function of the temperature dependent energy gap:

$$\frac{\alpha_s}{\alpha_n} = \frac{2}{\exp\,(\mathscr{E}/kT) + 1} \tag{33.2}$$

where k is Boltzmann's constant.

It should be emphasized, that Eq. (33.2) would not result from a simple energy gap theory unless the normal density of states were preserved. This is because

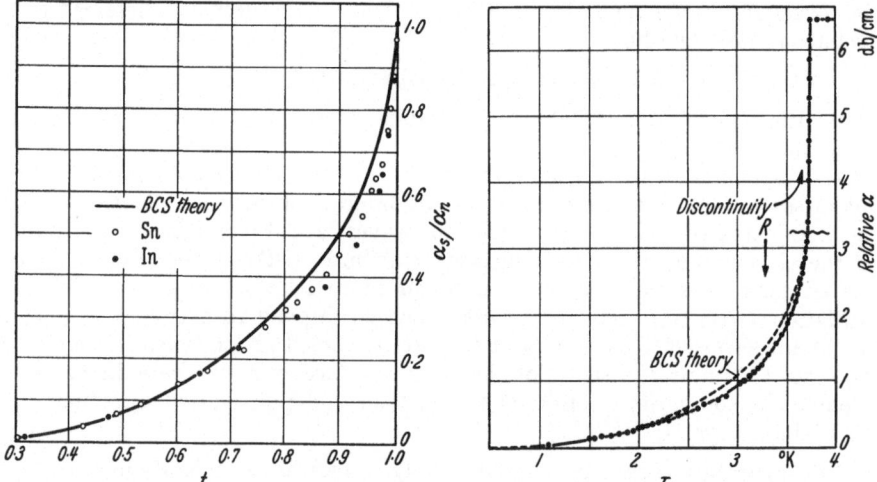

Fig. 24. Tin (54 Mc/sec) and indium (28.6 Mc/sec) longitudinal wave attenuation measurements compared with the BCS theory. After Morse [94].

Fig. 25. Observed shear wave attenuation in a polycrystalline sample of tin at 27.5 Mc/sec. The BCS theory is compared with that part of the attenuation labelled R. After Morse [94].

the factor $(1 - \mathscr{E}^2/E^2)$, used in deriving expression (33.2), exactly cancelled out the density of states modification. However, the use of any type of energy gap implies a change in the density of states, since the total number of states is conserved; n_s must, therefore, increase near the energy gap. Thus the BCS theory leads to a result that would be calculated for a simple energy gap model, if the normal density of states were preserved.

Fair agreement is found between experimental results and the theoretical predictions [Eq. (33.2)] in the case of longitudinal waves (see Fig. 24). The attenuation of transverse waves exhibits somewhat different features. For $q\,l_e > 1$ there is an initial discontinuous drop of α_s at the transition temperature, followed by a gradual drop upon further lowering of the temperature. The extent of the discontinuity decreases markedly with decreasing $q\,l_e$ and appears to tend to zero when $q l_e \to 0$. The decrease in α_s with decreasing temperature, after the discontinuity, agrees well with the prediction of the BCS theory (see Fig. 25). A qualitative explanation of this behavior was proposed by Morse [94], and may be summarized as follows. In an "ideal" metal (one obeying the free electron model) the interaction with transverse waves would be essentially electromagnetic in character and would not depend upon the actual change in the electron energy $E(\boldsymbol{k})$ with deformation of the lattice. In a "real" metal, however, there would also be a contribution from the deformation interaction for transverse waves. Thus the attenuation of transverse waves would be caused by a perturbing potential E_1 [see Eq. (30.6)] consisting of two terms: $E_1 = E_2 + E_3$ where E_2 refers to the electrodynamic interaction and E_3 is the "real" metal interaction, as discussed above. In the superconducting state E_2 should be effectively eliminated ("shorted-out"). The attenuation should thus change discontinuously, at the transition temperature, from a value determined by $E_2 + E_3$ to a value determined by E_3 alone. In the superconducting state the attenuation determined by E_3 should vary as $2f(\mathscr{E})$, as long as $q\,l_e > 1$, and $\alpha_s/\alpha_n = 2f(\mathscr{E})$ as for longitudinal waves; this is in agreement with experimental observations.

One can also rationalize the observed decrease of the discontinuity, when $q\,l_e \to 0$, on the same basis. For $q\,l_e \ll 1$, collisions are the dominant mechanism of interaction between electrons and the lattice, rather than electric and magnetic fields. One would thus expect the electrodynamic part of the decrease in attenuation at T_c to be small.

Perhaps the most remarkable agreement between theory and experiment, in superconductors, occurs in the ultrasonic measurement of the superconducting energy gap at absolute zero. This can be done by using the relation $\alpha_s/\alpha_n = 2f(\mathscr{E})$ and substituting measured values of α_s/α_n. Attenuation measurements carried out on tin by MORSE and co-workers [94] yielded a value for $2\mathscr{E}(0)$ (the energy gap at absolute zero) of $3.54\,kT_c$, where k is BOLTZMANN'S constant and T_c is the transition temperature. This compares with the value predicted by the BCS theory, $2\mathscr{E}(0) = 3.5\,kT_c$ for all metals (the maximum possible error in the value determined experimentally was shown to be $\pm 0.15\,kT_c$).

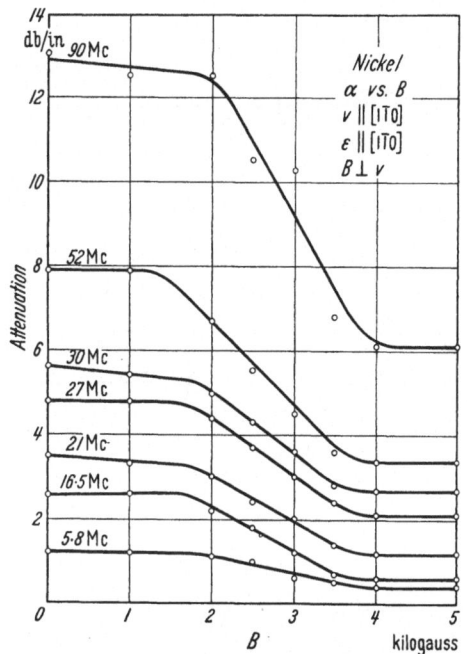

Fig. 26. Attenuation as a function of induction B in nickel for a compressional wave propagation in the [1T0] direction; B is perpendicular to the propagation direction. Frequencies are as indicated. After LEVY and TRUELL [112].

VI. Magneto-elastic interactions.

34. Stress wave interaction with magnetic domain walls. Experimental results. It is well known that internal strain affects Bloch wall motion in a ferromagnetic material. This coupling is responsible for a large proportion of the losses which occur when a stress wave propagates through a ferromagnetic material. The magnetomechanical losses have been studied [109] to [113] by ultrasonic methods, particularly in single crystals of iron silicon and nickel.

The accumulated experimental observations are outlined in what follows. Not all of these results have a theoretical interpretation. Both attenuation [112] and velocity [113] measurements have been made on crystals grown from 99.9% nickel and the experimental results for attenuation as a function of the magnetic induction are shown in Fig. 26 for ultrasonic frequencies from 5 to 90 Mc/sec and for compressional waves propagating in a $\langle 110 \rangle$ direction. Fig. 27 shows the results for the same type of measurements also made in a ⟨110⟩ direction. In this case, however, the magnetic field is parallel to the propagation direction

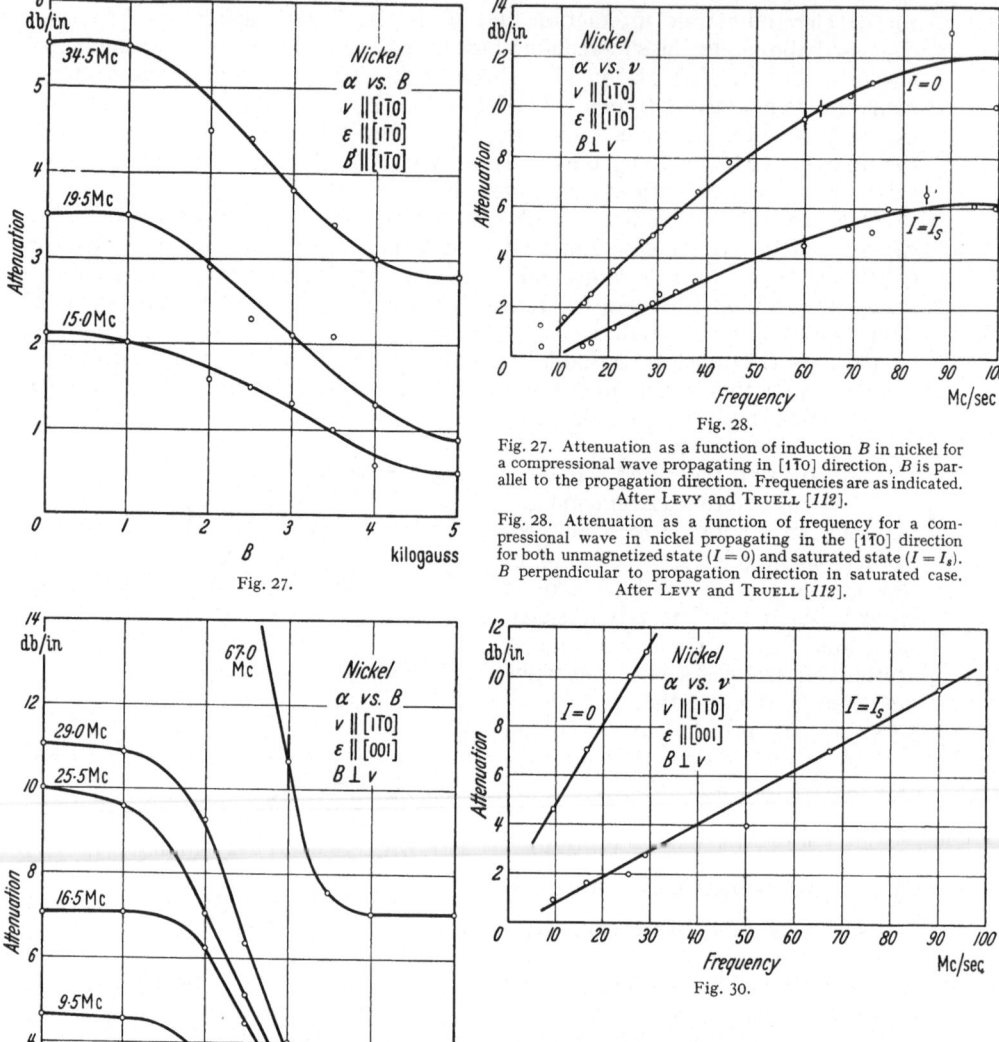

Fig. 27. Attenuation as a function of induction B in nickel for a compressional wave propagating in [1T0] direction, B is parallel to the propagation direction. Frequencies are as indicated. After LEVY and TRUELL [112].

Fig. 28. Attenuation as a function of frequency for a compressional wave in nickel propagating in the [1T0] direction for both unmagnetized state ($I = 0$) and saturated state ($I = I_s$). B perpendicular to propagation direction in saturated case. After LEVY and TRUELL [112].

Fig. 29. Attenuation as a function of induction B for a transverse wave in nickel propagating in the [1T0] direction and polarized in the [001] direction B is perpendicular to propagation direction. Frequencies as indicated. After LEVY and TRUELL [112].

Fig. 30. Attenuation as a function of frequency for a transverse wave in nickel propagating in the [1T0] direction and polarized in the [001] direction for both unmagnetized state ($I = 0$) and saturated state ($I = I_s$). B is perpendicular to propagation direction. After LEVY and TRUELL [112].

rather than perpendicular as in Fig. 26. Attenuation-frequency results for compressional waves are shown in Fig. 28 for both the unmagnetized state and the magnetically saturated state. Similar data taken with transverse waves are shown

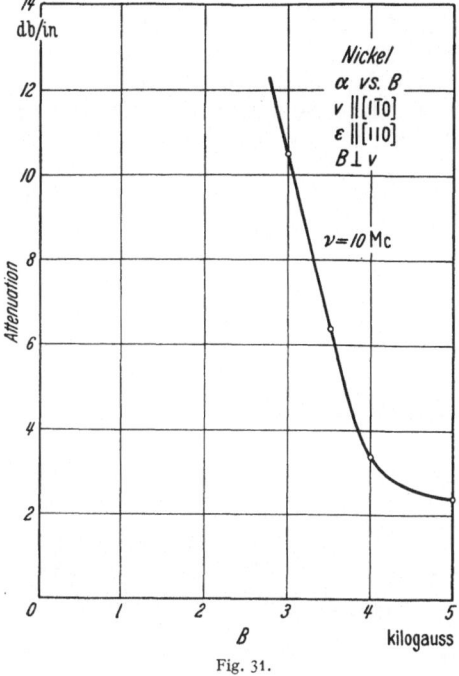

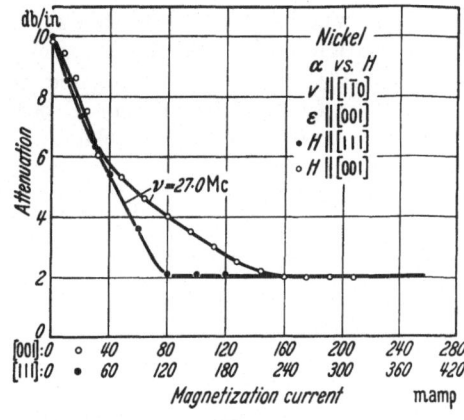

Fig. 32.

Fig. 31. Attenuation as a function of induction for a transverse wave in nickel propagating in the [1T0] direction and polarized in the [110] direction. B is perpendicular to the propagation direction. The frequency is 10 Mc/sec. After LEVY and TRUELL [112].

Fig. 32. Anisotropy of attenuation in nickel at 27 Mc/sec for shear wave propagating in [1T0] direction polarized in [001] direction. Anisotropy arises from application of magnetic field in [001] direction or in [111] direction. After LEVY and TRUELL [112].

in Figs. 29 and 30. In some directions the transverse wave attenuation in nickel is so high that it cannot be measured at zero magnetic field. Fig. 31 shows such a case.

Additional data relating the attenuation with the magnetizing current for an applied magnetic field in two different crystal directions perpendicular to the propagation direction has a form as shown in Fig. 32. These curves of Fig. 32 when inverted have a form and relationship to one another very similar to the corresponding magnetization curves for a single crystal of nickel [114]. In addition it is found that attenuation-magnetic field curves, taken over a cycle of the magnetic field, are similar to hysteresis loops.

Measurements of attenuation and velocity in

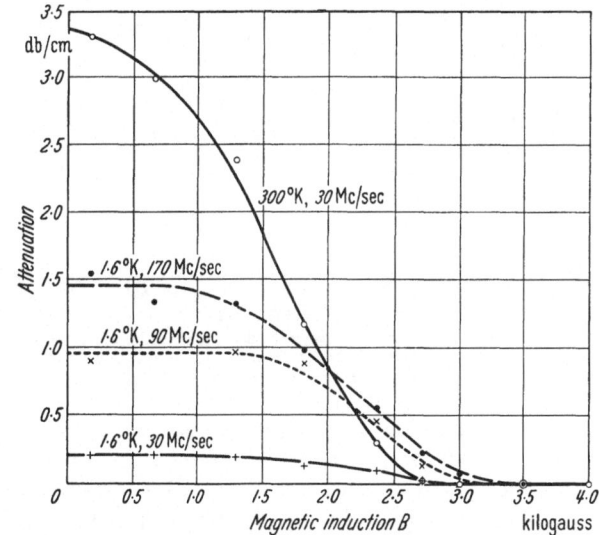

Fig. 33. Attenuation of compressional waves in nickel as a function of magnetic induction B at 300° K and at 1.6° K for various frequencies. Waves propagating along [1T0] direction, B along [111] direction.

nickel have been made as a function of temperature and as a function of magnetic field at low temperatures.

In general it can be said that as the magnetic field is increased from zero the attenuation decreases and the velocity increases.

Fig. 33[1] shows (for compressional waves) the change in attenuation as a function of magnetic induction B both at 300° K at 30 Mc/sec and at 1.6° K at 30 Mc/sec, 90 Mc/sec and 170 Mc/sec. It is seen that with no applied magnetic field the temperature change from 300 to 1.6° K causes an attenuation decrease of more than a factor of ten for a 30 Mc/sec frequency.

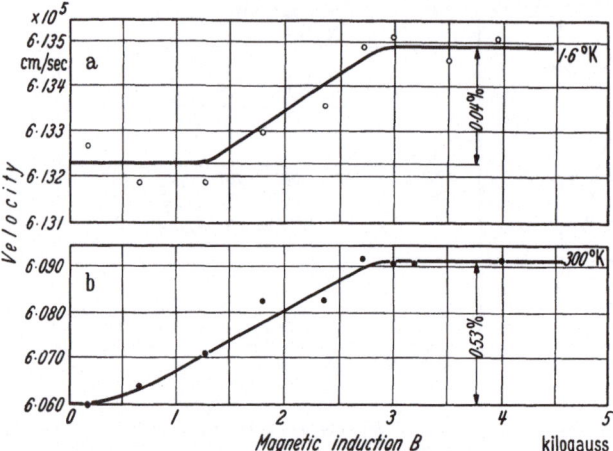

Fig. 34 a and b. Velocity of compressional waves in nickel as a function of magnetic induction B. Waves propagating along [1$\bar{1}$0] direction, B along [111] direction. a At 1.6° K. b At 300° K.

The velocity increase measured as a function of magnetic induction is shown in Fig. 34[1] for compressional waves at 30 Mc/sec. Fig. 34 a shows the increase measured at 1.6° K — a total change of only 0.04 %. Fig. 34 b shows the velocity increase as a function of magnetic induction at 300° K. In this case the total change was 0.53 %. In these two cases the measurement of the difference is more accurate than the velocity values themselves.

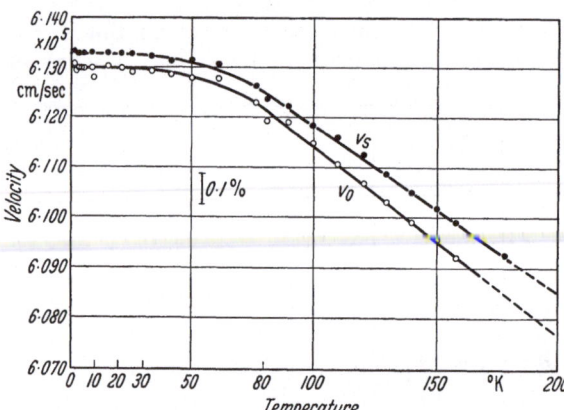

Fig. 35. Velocity of compressional waves in nickel as a function of temperature, for magnetized state v_s and for unmagnetized state v_0. Waves propagating along [1$\bar{1}$0] direction, B in the [111] direction for v_s.

Measurements of velocity as a function of temperature for magnetized (v_s) and unmagnetized states (v_0) are shown in Fig. 35[1] where it is seen that the velocity decreases with increasing temperature. The difference between v_0 and v_s increases with increasing temperature.

The difference between attenuation values for saturated and for unmagnetized nickel are shown in Fig. 36[1] where α_0 is the attenuation curve for no magnetization and α_s is the attenuation curve for a sample magnetically saturated.

A background value of 0.75 db/cm has been subtracted from both α_0 and α_s to obtain the curves shown.

The magnetomechanical losses associated with the propagation of a high frequency stress wave through a ferromagnetic crystal (not saturated magnetic-

[1] Unpublished data of E. Roland Dobbs, Bruce Chick and T. Fitzgerald (Brown University).

ally) arise from the fact that the spin walls vibrate under the influence of the passing stress wave. The vibrating spin wall constitutes a magnetic field changing with time; consequently an alternating electromotive force is induced which in turn induces electrical currents in the materials, hence losses.

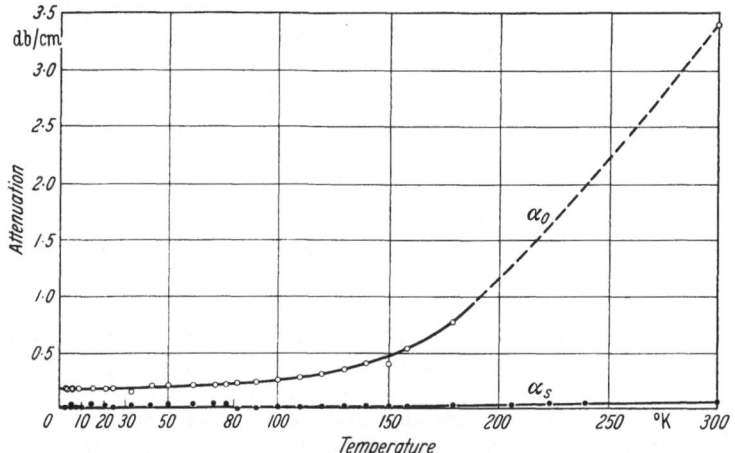

Fig. 36. Attenuation of compressional waves in nickel as a function of temperature for unmagnetized state α_0 and for magnetized state α_s. Waves propagating along [1$\bar{1}$0] direction, B along [111] direction for α_s.

35. Outline of an analytical analysis of the domain wall motion. A calculation of the losses for 180° spin walls has been given [*115*] on the basis of a classical electromagnetic picture in which the magnetization I is considered to be harmonic in time. The current density j is then calculated as well as the average power dissipated per unit area of wall for a wall of thickness δ. The motion of a spin wall is obviously connected with the magnitude of the internal strains but no complete description of this effect seems to be available. It is assumed that a plane ultrasonic stress wave of frequency ν causes a spin wall to vibrate with this frequency and that the phase and amplitude of the vibration depend only on the frequency and the strain. It is also assumed that since the current, caused by a moving wall, decays exponentially with the distance from the wall, the current in the region between two neighboring walls is that arising only from the motion of these two walls.

For spin walls, equally spaced D cm apart, the loss in energy flux $\mathbf{S}$ per unit length of ultrasonic path can be written

$$\frac{d S_x}{d x} = - \alpha \, S_x$$

or

$$S_x(x) = S_0 \, e^{-\alpha x},$$

hence one has an attenuation of the energy flux per unit length of path in the material. Finally α in units of db/cm turns out to be

$$\alpha \, (\text{db/cm}) = 4.343 \, (16\pi^3) \, I_s^2 \, \sigma^{\frac{1}{2}} \, D^{-1} \, \nu^{\frac{3}{2}} \, [k(\nu)]^2$$

where I_s is the saturation value of the magnetization, σ is the electrical conductivity, 1.28×10^{-4} for nickel, D is the distance between 180° spin walls, ν is the ultrasonic frequency, and $k(\nu)$ is the factor which relates the amplitude of vibration

15*

of a spin wall with the flux density of the ultrasonic wave causing the vibration; in other words, the amplitude u_0 of the vibration of the spin wall is related to the energy flux S of the stress wave by

$$u_0 = k(v)\,|S|^{\frac{1}{2}}.$$

$k(v)$ has been found from experiment to depend on frequency as $v^{-\frac{3}{8}}$ for nickel in the range 10 to 50 Mc/sec. The magnitude of $k(v)$ is connected with the magneto-striction coefficients.

36. Interaction of spin waves and ultrasonic waves in ferromagnetic crystals. The interaction of spin waves [116] (magnons) and ultrasonic waves in ferro-magnetic materials has been discussed [117] to [119] in view of the magneto-elastic coupling between spin waves and phonons. In particular KITTEL [117] has examined the resonance behavior to be expected when both the spin waves and the phonon wavelengths and frequencies are equal. It is shown that at high ultrasonic frequencies the resonance condition may provide a measure of the value of the exchange energy constant. KITTEL also showed that a ferro-magnetic material may have non-reciprocal acoustic properties. At resonance between spin waves and phonons the phonon or ultrasonic attenuation is expected to be large. KITTEL derives the dispersion relation for a spin wave and the phonons as well as the equations of motion for the phonons. It is shown that the attenuation of a transverse acoustic wave under resonance conditions with the spin system is approximately given by

$$\frac{k_2}{k_1} \approx \frac{\omega_s \tau\, b_2^2}{2G\,M_s^2} = \frac{\gamma\,\tau\,b_2^2}{2GM_s} \qquad k_2 \ll k_1, \qquad (36.1)$$

where $k = k_1 - i\,k_2$ is the phonon propagation factor, G is an elastic constant $\approx 10^{12}$ dynes/cm, τ is spin relaxation time $= 10^{-8}$ sec (ferrite), γ is the magneto-gyric ratio $= \dfrac{g\,e}{2m\,c} \approx 2 \times 10^7$ (oersted/sec)$^{-1}$, b_2 is a magnetoelastic coupling coeffi-cient $= 10^8$ ergs/cm^3 for nickel, and M_s is the saturation magnetization $= 500$ gauss. $\omega_s = \gamma\, M_s$.

The result of using these numerical values is $k_2/k_1 \approx 2$ which means that the condition $k_2 \ll k_1$ is not fullfilled and that the attenuation of the ultrasonic wave is large at the resonance frequency of the spin system. The damping falls off considerably as the frequency is moved away from resonance.

It should also be noted that the ultrasonic wave may have a circular polarization of either one sense or the other, and that the attenuation for circular polarization of one sense may be considerably less than that for the opposite sense. The attenuation for circular polarization of opposite sense to that given above is approximately

$$\frac{k_2}{k_1} = \frac{\gamma\, b_2^2}{8\omega^2\,\tau\,G\,M_s} \qquad (36.2)$$

or a value lower by a factor $\dfrac{1}{4\omega^2\,\tau^2}$ than the value for the sense of circular polariza-tion given above.

This unequal attenuation of the transverse ultrasonic waves of one sense or the other forms the basis for saying that non reciprocal acoustic elements are in principle possible.

Finally there is the feature of such magnetoelastic coupling that it is expected that a thin ferromagnetic crystal might be driven as a magnetostrictive oscillator at very high frequencies.

The experimental observation of the excitation of ultrasonic waves at microwave (1000 Mc/sec) frequencies by a ferromagnetic single crystal has been attained by Bömmel and Dransfeld [120], who observed the generation of transverse waves from a thin nickel film deposited on a quartz surface and excited to ferromagnetic resonance. In the same experiment it was also found that ferromagnetic resonance could be excited by incident ultrasonic waves. The arrangement used by Bömmel and Dransfeld is shown in Fig. 37 with the quartz sample placed between the two cavities and the nickel film (18000 Å) on one end of the quartz. The nickel film has the radiofrequency magnetic field parallel to it in the cavity on the left while the other end of the quartz rod has the radiofrequency electric field perpendicular to it. The d.c. magnetic field is parallel to the axis of the quartz rod. With transverse ultrasonic waves generated in cavity C_2 the nickel film served as an acoustic receiver. With the nickel film driven at ferromagnetic resonance in cavity C_1, transverse ultrasonic waves were generated in the nickel film and transmitted into the quartz rod and the end of the rod in cavity C_2 then served as the acoustic receiver.

At ferromagnetic resonance the total magnetic moment will precess around the axis of the constant magnetic field, and because of the magnetoelastic coupling, the surface of the nickel film will experience a rotational shear strain around this axis; the film surface will thereby generate circularly polarized transverse sound waves which propagate into the quartz.

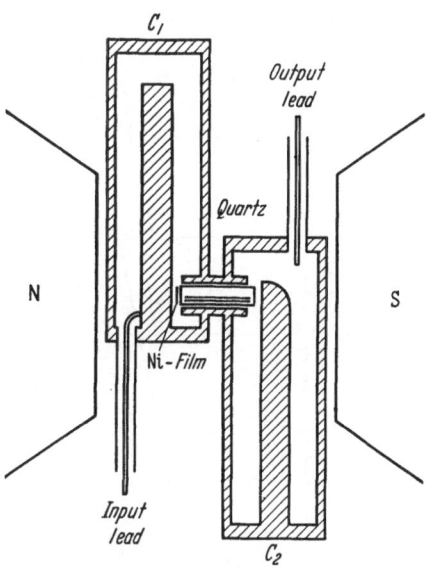

Fig. 37. Experimental arrangement for detection of high frequency ultrasonic waves generated in a nickel film by ferromagnetic resonance. After Bömmel and Dransfeld [120].

When the spin precession in the sample is uniform (i.e. spin wave of infinite wavelength) the magnetostrictive stress will also be uniform and, initially, a displacement may occur only at the surface; hence elastic waves are excited at the surface. If the spin precession is not uniform, then spin waves of finite wavelength may exist and the excitation of ultrasonic waves is accomplished as mentioned in the above discussion concerned with Kittel's paper [117].

In the particular experiment by Bömmel and Dransfeld the nickel film was sufficiently thin compared with the ultrasonic wavelength so that there was very little difference between the spin wave excitation with finite or infinite wavelength.

The theory of relaxation processes in ferromagnetics at low temperatures has been discussed by Akhieser [121] who treated spin-spin and spin-lattice effects and who discussed, among other things, the scattering of spin waves by lattice vibrations. Other discussions dealing with the interaction of spin waves and phonons with particular regard to sound absorption are those of Akhieser and Shishkin [118] and more recently Kaganov and Chikvashvili [119]. Other work pertinent to the general problem of spin-phonon interaction in ferromagnetic materials is given in Refs. [122] and [123].

VII. Stress wave interaction with thermal elastic waves.
Phonon-phonon interaction.

37. Description of the problem. The measurement of the attenuation and velocity of ultrasonic stress waves as a function of temperature in quartz [124] to [128] and germanium [129] to [131] has aroused interest in the mechanisms responsible for the observed behavior.

The experimental situation is simply that ultrasonic stress waves interact in a crystalline solid with thermal or lattice vibration stress waves despite the fact that there is an appreciable frequency difference between the two as well

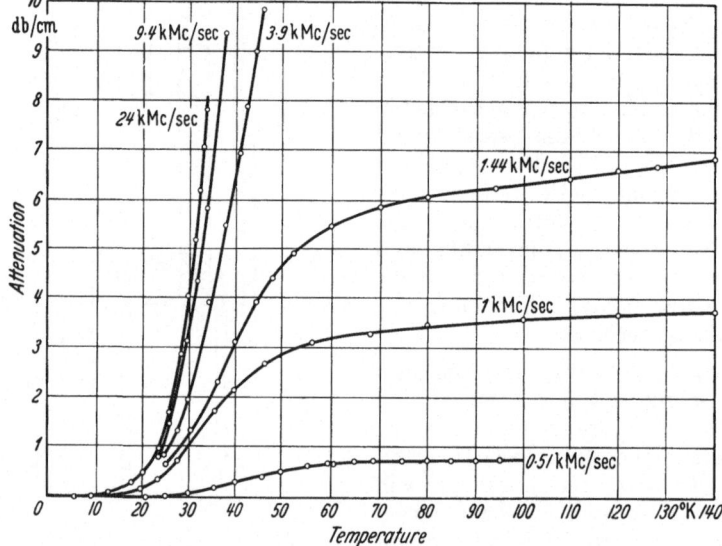

Fig. 38. Attenuation of compressional waves as a function of temperature of X-cut crystalline quartz for the frequencies indicated in kilomegacycles/sec.

as the fact that no such interaction should occur, even at equal frequencies, if the waves were linearly elastic. If the interatomic forces were harmonic there would be no interaction between the ultrasonic waves and the lattice waves. Corresponding to the propagation of waves in the harmonic approximation there is, in the quantum situation, the "free phonon" or "non interacting" phonon case. With anharmonic lattice forces, however, phonons may scatter other phonons. The anharmonic behavior may be approximated by means of the cubic terms in the strain energy function which means of course the third order elastic constants; higher terms may also be involved.

The phonon-phonon interaction may be thought of as the scattering of one wave by the other. One wave, propagating through the solid, causes displacements of the atoms from their normal or rest positions and may cause the elastic properties to be altered. Gradients in elastic properties are in general the cause of scattering of elastic stress waves, and if a second stress wave, also propagating through the lattice, encounters the effects of the first wave, a scattering of one wave by the other can occur providing that at least one of the waves can produce gradients in elastic properties as a consequence of the anharmonic interaction forces.

Another way of saying the same thing [132] is that sound waves on passing through a crystal disturb the equilibrium distribution of the thermal phonons and that the tendency to reestablish equilibrium in the phonon distribution leads to absorption of the ultrasonic waves. A measure of the connection among the thermal phonon modes as well as between thermal and ultrasonic modes (or coupling among the modes) is given by the third order elastic coefficients of a solid and for this reason these coefficients are of considerable interest and importance in the phonon-phonon interactions. Consequently the third order coefficients are related to the relaxation times for normal phononphonon collisions and umklapp processes, both of which may govern the relaxation of a disturbed phonon distribution toward an equilibrium distribution.

It should, of course, be noted that there is, in any solid crystalline lattice, the spectrum of lattice vibration waves extending over a frequency range from about 10^{11} to 10^{13} cycles per second. On the other hand, the ultrasonic wave is essentially monochromatic at whatever frequency is chosen.

38. Experimental data for quartz and germanium. Experiments on phonon-phonon effects have been carried out in the frequency range from 5×10^8 to 2.5×10^{10} cycles/sec. The attenuation temperature behavior for crystalline quartz (Fig. 38) includes data from three sources [124], [125], [127], [128] and covers the frequency range mentioned. Similar data [129] to [131] in the case of germanium are shown in Fig. 39.

The quartz data show a relatively rapid drop in attenuation with decreasing temperature at about 40° K whereas the germanium data of Fig. 39 have a similar attenuation decrease at a slightly higher temperature around 50° K. Unfortunately the data of Fig. 38 do not permit one to decide whether the location (on the temperature scale) of the attenuation change is frequency dependent or not.

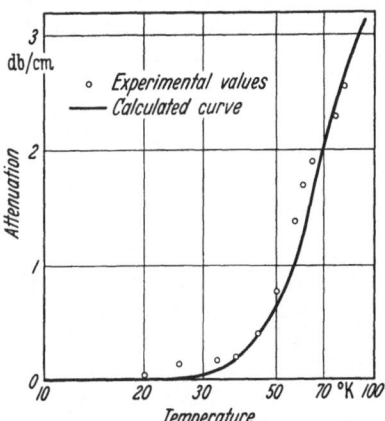

Fig. 39. Attenuation of a compressional wave as a function of temperature for germanium in [100] direction at 508 Mc/sec. After VERMA and JOSHI [131].

39. Calculation of attenuation due to phonon-phonon interaction. A number of attempts have been made [125], [132] to [135] to calculate the dependence of sound absorption on frequency and temperature. None of these has been fully satisfactory in accounting for the observed behavior.

Two approaches leading to expressions for ultrasonic attenuation as a function of temperature have been made, one by BÖMMEL and DRANSFELD [125] and the other by WOODRUFF and EHRENREICH [135], both based on the work of AKHIESER [132].

α) *Method of Bömmel and Dransfeld.* The first approach makes direct use of the idea of AKHIESER [132], mentioned earlier, that the ultrasonic wave passing through the crystal produces a perturbation of the thermal phonon distribution with the result that the thermal phonon distribution is no longer an equilibrium one and no longer a Planck distribution. The reestablishment of equilibrium in the phonon distribution involves an increase in entropy of the system which in turn involves an absorption of ultrasonic energy.

In the region of present interest the wavelength of a thermal phonon is always shorter than the ultrasonic wavelength.

However, because of the previously mentioned anharmonic effects, both sound waves and thermal phonons can create gradients in the elastic properties of the lattice; these gradients are "seen" by both sets of waves and are thereby scattered.

The thermal phonon velocity is altered by the scattering process by amounts which differ for phonons of different polarization j and different propagation vector k.

The polarization vector j and propagation vector k determine a "branch" of the phonon spectrum. The relation between the fractional velocity change $(\Delta v/v)_{j,k}$ for a branch (j, k) and the relative density change $(\Delta \varrho/\varrho)$ is given by $\gamma_{j,k}$ the Grüneisen constant

$$\left(\frac{\Delta v}{v}\right)_{j,k} = \gamma_{j,k}\left(\frac{\Delta \varrho}{\varrho}\right). \tag{39.1}$$

Unfortunately the Grüneisen constant varies considerably from branch to branch as well as with frequency and temperature.

Bömmel and Dransfeld assumed that dispersion can be neglected and that phonons of all frequencies in a single branch have the same velocity change for a given strain. The temperature of the phonons in a given branch is then changed in a definite way by a compressional strain

$$\left(\frac{\Delta T}{T}\right)_{j,k} = \left(\frac{\Delta v}{v}\right)_{j,k} = \gamma_{j,k}\left(\frac{\Delta \varrho}{\varrho}\right). \tag{39.2}$$

After a deformation each branch has in general a different temperature. In particular γ can have large differences among different branches and may even be positive or negative so that some branches of the phonon spectrum may be cooled while others are being heated. The main difficulty with this approach to a quantitative handling of the problem seems to be the lack of sufficient information about the $\gamma_{j,k}$ for the many branches.

The simplification used by Bömmel and Dransfeld was that of dividing all of the branches into two groups with different average temperatures and comparable specific heats. One group is assumed to contain all branches having undergone appreciable positive temperature changes, and this temperature change is labeled $(\Delta T/T)_1$. The second group consists of all of the remaining branches having small or negative temperature changes $(\Delta T/T)_2$. The average fractional temperature change between the two groups assumed is then

$$\left(\frac{\Delta T}{T}\right)_{\text{Ave}} = \left(\frac{\Delta T}{T}\right)_1 - \left(\frac{\Delta T}{T}\right)_2 \tag{39.3}$$

which is next used to define an average γ for the entire system of all branches

$$\left(\frac{\Delta T}{T}\right)_{\text{Ave}} = \gamma_{\text{Ave}}\left(\frac{\Delta \varrho}{\varrho}\right). \tag{39.4}$$

A compressional wave propagating through the lattice may produce a time dependent periodic temperature difference between the two groups of phonon branches assumed. A relaxation time τ is assumed for the heat exchange which takes place between the two groups; this exchange in time leads to an increase in entropy S corresponding to the absorption of the sound wave.

The attenuation in db/inch is calculated to be

$$\alpha = 4.34 \frac{v \int_0^{t_0} T \, dS}{I_0} = 1.1 \frac{C \, T \, \gamma_{\text{Ave}}^2}{\varrho \, v^3} \frac{\omega_s^2 \tau}{1 + (\omega_s \tau)^2} \tag{39.5}$$

where t_0 is the period of the ultrasonic wave, v is the compressional wave velocity, I_0 is the energy flux or intensity of the ultrasound, C is the specific heat per cubic centimeter, and ϱ is the material density.

The relaxation time τ is the mean time between phonon collisions. It was assumed by BÖMMEL and DRANSFELD that the energy exchange between phonons occurs by means of the phonon-phonon umklapp processes.

Using the thermal conductivity expression

$$\varkappa = \tfrac{1}{3} C v \Lambda$$

where C is the specific heat, v is the average phonon velocity, and Λ is the phonon mean free path between collisions, i.e. $\Lambda = v \tau$, one obtains

$$\varkappa = \tfrac{1}{3} C v^2 \tau. \tag{39.6}$$

From the manner in which the specific heat and sound velocity depend (quartz) on temperature around 40° K it can be argued that τ varies as $1/T^2$. At higher temperatures, $\tau \ll t_0$ (period of sound wave) so that $\omega_s \tau \ll 1$ and the attenuation expression becomes

$$\alpha = 1.1 \, C \, T \, \gamma_{\text{Ave}}^2 \, \frac{\omega_s^2 \tau}{\varrho \, v^3}. \tag{39.7}$$

It turns out that this expression is practically temperature independent above about 60° K in agreement with the data on quartz. It also turns out, from measurements, on thermal conductivity and specific heat, that $\omega_s \tau \cong 1$ at 40° K for an ultrasonic frequency of 1000 Mc/sec; the attenuation value α at this point agrees well with the measured value of 5 db/inch. A value of $\gamma_{\text{Ave}} = 2$ was used for this calculation.

Expression (39.5) has been used to calculate ultrasonic attenuation [131] as a function of temperature for germanium, and the calculated values have been compared with measured values [129], [130]. The solid curve of Fig. 39 has been calculated by VERMA and JOSHI [131]. The relaxation time τ was calculated for low temperatures using (39.6) together with low temperature values of thermal conductivity [151], specific heat and the elastic constant ε_{11} [153]. The value of γ_{ave} used in (39.5) was obtained by matching the calculated attenuation value with the experimental value at 70° K. The agreement between the experimental values (points) and the calculated values (solid curve) of Fig. 39 is rather good.

β) Method of Woodruff and Ehrenreich. The second of the two approaches, mentioned above, leading to expressions for ultrasonic attenuation as a function of temperature is that of WOODRUFF and EHRENREICH [135] who extended the work of AKHIESER. The point of view taken is that the sound wave modulates the frequencies of all the thermal phonons and that this changes the energy distribution of the thermal phonons. The Boltzmann equation is used to determine the modification made by the sound in the thermal phonon distribution. A perturbed frequency and temperature are calculated for the modified or perturbed phonon distribution and relaxation times are assumed for normal phonon scattering processes as well as for Umklapp processes which do not conserve the phonon wave vector. Finally the average time rate at which energy is transferred from the sound wave to the thermal phonon is calculated and the resulting attenuation expression obtained by WOODRUFF and EHRENREICH is

$$\alpha \, (\text{db/inch}) = \frac{C_v \, T \gamma^2 \, \omega_s}{2 \varrho \, v^3} \, \text{Im} \left[\frac{(I_{00}^*)^2}{I_{01}} + i \, \omega_s \, \tau \, I_{00}^* \right]. \tag{39.8}$$

ω_s is the sound wave frequency, C_v is the specific heat, T is the temperature, and ϱ is the density of the material, v is the sound wave velocity, γ is the Grüncisen constant, $I_{mn}(\omega_s \tau) \equiv \int\limits_{-1}^{+1} \dfrac{\mu^m(1-\mu)^n}{1 - i\,\omega_s\,\tau\,(1-\mu)}\,d\mu$, and τ is the combined relaxation time for normal and umklapp processes;

$$\frac{1}{\tau} = \frac{1}{\tau_u} + \frac{1}{\tau_N}.$$

In this approximation it has been assumed that $(\tau_N/\tau_u) \gg 1$ and that $\tau_u \approx \tau$ may be associated with the thermal conductivity relaxation time.

In the course of the derivation of the above expression it has been assumed that various quantities, such as $\omega_0(k)$ the frequency of phonon of propagation vector k in the unstrained crystal, are independent of the direction of k although they may still depend on the magnitude $|k|$ and on the mode j. In addition, in order to obtain tractable expressions, it has been assumed that τ_u and τ_N are independent of k altogether. The last assumption seems to be more serious than the preceeding ones. In the case $\omega_s \tau \ll 1$ expression (39.8) becomes

$$\alpha = \frac{C_v\,T\,\gamma^2\,\tau\,\omega_s^2}{3\,\varrho\,v^3} \tag{39.9}$$

or in terms of the thermal conductivity $\varkappa \equiv \frac{1}{3} C_v\,v^2\,\tau$ it becomes

$$\alpha = \frac{\gamma^2\,\omega_s^2\,\varkappa\,T}{\varrho\,v^5} \tag{39.10}$$

and, at temperatures greater than the Debye temperature Θ_D of the solid, $\varkappa \sim 1/T$, hence α becomes independent of temperature and this is experimentally observed in quartz and germanium. In the case $\omega_s \tau \gg 1$ expression (39.8) becomes

$$\alpha = \frac{\pi\,\gamma^2\,\omega_s\,C_v\,T}{4\,\varrho\,v^3}. \tag{39.11}$$

It may be noted that although the analysis used by Woodruff and Ehrenreich does not really apply in the last case $\omega_s \tau \gg 1$ (because quantum effects begin to become important) there is some similarity to a quantum mechanical result of Landau and Rumer [133] in that the dependence on T and on ω_s is the same in both cases.

In this analysis it has been assumed that the relaxation of the phonon distribution occurs toward the modified or perturbed phonon distribution associated with a temperature T'. Another approximation for α is possible, however, if it is considered that the relaxation of the phonon distribution may occur toward the unperturbed thermal equilibrium distribution for $\omega_s \tau > 1$ because in this case a thermal phonon will travel many sound wavelengths between collisions. Under these circumstances collisions should not occur with sufficient frequency in any local region to alter the distribution toward which relaxation occurs. In this particular case the attenuation results may be written

$$\alpha = \frac{3\gamma^2\,\omega_s^2\,\varkappa\,T}{\varrho\,v^5} \left(\frac{1}{2\omega_s\,\tau}\right) \text{arc tan } (2\omega_s\,\tau). \tag{39.12}$$

This expression is similar to (39.8) except that the first term in the brackets of (39.8) is missing. With $\omega_s \tau \ll 1$ expression (39.12) yields a value of α three times larger than that given by (39.10), but for $\omega_s \tau \gg 1$ it leads to the same value as that given by (39.11). Comparison of these results with attenuation measurements made on quartz at 1000 and at 3900 Mc/sec is shown in Figs. 40 and 41. The Grüneisen constant γ was treated as an adjustable parameter and was determined by matching the experimental and calculated attenuation at 60° K for 1000 Mc/sec and 40° K for 3900 Mc/sec. The thermal conductivity relaxation time was obtained from experimental values of $\varkappa$ and C_v.

The magnitude of the attenuation seems to be represented well by relation (39.8) but the shape of the curve is given somewhat better by Eq. (39.12). The results of this calculation seem to yield correct qualitative features as well as correct quantitative results in the range $\omega_s \tau \ll 1$.

A summary of some of the general background in phonons and phonon-phonon processes is to be found in "Electrons and Phonons" by ZIMAN [136].

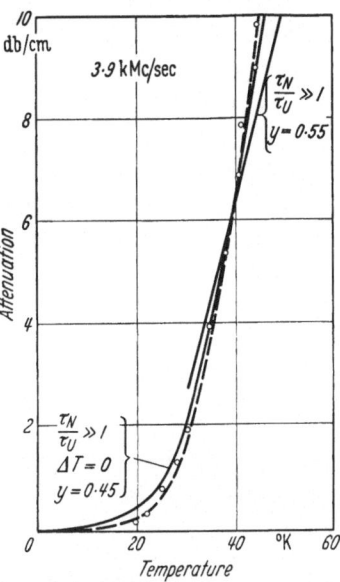

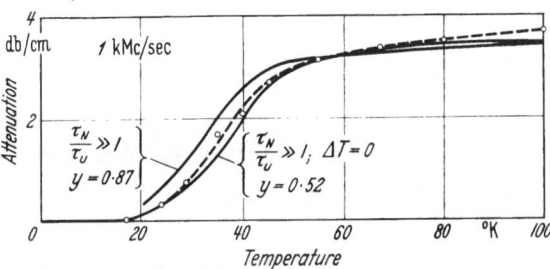

Fig. 40. Attenuation of compressional waves as a function of temperature at 1000 Mc/sec. Calculated solid curves compared with experimental dashed curve taken from Ref. [125]. After WOODRUFF and EHRENREICH [135].

Fig. 41. Attenuation of compressional waves as a function of temperature at 3900 Mc/sec. Dashed curve and experimental data taken from Ref. [125]. Solid curves are calculated. After WOODRUFF and EHRENREICH [135].

VIII. Stress wave interaction with nuclear spin systems.

40. Conditions for interaction with nuclear spin systems. Ultrasonic waves are coupled to nuclear spins[1] through the interaction of the quadrupolar moment of the nucleus with the time varying electric field gradient generated by the ultrasonic wave. When the lattice is distorted by an ultrasonic wave, the electric field gradient at the site of a nucleus varies periodically with the frequency of the wave. The variation of the electric field gradient produces the necessary time-dependent perturbation of the spin energy levels to bring about transitions. The frequency of the ultrasonic wave must correspond to the frequency of $\Delta m = \pm 1$ or $\Delta m = \pm 2$ transitions, where the quantum number m labels the allowed values of the nuclear spin.

The occurrence of such transitions induced by ultrasonic waves was first demonstrated experimentally by PROCTOR and co-workers [137] to [139]. The basic purpose of these experiments is to obtain information on the strength and nature of the coupling between lattice waves at ultrasonic frequencies, and the nuclear quadrupole moments.

The electrostatic interaction energy H of a nucleus with the charges in its environment can be expressed by:

$$H = \int \varrho(x) V(x) d^3x \qquad (40.1)$$

where the integration is taken over the nuclear volume, $\varrho(x)$ is the nuclear charge density and $V(x)$ is the electrostatic potential arising from all charges other than

---

[1] Only nuclei of spin 1, or greater, are considered in this coupling.

those of the nucleus under consideration. The potential $V(x)$ can be expanded in a power series about the center of mass of the nucleus and Eq. (40.1) can be rewritten in the form:

$$H = \int \varrho(x)\, d^3x \left\{ V_0 + \sum_i \left(\frac{\partial V}{\partial x_i}\right) x_i + \frac{1}{2} \sum_{i,j} \left(\frac{\partial^2 V}{\partial x_i \partial x_j}\right) x_i x_j + \cdots = Z\, e\, V_0 + \right.$$
$$\left. + \sum_i P_i \left(\frac{\partial V}{\partial x_i}\right) + \frac{1}{2} \sum_{i,j} Q_i \left(\frac{\partial^2 V}{\partial x_i \partial x_j}\right) + \cdots \right. \tag{40.2}$$

The interaction energy of interest here is the one arising from the quadrupole term and this can be written as:

$$H = \frac{1}{2} \sum_{ij} Q_{ij} V_{,ij} \tag{40.3}$$

where

$$Q_{ij} = \int \varrho(x)\, x_i\, x_j\, d^3x$$

is the electric quadrupole tensor and

$$V_{,ij} = \frac{\partial^2 V}{\partial x_i \partial x_j}.$$

Both Q_{ij} and $V_{,ij}$ are second rank symmetric tensors, moreover, $V_{,ij}$ can be expressed as

$$V_{,ij} = -\frac{E_j}{x_i}$$

where the E_j are the components of the electric field at the nucleus. Thus $V_{,ij}$ represents the electric field gradient which interacts with the nuclear electric quadrupole moment.

It must be pointed out that Eq. (40.3) is qualitatively correct, but in order to obtain a quantitative evaluation of the interaction energy, additional effects must be taken into account. The most general of these effects is that the potential arising from a quadrupolar nucleus can induce a quadrupole moment in the electron charge cloud surrounding the nucleus. Depending on the state of the core electrons there is an increase (antishielding) or decrease (shielding) of the nuclear quadrupole moment brought about by this polarization [140]. In general, the antishielding effect is larger, often by a considerable amount, than the nuclear quadrupole moment; the shielding effect is usually small and often negligible.

There are also the so-called covalency effects which have been discussed in detail by Yosida and Moriya [141]. These effects arise from mixing of p- or d-like orbitals into s-like wave functions, when the lattice is deformed by vibrations. The electrons in these orbitals give rise to an electric field gradient at the nucleus which interacts with the nuclear quadrupole moment. The resulting quadrupole spin-lattice coupling can be much larger than with the direct field alone.

41. The probability of transitions between nuclear spin levels. The probability per unit time that an ultrasonic wave will produce a transition of nuclear spin between its Zeeman levels was evaluated by Kraus and Tantilla [142]; these authors considered the quadrupolar interaction as a perturbation of the Zeeman energy. In this treatment the quadrupole term of the electrostatic potential at a position r_n from the center of mass of the nucleus, where ϱ_n is the nuclear charge density due to a charge q_a, external to the nucleus, after displacement, is given by

$$V_\alpha = \frac{q_\alpha}{6\,(r^{(\alpha)})^5} \, A : B \tag{41.1}$$

where A and B are second rank tensors[1]. After suitable transformations, Eq. (41.1) is rewritten in the explicit form

$$V_\alpha = \frac{q_\alpha}{(r^{(\alpha)})^5} \sum_{i=-2}^{2} C_i A_i B_{-i} \tag{41.2}$$

where C_i are numerical constants. The contribution from the charge q_α to the nuclear quadrupole interaction energy is found by multiplying V_α with the nuclear charge density and integrating over the region occupied by the central nucleus:

$$H_\alpha = \int V_\alpha \varrho_n(r)\, d\tau_n \tag{41.3}$$

where H_α is the contribution to the Hamiltonian from the charge q_α. Substitution of V_α from Eq. (41.2) yields

$$H_\alpha = \frac{q_\alpha}{(r^{(\alpha)})^5} \sum_{i=-2}^{2} C_i Q_i B^{(\alpha)}_{-i} \tag{41.4}$$

where Q_i are the components of the nuclear quadrupole moments, defined as

$$Q_i \equiv \int \varrho_n(r) A_i\, d\tau_n \quad (i=0, \pm 1, +2) \tag{41.5}$$

and A_i are functions of the cartesian coordinates of an element of the nuclear charge. $B^{(\alpha)}_{-i}$ are functions of the relative displacement of the charge q_α with respect to the displacement of the central nucleus.

If the quantum number m designates the allowed values of the nuclear spin operator I_z, only processes for which $\Delta m = \pm 2$ are considered. Consequently, the matrix elements of the operator H, of interest here are

$$(I_m |H| I_{m+2}) = (I_m |Q_{-2}| I_{m+2}) \sum_\alpha \frac{q_\alpha}{(r^{(\alpha)})^5} C_2 B^{(\alpha)}_{+2} \tag{41.6}$$

$$(I_m |H| I_{m-2}) = (I_m |Q_{+2}| I_{m-2}) \sum_\alpha \frac{q_\alpha}{(r^{(\alpha)})^5} C_2 B^{(\alpha)}_{-2} \tag{41.7}$$

and

$$(I_m |Q_{+2}| I_{m-2}) = \frac{3eQ}{I(2I-1)} [(I-m+2)(I+m-1)(I-m+1)(I+m)]^{\frac{1}{2}}$$

$$(I_m |Q_{-2}| I_{m+2}) = \frac{3eQ}{I(2I-1)} [(I+m+2)(I-m-1)(I+m+1)(I-m)]^{\frac{1}{2}}.$$

The presence of an ultrasonic wave is taken into account through the displacements of the lattice points. These displacements are described by:

$$s = s_0 \cos(kx - \delta) \cos \omega t, \tag{41.8}$$

where $k = 2\pi/\lambda$, λ is the wavelength of the ultrasonic wave, s_0 its amplitude and δ is an arbitrary phase factor. The displacement of the nucleus is taken as:

$$s_1^{(1)} = s_{01} \cos(ka - \delta) \cos \omega t, \tag{41.9}$$

where a is the equilibrium lattice constant. After expanding $B^{(\alpha)}_{\pm 2}/(r^{(\alpha)})^5$ in the relative displacements of the charge q_α and retaining the appropriate terms, the matrix

---

[1] A and B have five linearly independent components of the form A_i where $i = 0, \pm 1, \pm 2$; $A_0 = r^2 Y_2^0$; $A_{\pm 1} = r^2 Y_2^{\pm 1}$; $A_{\pm 2} = r^2 Y_2^{\pm 2}$, where Y_2^m are the unnormalized spherical harmonics of degree two.

elements (41.6) and (41.7) become:

$$(I_m | H | I_{m+2})$$
$$= \frac{-e^2 Q \gamma}{8 I (2I-1)} [(I+m+2)(I-m-1)(I+m+1)(I-m)]^{\frac{1}{2}} \frac{18}{a^3} s_{01} k \sin \delta \cos \omega t, \Big\} \quad (41.10)$$

$$(I_m | H | I_{m-2}) =$$
$$- \frac{e^2 Q \gamma}{8 I (2I-1)} [(I-m+2)(I+m-1)(I-m+1)(I+m)]^{\frac{1}{2}} s_{01} k \sin \delta \cos \omega t \Big\} \quad (41.11)$$

where the substitution $q_\alpha = \gamma e$ was made for all α; γ is a constant and e is the electronic charge.

If the nucleus is initially in the state m, the probability that in time t the nucleus will make a transition to one of the states k, is given from first-order perturbation theory as:

$$W_{km} = \frac{|H^0|^2}{\hbar} g(\omega) \frac{\pi}{2} t \qquad (41.12)$$

where H^0 is time independent and $g(\omega)$ is the normalized shape function of the nuclear resonance line. Averaging over the frequency spread $\delta \omega$ of the resonance line, the average transition probability per unit time will be given by

$$W = \left(\frac{W_{km}}{t} \right)_{\text{Ave}} = \frac{|H^0_{km}|^2}{4 \hbar^2 \delta \nu} . \qquad (41.13)$$

Finally, after substitution of the matrix element H^0_{km} from Eqs. (41.10) and (41.11), the transition probabilities per unit time for $\Delta m = \pm 2$ are given respectively by:

$$W_{m,m-2} = \frac{e^4 Q^2 \gamma^2}{I^2 (2I-1)^2} (I-m+2)(I-m-1)(I+m+1)(I-m) \frac{(s_{01})^2 k^2}{a^6} \sin^2 \delta, \quad (41.14)$$

$$W_{m,m+2} = \frac{81}{64 \hbar^2} \frac{e^4 Q^2 \gamma^2}{I^2 (2I-1)^2} (I-m+2)(I-m-1)(I-m+1)(I+m) \frac{(s_{01})^2 k^2}{a^6} \sin^2 \delta. \quad (41.15)$$

42. Changes in nuclear magnetization. Kraus and Tantilla [142] have also considered the equilibrium magnetization of nuclei, approached in the presence of ultrasonic vibrations at the frequency of the spin transitions; this magnetization is less than the magnetization in the presence of thermal vibrations alone [138]. This treatment will now be summarized. The total time rate of change of the m_z component of the macroscopic magnetization due to the nuclear spin system can be expressed by:

$$\frac{dm_z}{dt} = \frac{dm_z}{dt}\Big|_{\Delta m = \pm 1} + \frac{dm_z}{dt}\Big|_{\Delta m = \pm 2} + \frac{dm_z}{dt}\Big|_{\text{ultrasonic}} \qquad (42.1)$$

where

$$\frac{dm_z}{dt}\Big|_{\Delta m = +1} = \frac{6(m_{z_0} - m_z)}{I(I+1)(2I+1)} \sum_{m=I}^{-I+1} \omega^m_{m-1}. \qquad (42.2)$$

Here $m_{z_0} = \frac{1}{3} \gamma \hbar I(I+1) n (\hbar \omega / kT_l)$ represents the value of m_z when the spin system is in equilibrium with the lattice (T_l is the lattice temperature and k is Boltzmann's constant).

ω^m_{m-1} is the probability per unit time of a thermally induced transition from the state m to the state $m-1$;

$$\omega^m_{m-1} = \frac{9 e^2 Q^2}{4 I^2 (2I-1)^2} (2m-1)^2 (I+m)(I-m+1) F(l), \qquad (42.3)$$

where $F(l)$ is a function of the lattice coordinates only.

A similar expression is obtained for $\dfrac{d m_z}{dt}\bigg|_{\Delta m=\pm 2}$

$$\frac{d m_z}{dt}\bigg|_{\Delta m=\pm 2} = \frac{24\,(m_{z_0} - m_z)}{I(I+1)(2I+1)} \sum_{m=I}^{-I+2} \omega_{m-2}^m, \tag{42.4}$$

where

$$\omega_{m-2}^m = \frac{9 e^2\,Q^2}{I^2\,(2I-1)^2}\,(I+m)\,(I-m+1)\,(I+m-1)\,(I-m+2)\,G(l). \tag{42.5}$$

Here again $G(l)$ is a function of lattice coordinates only.

Assuming that the ultrasonic frequency is that corresponding to $\Delta m \pm 2$ transitions, one obtains:

$$\frac{d m_z}{dt}\bigg|_{\text{ultrasonic}} = \frac{-24\,m_z}{I(I+1)(2I+1)} \sum_{m=I}^{-I+2} W_{m-2}^m \tag{42.6}$$

where W_{m-2}^m is given by Eq. (41.14).

The time rate of change of m_z can also be expressed in terms of relaxation times T', T'' and T_u, as follows:

$$\frac{1}{T'} = \frac{6}{I(I+1)(2I+1)} \sum_{m=1}^{-I} \omega_{m-1}^m, \tag{42.7}$$

$$\frac{1}{T''} = \frac{24}{I(I+1)(2I+1)} \sum_{m=I}^{-I+2} \omega_{m-2}^m, \tag{42.8}$$

$$\frac{1}{T_u} = \frac{24}{I(I+1)(2I+1)} \sum_{m=I}^{-I+2} W_{m-2}^m, \tag{42.9}$$

$$\frac{1}{T_1} = \frac{1}{T'} + \frac{1}{T''}. \tag{42.10}$$

Following these definitions, Eq. (42.1) may be written:

$$\frac{d m_z}{dt} = \frac{m_0 - m_z}{T_1} - \frac{m_z}{T_u} \tag{42.11}$$

where T_1 is the observed spin-lattice relaxation time. The ultrasonic relaxation time can be computed from Eqs. (41.14) and (42.9).

If a constant ultrasonic excitation is applied for a period of time much longer than T_1, the magnetization density is constant in a region where the phase δ is constant, that is $d m_z/dt = 0$ and

$$m_z = \frac{m_{z_0}}{1 + (T_1/T_u)} = \frac{m_{z_0}}{1 + \beta \sin^2 \delta} \tag{42.12}$$

where

$$\beta = \frac{T_1}{\sin^2 \delta}\,\frac{24}{I(I+1)(2I+1)} \sum_{m=I}^{-I+2} W_{m-2}^m. \tag{42.13}$$

The total magnetization is obtained by integrating Eq. (42.13) over the volume of the crystal; for a uniform cylinder with a wave in the x direction $\delta = \dfrac{2\pi x}{\lambda}$, and:

$$M_z = \int m_z\,d\tau = m_{z_0} \int \frac{A\,dx}{1 + \sin^2\left(\dfrac{2\pi x}{\lambda}\right)} = \frac{M_{z_0}}{(1+\beta)^{\frac{1}{2}}} \tag{42.14}$$

where A is the cross-sectional area of the cylinder and M_{z_0} is the total magnetization of the nuclei in the absence of ultrasonic radiation.

43. Experimental approach. The experimental technique used by Proctor and co-workers [137] to [139] will be described briefly[1]. This technique is based on a pulsed nuclear induction method of measuring nuclear magnetization.

When a short duration pulse of magnetic field is applied to a crystal at the transition frequency, the transient nuclear induction signal is proportional to the static magnetization of the specimen, which is a function of the population of the various nuclear states. Thus the amplitude of the transient nuclear induction signal A_1 (see Fig. 42), following the pulse of high frequency flux, is proportional to the population difference between the quadrupole states in thermal equilibrium at the temperature of the specimen. Following the pulse, the newly created population difference Δ will attempt to return to its equilibrium value Δ_e, at a rate determined by the relaxation time T_1 which characterizes the spin-lattice interactions. If the time interval between successive pulses is longer than T_1, and the population difference Δ is zero following the application of the first pulse[2], $\Delta(0)=0$ the rate of approach of Δ to its equilibrium value Δ_e may be described by the relation:

$$\frac{\Delta(t)}{\Delta_e} = (1 - e^{-t/T_1}). \qquad (43.1)$$

If A_2 is the amplitude of the transient nuclear induction signal after a second pulse of high frequency flux (again of duration τ)[2] is applied after a time t (see Fig. 42), then:

$$\frac{A_2}{A_1} = (1 - e^{-t/T_1}). \qquad (43.2)$$

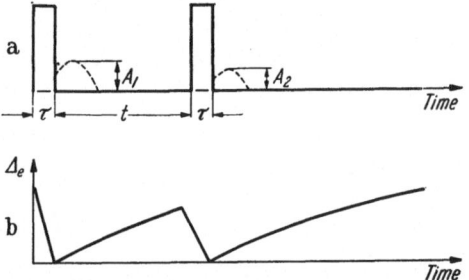

Fig. 42a and b. (a) Following the first strong radio frequency pulse of duration τ there is a transient nuclear induction signal of amplitude A_1 proportional to the population difference between two states at thermal equilibrium. After a time t a second pulse of the same duration is applied and the induction signal of amplitude A_2 appears. (b) The instantaneous population difference over the same interval as (a). Δ_e indicates the equilibrium population difference at the beginning of the radio frequency pulse. After Proctor and Tantilla [138].

The measured ratio A_2/A_1, as a function of the time interval t, yields the value of T_1 from a plot of log $[1-(A_2/A_1)]$ as a function of t.

The next step consists of measuring the influence of an ultrasonic excitation applied at the appropriate transition frequency. If ξ_0 and ξ represent the amplitudes of the nuclear induction signal, respectively in the absence and in the presence of ultrasonic excitation, their ratio can be represented by:

$$\left(\frac{\xi_0}{\xi}\right)^2 = 1 + k_1 V^2 \qquad (43.3)$$

where V is the peak voltage applied to the transducer.

The value of k_1 can be determined experimentally from the slope of the curve of $(\xi_0/\xi)^2$ versus V^2. The product $k_1 V^2$ can be expressed by putting $M_{z_0}=\xi_0$, $M_z=\xi$ and $dM_z/dt=0$ in Eq. (42.14) and combining with Eqs. (41.14), (41.15), and (42.13), the product $k_1 V^2$ can be expressed as follows:

$$k_1 V^2 = \frac{243}{8 a^6 \hbar^2 \delta\nu} \frac{e^4 Q^2 \gamma^2 A^2 k^2}{I^3 (I+1)(2I+1)(2I-1)^2} T_1 (I-m+2)(I+m-1)(I-m+1)(I+m). \qquad (43.4)$$

Here A is the amplitude of ultrasonic vibration and $\delta\nu$ is the ultrasonic line width (the other terms were defined before). From expression (43.4) the para-

---

[1] The results obtained by Proctor and co-workers have been discussed by Verma [143].
[2] The pulse duration τ is adjusted to produce the maximum induction transient amplitude, which occurs when the expectation value of the population excess, Δ, is zero.

meter γ can be determined, provided the amplitude of the ultrasonic vibration is known. This amplitude can be evaluated, in principle, from the power used to drive the ultrasonic transducer, and the attenuation coefficient of ultrasonic waves in the specimen. Unfortunately such evaluations are always difficult and frequently unreliable. This difficulty was circumvented by JENNINGS, TANTILLA and KRAUS [144] in their experiments on sodium iodide, at the cost of obtaining information about the absolute value of quadrupole coupling between a sodium or iodine nucleus and the lattice. These authors determined separately the magnetization for sodium and iodine nuclei, in the same specimen, as a function of ultrasonic power transmitted to the crystal. Thus, they found a value for γ_I/γ_{Na}; this ratio does not depend on an explicit knowledge of the ultrasonic energy density, which is the same for measurements on both nuclei.

IX. The effect of ultrasonic stress waves on electron paramagnetic resonance in solids.

44. Stress waves and electron spin level transitions. The use of paramagnetic centers as detectors of ultrasonic radiation at microwave frequencies has been discussed by C. KITTEL [145]. Experimental evidence for the interaction between high frequency ultrasonic waves and electron spins has been shown by observing changes induced in an electron spin resonance line [127], [146], [147].

Paramagnetic resonance is observed as the result of the transition of electrons, induced by a microwave field, from one energy level to another where the levels in question arise from splitting in the presence of a constant magnetic field.

In a magnetic field H the energy levels are raised and lowered in energy by an amount $\mu \cdot H$ where μ is the magnetic moment. The total separation is then, in the elementary picture, twice this amount or $\hbar \omega = \Delta E = g\left(\dfrac{e\hbar}{2mc}\right)H$ where for an electron spin the g factor is equal to 2 and $\dfrac{e\hbar}{2mc} = -9.27 \times 10^{-20}$ ergs/gauss. One may observe the microwave energy absorption during the energy level transition when the resonance condition is satisfied. Such transitions can be induced either by ultrasonic waves or electromagnetic waves of proper frequency. This microwave absorption is altered by the presence of ultrasonic waves in the solid.

Where only two levels are considered the population of electrons in thermal equilibrium in each level is given by [148]

$$\frac{N_1^0}{N} = \frac{e^{\mu H/kT}}{e^{\mu H/kT} + e^{-\mu H/kT}}, \tag{44.1}$$

$$\frac{N_2^0}{N} = \frac{e^{-\mu H/kT}}{e^{\mu H/kT} + e^{-\mu H/kT}}, \tag{44.2}$$

where N_1^0 and N_2^0 are the lower and upper level populations and $N = N_1^0 + N_2^0$.

If $\Delta N = N_2 - N_1$ is the excess spin population of level 2 over level 1 at any time, and ΔN_0 is the difference for thermal equilibrium, the transition rate may be expressed [145] as

$$\frac{d(\Delta N)}{dt} = \frac{\Delta N^0 - \Delta N}{\tau_1} - W \Delta N \tag{44.3}$$

where τ_1 is spin lattice relaxation time and where W is the transition rate from level 1 to level 2 for phonon induced transitions. An expression for W is given by KITTEL [145]

$$W = \frac{2\pi}{h^2} |G|^2 \varepsilon_{ij}^2 \tau_2. \tag{44.4}$$

τ_2 is the spin-spin relaxation time and G is a coupling constant in the strain part of the Hamiltonian $H^1 = G\, S_i\, S_j\, \varepsilon_{ij}$ where S is, the spin of a given paramagnetic center in units of $\hbar$; ε_{ij} is a strain component.

Where $h\nu \ll kT$ the strain-spin interaction term H^1 leads to the spin-lattice (direct) relaxation time τ_1 assuming thermal equilibrium in the phonon distribution

$$\tau_1 \cong \frac{\varrho\, v^5\, \hbar^2}{|G|^2 \omega^2\, kT} \approx \frac{1}{10^{32}\, |G|^2\, T}\,. \tag{44.5}$$

T is the temperature, v is the sound velocity and ϱ is the material density. From this expression it can be estimated that G must be of the order of 3×10^{-14} ergs. From relation (44.3) one can obtain [145] a steady state solution of the form

$$\Delta N = \Delta N^0 \left(1 + \frac{2\pi |G|^2\, \varepsilon_{ij}^2\, \tau_1\, \tau_2}{\hbar^2}\right)^{-1} \tag{44.6}$$

which shows that saturation effects become important when

$$\varepsilon_{ij}^2 \approx \frac{\hbar^2}{2\pi\, \tau_1\, \tau_2\, |G|^2} \tag{44.7}$$

using $\tau_1 = 10^{-5}$, $\tau_2 = 10^{-6}$, $|G|^2 \approx 10^{-27}$ one finds that $\varepsilon_{ij} \approx 3 \times 10^{-9}$ and this value of strain corresponds to a phonon energy flux of about 0.1 μwatts/cm².

45. Experimental results. The effect of 9200 Mc/sec ultrasonic stress waves on electron spin resonance of manganese impurity in a quartz [146], [147] single crystal has been shown by Jacobsen, Shiren and Tucker.

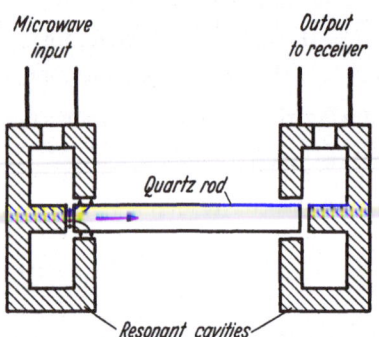

Fig. 43. Two cavity arrangement for microwave ultrasonic experiments. After Jacobsen [127] and Bömmel and Dransfeld [124], [125].

The paramagnetic resonance absorption line was observed at one end of a quartz rod. Ultrasonic stress waves were introduced at the other end and the effect on the resonance line was measured as a function of the power and frequency of the ultrasonic waves.

The experiments were carried out in the neighbourhood of 2° K. Two resonant cavities and the quartz sample were arranged roughly as shown in Fig. 43. The quartz crystal was 0.3 cm in diameter and 3 cm long and extended between the two cavities one of which was the driving cavity; the other was the receiving cavity which formed part of the paramagnetic resonance spectrometer.

Ultrasonic waves, generated at the end of the quartz rod in the driving cavity [125], [149], [150], interact with the spins of the manganese impurities and this interaction is defected in the second cavity (the spectrometer cavity).

Fig. 44 shows some of the results [146] of Jacobsen, Shiren and Tucker. Two effects are dominant. The effect of increasing ultrasonic power, with other quantities held constant, is that of decreasing the spin resonance signal amplitude. The second effect is that of a shift in the position of the peak of Fig. 44 with increasing ultrasonic power. The peak shifts toward higher spectrometer power with increasing ultrasonic power.

By varying the ultrasonic frequency, with the spectrometer frequency held constant, the bandwidth was obtained for the interaction of the ultrasonic waves with the electron spins. A result for one of the manganese resonance lines is shown in Fig. 45 from [*146*] JACOBSEN, SHIREN and TUCKER. The ultrasonic line width varied from line to line and was found to be less than the microwave line width for the same line.

The results were somewhat different with paramagnetic centers obtained from the irradiation of quartz by neutrons, electrons, or x rays. As with the manganese centers increasing the ultrasonic power decreased the signal amplitude, but the observed ultrasonic bandwidth was considerably wider (more than ± 500 Mc/sec) than in the case with the manganese centers. At present there appears to be no satisfactory way of interpreting these

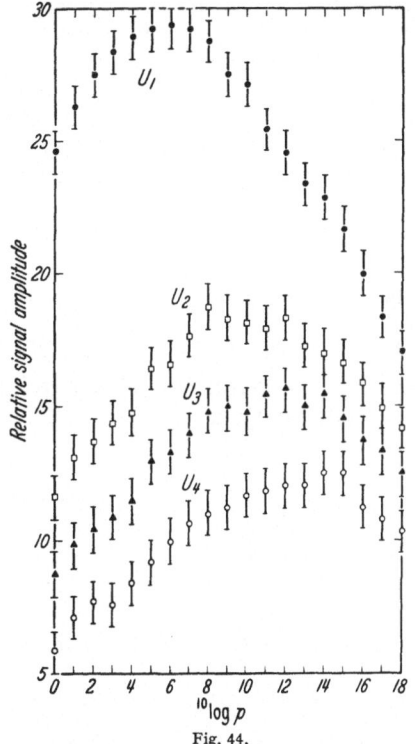

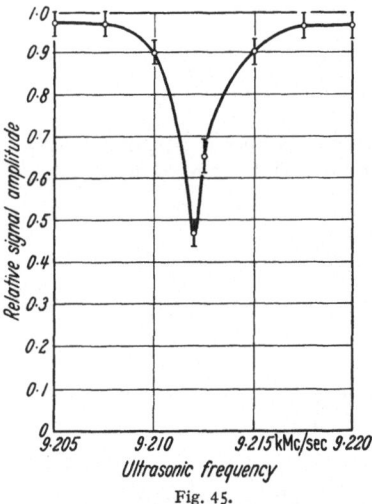

Fig. 44. Derivative of absorption as a function of the logarithm of spectrometer power P for several relative values of ultrasonic power U for a line due to manganese in quartz. $u_4 > u_3 > u_2 > u_1$. After JACOBSEN, SHIREN and TUCKER [*146*].

Fig. 45. Amplitude of derivative of absorption as a function of ultrasonic frequency with spectrometer frequency, magnet field ultrasonic power, and spectrometer power all held constant. The data are for the same line of manganese in quartz as in Fig. 42. After JACOBSEN, SHIREN and TUCKER [*146*].

bandwidths generally, and in particular the large ultrasonic bandwidths (>1000 Mc/sec) of the radiation induced centers are not understood at all.

X. Stress waves and electrical phenomena in piezoelectric crystals.

46. Phenomenological treatment. In any piezoelectric solid electromagnetic waves accompany stress waves and vice versa. A stress wave propagating in a piezoelectric crystal may be accompanied by oscillating electric fields, which lead to an increase in the effective values of the relevant elastic constants; this effect may be described as a "stiffening" of the crystal. In general, the coupling between the electromagnetic and elastic modes will cause three stress waves and two electromagnetic waves to propagate.

When a crystal is both piezoelectric and semiconducting (for example CdS), attenuation and dispersion of the stress waves may result from their interaction with the currents and space charge produced by the internal electric fields mentioned above. The contribution of the effects of these electric fields to the value of the elastic constants in a piezoelectric solid has been treated by Kyame [154], Koga et al. [155], and Pailloux [156].

Analysis of wave propagation for piezoelectric materials involves the combined use of the elastic wave equations and Maxwell's equations. One can begin such an analysis by considering the thermodynamic relations for the energy of a crystal under stress. The stress-strain relation contains an additional stress term caused by the electric field E arising from the applied stress [1]

$$\sigma_{ij} = c_{ijkl}\,\varepsilon_{kl} - e_{ijr}\,E_r. \tag{46.1}$$

Where the e_{ijr} are piezoelectric constants relating electric field to stress. The expression for the electric displacement in terms of strain and electric field is

$$D_j = e_{jkl}\,\varepsilon_{kl} + p_{jr}\,E_r, \tag{46.2}$$

where the p_{rj} are the components of the dielectric tensor. In addition the magnetic induction $\boldsymbol{B}$ is assumed to be related to the magnetic field H

$$\boldsymbol{B} = \mu\,\boldsymbol{H} \tag{46.3}$$

assuming isotropic magnetic behavior. The current density is related to the electric field by the components b_{ki} of the conductivity tensor

$$J_k = b_{ki}\,E_i. \tag{46.4}$$

Using Eq. (3.1)

$$\sigma_{ij,j} = \varrho\,\ddot{s}_i$$

together with Eq. (46.1) one finds the equation of motion,

$$\sigma_{ij,j} = c_{ijkl}\,\varepsilon_{kl,j} - e_{ijr}\,E_{r,j} = \varrho\,\ddot{s}_i, \tag{46.5}$$

and in the manner used to obtain the Eq. (1.11) one finds, in terms of the displacements,

$$c_{ijkl}\,s_{l,kj} - e_{ijr}\,E_{r,j} = \varrho\,\ddot{s}_i. \tag{46.6}$$

From the electromagnetic field equations one obtains the wave equation

$$\nabla^2\,\boldsymbol{E} = \mu\,\ddot{\boldsymbol{D}} + \mu\,\dot{\boldsymbol{J}}. \tag{46.7}$$

Consider a plane wave propagating in the x_1 direction so that

$$\frac{\partial}{\partial x_2} = \frac{\partial}{\partial x_3} = 0.$$

Then from (46.7), (46.4) and (46.2)

$$\frac{\partial^2 E_q}{\partial x_1^2} = \mu\left[e_{qkl}\,\ddot{\varepsilon}_{kl} + p_{qr}\,\ddot{E}_r + b_{qr}\,\dot{E}_r\right]$$

or

$$\frac{\partial^2 E_q}{\partial x_1^2} = \mu\left[e_{qkl}\,\frac{\partial^3 s_l}{\partial x_k\,\partial t^2} + p_{qr}\,\ddot{E}_r + b_{qr}\,\dot{E}_r\right]. \tag{46.8}$$

For the special plane wave case Eqs. (46.6) and (46.8) become

$$\varrho\,\ddot{s}_i = c_{i11l}\,\frac{\partial^2 s_l}{\partial x_1^2} - e_{i1r}\,\frac{\partial E_r}{\partial x_1}, \tag{46.9}$$

$$\frac{\partial^2 E_q}{\partial x_1^2} = \mu\left[e_{q1l}\,\frac{\partial^3 s_l}{\partial x_1\,\partial t^2} + p_{qr}\,\ddot{E}_r + b_{qr}\,\dot{E}_r\right]. \tag{46.10}$$

Now

$$(\operatorname{curl} H)_1 = 0 = \left(\frac{\partial H_3}{\partial x_2} - \frac{\partial H_2}{\partial x_3}\right) = \dot{D}_1 + \dot{J}_1.$$

Hence,

$$e_{11l}\frac{\partial^2 s_l}{\partial x_1 \partial t} + p_{1r}\dot{E}_r + b_{1r}E_r = 0, \tag{46.11}$$

and this relation will be used to eliminate E_1 from Eqs. (46.9) and (46.10). Plane wave solutions of the form

$$s_l = s_{0l}\, e^{i(\omega t - k\, x_1)}, \tag{46.12}$$

$$E_l = E_{0l}\, e^{i(\omega t - k\, x_1)} \tag{46.13}$$

are assumed, in which case Eq. (46.6) becomes

$$\omega^2\varrho\, s_i = k^2\, c_{i11l}\, s_l - i\, k\, e_{i1l}\, E_l \tag{46.14}$$

and Eq. (46.10) yields

$$k^2\, E_q = -i\,\mu\,\omega^2\, k\, e_{q1l}\, s_l + \mu\,\omega^2\, p_{ql}\, E_l - i\,\omega\,\mu\, b_{ql}\, E_l. \tag{46.15}$$

From Eq. (46.11) it is found that

$$E_1 = -\left\{\frac{\omega\, k\, e_{11k}\, s_k + (b_{1l} + i\,\omega\, p_{1l})\, E_l}{(b_{11} + i\,\omega\, p_{11})}\right\} \tag{46.16}$$

where

$$k = 1, 2, 3 \quad \text{and} \quad l = 2, 3.$$

Substitution of (46.16) in (46.14) yields the result

$$\omega^2\varrho\, S_i = k^2\left[c_{i11k} + \frac{e_{i11}\, e_{11k}}{\left(p_{11} - i\,\dfrac{b_{11}}{\omega}\right)}\right] S_k - i\, k\left[e_{i1l} - \frac{e_{i11}\left(p_{1l} - i\,\dfrac{b_{1l}}{\omega}\right)}{\left(p_{11} - i\,\dfrac{b_{11}}{\omega}\right)}\right] E_l \tag{46.17}$$

or

$$\omega^2\varrho\, s_i = k^2\, c^1_{i11k}\, S_k - i\, k\, e^1_{i1l}\, E_l \qquad i,\, k = 1, 2, 3 \quad \text{and} \quad l = 2, 3 \tag{46.18}$$

where c^1_{i11k} and e^1_{i1l} are given by identification of (46.18) with (46.17).

Substitution of (46.16) in (46.15) leads to

$$k^2\, E_q = -i\,\mu\,\omega^2\, k\left[e_{q1k} - \frac{e_{11k}\left(p_{q1} - i\,\dfrac{b_{q1}}{\omega}\right)}{\left(p_{11} - i\,\dfrac{b_{11}}{\omega}\right)}\right] s_k +$$

$$+ \mu\,\omega^2\left[p_{ql} - \frac{p_{q1}\left(p_{1l} - i\,\dfrac{b_{1l}}{\omega}\right)}{\left(p_{11} - i\,\dfrac{b_{11}}{\omega}\right)}\right] E_l -$$

$$- i\,\mu\,\omega\left[b_{ql} - \frac{b_{q1}\left(p_{1l} - i\,\dfrac{b_{1l}}{\omega}\right)}{\left(p_{11} - i\,\dfrac{b_{11}}{\omega}\right)}\right] E_l \tag{46.19}$$

or

$$k^2\, E_q = -i\,\mu\,\omega^2\, k\, e^1_{q1k}\, s_k + \mu\,\omega^2\left(p^1_{ql} - i\,\frac{b^1_{ql}}{\omega}\right) E_l \tag{46.20}$$

where expressions for e^1_{q1k}, p^1_{ql} and b^1_{ql} are found by comparison of (46.20) with (46.19).

Eqs. (46.20) and (46.18) specify the combined electromagnetic-elastic behavior of the piezoelectric material. The existence of the solutions of the five equations [three from (46.18) and two from (46.20)] depends on the vanishing of the determinant of the coefficients

$$
\begin{vmatrix}
(\omega^2 \varrho - k^2 c_{1111}^1) & -k^2 c_{1112}^1 & -k^2 c_{1113}^1 & i k e_{112}^1 & i k e_{113}^1 \\
-k^2 c_{2111}^1 & (\omega^2 \varrho - k^2 c_{2112}^1) & -k^2 c_{2113}^1 & i k e_{212}^1 & i k e_{213}^1 \\
-k^2 c_{3111}^1 & -k^2 c_{3112}^1 & (\omega^2 \varrho - k^2 c_{3113}^1) & i k e_{312}^1 & i k e_{313}^1 \\
i \mu \omega^2 k e_{211}^1 & i \mu \omega^2 k e_{212}^1 & i \mu \omega^2 k e_{213}^1 & (k^2 - \mu \omega^2 p_{22}^1 + i \mu \omega b_{22}^1) & -(\mu \omega^2 p_{23}^1 - i \mu \omega b_{23}^1) \\
i \mu \omega^2 k e_{311}^1 & i \mu \omega^2 k e_{312}^1 & i \mu \omega^2 k e_{313}^1 & -(\mu \omega^2 p_{32}^1 - i \mu \omega b_{32}^1) & (k^2 - \mu \omega^2 p_{33}^1 + i \mu \omega b_{33}^1)
\end{vmatrix}
\qquad (46.21)
$$

The vanishing of this determinant leads to a fifth degree equation in k^2 or five possible values of the phase velocities for the propagation of the two electromagnetic waves and the three elastic waves. The coefficients in the determinant are complex and frequency dependent; consequently the values of k or the corresponding phase velocities will also be complex and frequency dependent. In other words the elastic and electromagnetic waves will propagate with attenuation and dispersion. The attenuation and velocity effects are large under conditions such as those which exist for example in cadmium sulphide.

47. Wave propagation in piezoelectric semiconductors. On the basis of the preceding treatment, Hutson and White [157] derived relations for the attenuation and dispersion of a plane stress wave in a piezoelectric semiconductor; these relations are based on a one dimensional approximation where E produces an electric field in the x_1 direction. A summary of their results follows.

The equations of state corresponding to the one-dimensional problem are

$$
\sigma = c \varepsilon - e E, \qquad (47.1)
$$

$$
D = e \varepsilon + p E. \qquad (47.2)
$$

Expressing E in terms of ε and differentiating (47.1) with respect to x leads to a wave equation. If D is assumed constant this equation takes the form

$$
\varrho \frac{\partial^2 s}{\partial t^2} = c \left(1 + \frac{e^2}{c p} \right) \frac{\partial^2 s}{\partial x^2} \qquad (47.3)
$$

where s is the displacement. The change in c due to the presence of the electric field is thus obvious. The condition of constant D further leads to a zero space charge density through Poisson's equation

$$
\frac{\partial D}{\partial x} = Q \qquad (47.4)
$$

while the continuity relation

$$
\frac{\partial J}{\partial x} = - \frac{\partial Q}{\partial t} \qquad (47.5)
$$

where J is the current, indicates that in this case the varying current density due to the piezoelectric fields is zero, which corresponds to a very low conductivity in the medium.

In the case of very high conductivity the field E accompanying the wave will be zero and the elastic constant c will remain unaffected [Eq. (47.1)] while the stress wave will be accompanied by D fields, currents and varying space charge.

The case of specific interest here is that corresponding to intermediate values of conductivity, in the range encountered in semiconductors. In this range, Eqs. (47.4) and (47.5) are used, together with an appropriate expression for J, to obtain values of D and E. These, in turn, permit one to eliminate E from the wave equation.

For an extrinsic semiconductor (assumed to be n type), the current density may be expressed by:

$$J = q\,(n + f\,n_s)\,\mu\,E + (\mu\,kT)\,f\,\frac{\partial n_s}{\partial x}$$ (47.6)

where the first term is due to drift and the second term is due to diffusion; q is the electronic charge, k is BOLTZMANN'S constant, T is the temperature, n is the mean density of electrons in the conduction band and f is the fraction of acoustically produced space charge density n_s which is mobile[1]. Thus $(n + f\,n_s)$ is the instantaneous local density of electrons in the conduction band. Eqs. (47.4) to (47.6) combined with plane wave representations of D and E, Eqs. (46.12) and (46.13), lead to the following expression for D:

$$D = \frac{-i\left(\dfrac{nq\mu}{\omega}\right)E}{1 + i\omega\left(\dfrac{k}{\omega}\right)^2\dfrac{\mu f kT}{q}}\,.$$ (47.7)

In the case of small conductivity modulation $(f\,n_s \ll n)$ Eq. (47.7) may be further simplified and written in the form

$$D = \frac{-i\left(\dfrac{b}{\omega}\right)E}{\left[1 + i\omega\left(\dfrac{k}{\omega}\right)^2\dfrac{\mu f kT}{q}\right]}$$ (47.8)

where $b = n\,q\,\mu$ represents the average conductivity.

The condition of small conductivity modulation $(f\,n_s \ll n)$ is satisfied when the effective drift velocity of the carriers in the piezoelectric field $f\,\mu\,E$, is much less than the velocity of the stress wave v. This imposes a limitation on the strain value:

$$\varepsilon \ll \frac{p\,v}{f\,\mu\,e}\,.$$

In order to determine the attenuation and dispersion of stress waves, use is made of the conductivity frequency, defined by $\omega_C = b/p$ and the diffusion frequency, defined by[2]

$$\omega_D = \frac{q}{f\,\mu\,kT}\left(\frac{\omega}{k}\right)^2 \approx \left(\frac{q}{f\,\mu\,kT}\right)v^2\,.$$

From Eqs. (47.1), (47.2) and (47.8) one obtains:

$$E = \frac{-e\,\varepsilon}{p}\left[\frac{1 + i\left(\dfrac{\omega}{\omega_D}\right)}{1 + i\left(\dfrac{\omega}{\omega_D}\right) + i\left(\dfrac{\omega_C}{\omega}\right)}\right]\,.$$ (47.9)

---

[1] For an extrinsic semiconductor in thermal equilibrium, the total space charge density Q may be expressed in terms of the energy levels and densities of the impurity states in the forbidden band, and the concentration of electrons in the conduction band. The condition of electrical neutrality corresponds to $Q = 0$, and the acoustically produced space charge is the periodic variation of Q about zero.

[2] ω_D is the frequency above which the wave length is sufficiently short for diffusion to smooth out carrier density fluctuations with the periodicity of the stress wave.

In the case of negligible charge carrier diffusion ($\omega_D \gg \omega_C, \omega$), Eq. (47.9) may be simplified to:

$$E = \frac{-e\,\varepsilon}{p} \, \frac{1 - i\left(\dfrac{\omega_C}{\omega}\right)}{1 + \left(\dfrac{\omega_C}{\omega}\right)^2} \tag{47.10}$$

and the effective elastic constant is obtained by substitution into (47.1)

$$\sigma = c\left[1 + \frac{e^2}{c\,p} \, \frac{1 - i\left(\dfrac{\omega_C}{\omega}\right)}{1 + \left(\dfrac{\omega_C}{\omega}\right)^2}\right]\varepsilon. \tag{47.11}$$

The velocity and attenuation are obtained in terms of the real and imaginary parts of the complex elastic constant[1]

$$v = v_0\left[1 + \frac{\dfrac{e^2}{2c\,p}}{1 + \left(\dfrac{\omega_C}{\omega}\right)^2}\right], \tag{47.12}$$

$$\alpha = \frac{\omega}{v_0} \, \frac{e^2}{2c\,p}\left[\frac{\dfrac{\omega_C}{\omega}}{1 + \left(\dfrac{\omega_C}{\omega}\right)^2}\right]. \tag{47.13}$$

These expressions show that at very low frequencies v tends to v_0 and α tends to zero, while in the high frequency limit they become:

$$v = v_\infty \approx v_0\left[1 + \frac{e^2}{2c\,p}\right], \tag{47.14}$$

$$\alpha = \alpha_\infty = \frac{\omega_C}{v_0} \, \frac{e^2}{2c\,p}. \tag{47.15}$$

In the vicinity of $\omega = \omega_C$ a simple relaxation type dispersion occurs. It should be emphasized that the relaxation frequency is given by the conductivity of the material.

When carrier diffusion is taken into account, the complete expressions for velocity and attenuation become:

$$v = v_0\left[1 + \frac{e^2}{2c\,p} \, \frac{1 + \dfrac{\omega_C}{\omega_D} + \left(\dfrac{\omega}{\omega_D}\right)^2}{1 + 2\dfrac{\omega_C}{\omega_D} + \left(\dfrac{\omega}{\omega_D}\right)^2 + \left(\dfrac{\omega_C}{\omega}\right)^2}\right], \tag{47.16}$$

$$\alpha = \frac{\omega}{v_0} \, \frac{e^2}{2c\,p}\left[\frac{\dfrac{\omega_C}{\omega}}{1 + 2\dfrac{\omega_C}{\omega_D} + \left(\dfrac{\omega}{\omega_D}\right)^2 + \left(\dfrac{\omega_C}{\omega}\right)^2}\right]. \tag{47.17}$$

In this case, for $\omega_D \gg \omega_C$, expressions (47.12) and (47.13) retain their validity for all frequencies, except that α approaches a constant value $(\omega_C/v_0)\,(e^2/2c\,p)$ in the frequency range between ω_C and ω_D, and drops to zero as ω becomes larger than ω_D. For $\omega_C \gg \omega_D$ the velocity and attenuation may be expressed by

$$v = v_0\left[1 + \frac{e^2}{2c\,p}\left(\frac{1 + \dfrac{\omega^2}{\omega_D\,\omega_C}}{2 + \dfrac{\omega^2}{\omega_D\,\omega_C} + \dfrac{\omega_D\,\omega_C}{\omega^2}}\right)\right] \tag{47.18}$$

---

[1] Expressions (47.11) and (47.12) are obtained on the assumption that $e^2/c\,p$ is small.

and

$$\alpha = \frac{\omega}{v_0} \frac{e^2}{2cp} \left(\frac{\dfrac{\omega_D}{\omega}}{2 + \dfrac{\omega^2}{\omega_D \omega_C} + \dfrac{\omega_D \omega_C}{\omega^2}} \right). \qquad (47.19)$$

In this case the maximum velocity change occurs at the frequency $\omega = (\omega_D \omega_C)^{\frac{1}{2}}$ whereas the frequency corresponding to maximum attenuation is $\omega = (\omega_D \omega_C / 3)^{\frac{1}{2}}$. Similar, but somewhat more complicated expressions can be derived for intrinsic semiconductors. The additional complication arises primarily form the necessity to deal with both electron and hole currents.

48. Light-sensitive ultrasonic attenuation in CdS. The combination of high photo-electric sensitivity and high piezoelectric sensitivity is possessed by a few materials such as cadmium sulphide. Ultrasonic attenuation changes in cadmium sulphide have been studied by GOBRECHT and BARTSCHAT [158]. These changes were studied as a function of concentration of various activators, as a function of the electromagnetic radiation absorbed, and as a function of temperature. Observations were made over the temperature range from 20 to $-180°$ C. The attenuation was measured at frequencies from about 1 to 7 Mc/sec. In general the larger the photoelectric effect the greater the attenuation encountered; this was determined by increasing the photosensitivity through the use of activators as well as simply going from the condition with light to that with no light incident on the crystal.

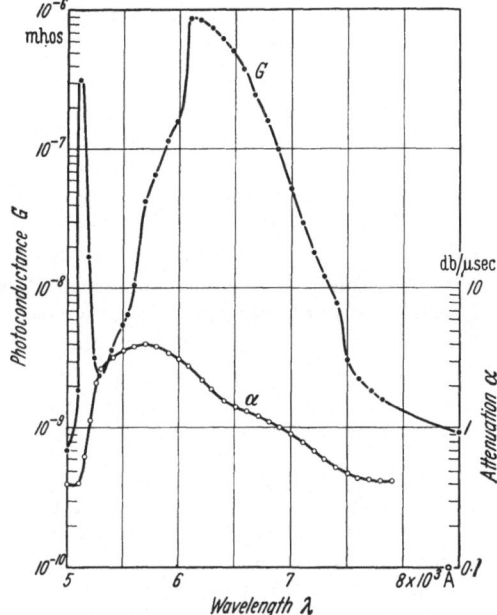

Fig. 46. Photoconductance G and ultrasonic attenuation α of a CdS crystal (type B) as a function of light wavelength used for illumination. Measurements made at an ultrasonic frequency of 45 Mc/sec; wavelength in angstroms. After NINE [159].

The attenuation increased with temperature, with the crystal in either light or dark condition. The attenuation also increased with larger concentration of activators; the activators have a large influence through their effect on increasing the photosensitivity. Exposure to infrared irradiation reduces the attenuation and tends to return the crystal to the dark condition (no light incident on the crystal). GOBRECHT and BARTSCHAT also showed that elastic stress waves affect the electrical conductivity of the crystal.

The results of attenuation and conductivity measurements [159], [160] on cadmium sulphide, at 45 Mc/sec, using illumination over a range of optical wavelengths from 5000 to 8500 Å, have the form shown in Fig. 46 (called type B behavior). In some instances however, instead of the large increase in attenuation as shown in Fig. 46, there is a small decrease in attenuation in the wave-length range shown (called: type A behavior). With type B behavior the magnitude of the attenuation change is much more than a factor of ten, in some cases, and when very intense light is used (room temperature) the attenuation change may be too large to be measurable with presently available techniques. There are also intermediate magnitudes of these effects between the two types of behavior indicated. The purity of the material is very important as indicated by the effect of the addition of activators. The remaining remarks have to do with crystals showing type B behavior.

In addition to the large attenuation changes between dark and light condition and the quenching effect of infra red radiation already mentioned, there is some interesting detail in the attenuation temperature behavior. The electrical conductivity was measured as well as attenuation, and there are peaks in the attenuation measurements which coincide rather well in position with peaks which also occur in the conductivity measurements.

Attenuation-temperature peaks agree in position with those from conductivity measurements. The amplitudes of the various peaks in the attenuation-temperature data depend strongly on the light excitation history of the sample.

The time rate of change of attenuation in going from the dark to the light condition of the sample is slower when the temperature is lower. The same is true of recovery from light condition back to dark condition when the light excitation is removed. Below 60° K these changes occur so slowly as to be almost undetectable. This behavior indicates the presence of a thermally activated process, the details of which have not yet been studied.

The values of attenuation measured in different crystallographic directions in cadmium sulphide have been shown to correlate with the magnitude of polarization effects to be expected from the piezoelectric properties of this compound.

It was suggested by Hutson [160] that these effects result from the relaxation interaction of mobile charge carriers with the longitudinal electric fields of piezoelectric origin, which accompany stress waves in CdS.

49. Ultrasonic amplification in CdS. On the basis of the considerations presented in Sect. 47, the possibility was pointed out [161] of achieving acoustic gain in a piezoelectric semiconductor, by applying a d.c. electric field pulse which causes the interacting charge carriers to drift, in the direction of wave propagation, faster than the stress wave velocity.

This prediction was confirmed experimentally by Hutson, McFee and White [101] who studied ultrasonic attenuation in CdS, in the presence of an external electric field pulse, with and without illumination of the crystal. Specifically, these investigators found that in the absence of illumination the external drift field pulse has no effect on the ultrasonic attenuation, whereas without a drift field, increasing illumination yields increasing attenuation and conductivity. (This behavior is consistent with the results of studies discussed in Sect. 48.)

When a drift field pulse, of suitable duration, is applied to an illuminated sample, variation of the drift field strength produces changes in the ultrasonic attenuation. For a certain range of field values the attenuation decreases with increasing field and for a sufficiently large field the attenuation drops to zero. Subsequent increases of the field produce "negative attenuation", which means amplification of the stress wave.

The appearance of negative attenuation (gain) is found to occur at the value of electric field such that the velocity of charge carriers (electrons, in this case) is equal to the velocity of the shear waves used in the experiment. The charge carrier mobility required for this coincidence to happen is approximately equal to the mobility deduced from independent Hall effect measurements at the same temperature. These observations, therefore, give further support to the suggested origin of the amplification effect.

An approximate theory of this effect can be formulated by adding a constant drift field, E, to the equation for current density [Eq. (47.6)]. In the case of one type of charge carrier this leads to a modified expression for α [see Eq. (47.17)]

$$\alpha = \frac{\omega\,e^2}{v\,2c\,p}\left[\frac{\dfrac{\omega_C}{\gamma\,\omega}}{1+\left(\dfrac{\omega_C}{\gamma\,\omega}\right)\left(1+\dfrac{\omega^2}{\omega_C\,\omega_D}\right)^2}\right] \qquad (49.1)$$

where $\gamma = \left(1 - \dfrac{v_d}{v}\right)$ and v_d is the drift velocity of the charge carriers. The experimental results are shown in Fig. 47. The dependence of attenuation on electric

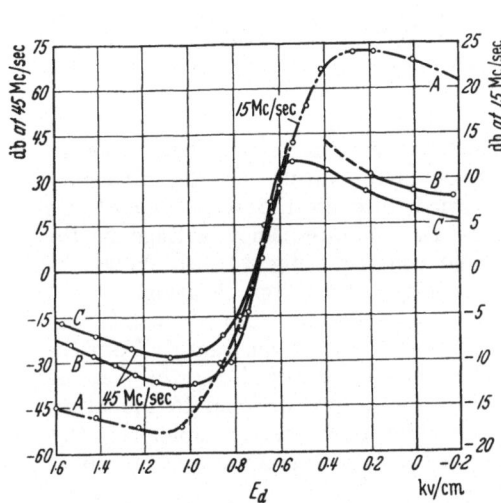

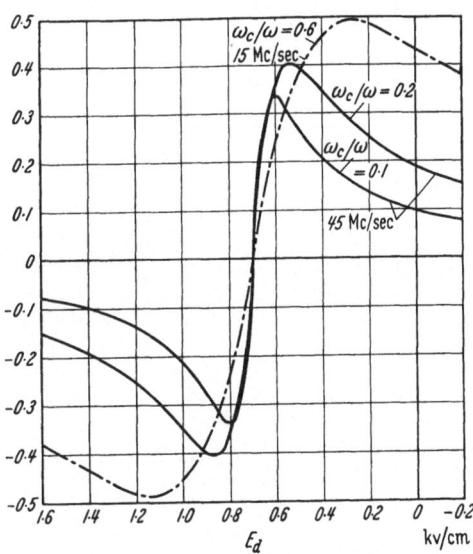

Fig. 47. Observed attenuation as a function of drift field E_d in CdS crystals. Zero db represents attenuation in unilluminated sample. Curve A: 15 Mc/sec, $\omega_c/\omega = 1.2$. Curve B: 45 Mc/sec, $\omega_c/\omega = 0.24$. Curve C: 45 Mc/sec, $\omega_c/\omega = 0.21$. The ω_c/ω values are averages for the whole sample computed from measured conductivities. After Hutson, McFee and White [162].

Fig. 48. Plots of the term in square brackets of Eq. (49.1). The numerical value of the term $\dfrac{\omega\,e^2}{v\,2c\,p}$, in this equation, is ≈ 150 db at 45 Mc/sec and ≈ 50 db at 15 Mc/sec, for the samples studied; thus Figs. 47 and 48 can be compared directly. After Hutson, McFee and White [162].

drift field, calculated on the basis of this theory, is in semiquantitative agreement with these observations, as shown in Fig. 48.

XI. Acoustoelectric effect.

50. Description of effect. A stress wave propagating through a semi conductor can, under certain circumstances, produce a d.c. potential difference across the ends of the semiconductor. This effect is called the acoustoelectric effect, and it is produced as the result of bunching of electrons and holes in the semi-conductor under the action of the deformation potential produced by the travelling stress wave [1].

The acoustoelectrical effect was first discussed by PARMENTER [163] who predicted an effect in metals. It has been shown, however, by WEINREICH [164] that the effect should vanish in metals although not in semi conductors, where the simultaneous bunching of electrons and holes is possible under the action of the deformation potential. With only one type of carrier the space charge effect (repulsion forces) should prevent appreciable bunching, making the acoustoelectric effect small. In a semiconductor, however, charge carriers of both signs when present make it possible to produce appreciable bunching of the charge without introducing large space charge forces of repulsion.

It has been pointed out by Holstein that it should be possible to produce an acoustoelectric effect with a single majority carrier if such carriers belong to complicated bands such as a multivalley band having minima in the energy surface. The deformation potentials seen by electrons, in different valleys, when a stress wave propagates through the lattice, will be different and it is then possible for various groups of electrons to bunch without upsetting seriously the electrical neutrality of the lattice. In this case intervalley scattering processes, and an associated relaxation time, are important rather than the presence of carriers of two signs.

Weinreich and White [165], [166] have observed the acoustoelectric effect in n type germanium using a transverse wave at a frequency of approximately 60 Mc/sec.

Fig. 49 shows the acoustoelectric field observed at liquid nitrogen temperature as the ultrasonic frequency is varied through the resonance of the transducer while the radiofrequency voltage of the transducer is held constant. Fig. 49 also shows the thermoelectric voltage arising from heating, by the ultrasonic power, of an indium absorber which was located at the opposite end of the sample from the transducer. The voltage produced by the heating of the indium absorber is proportional to the ultrasonic power travelling down the sample. In this case the estimated power (maximum) was about 0.9 watts/cm².

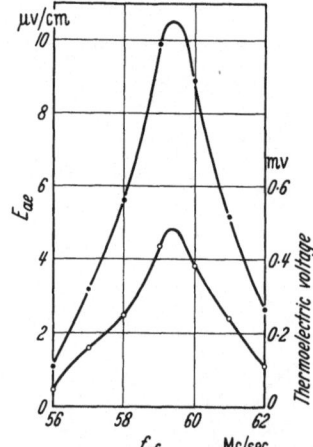

Fig. 49. Acoustoelectric effect (closed circles) and thermoelectric voltage developed by absorber (open circles) as rf frequency is varied through transducer resonance. After Weinreich and White [165].

For the case of a transverse wave travelling in the [100] direction and polarized in the [010] direction an approximate expression for the acoustoelectric field is provided by Weinreich [165].

$$E_{ae} = \frac{6\pi^2 q^2 \tau S}{e \lambda^2 kT} \tag{50.1}$$

where q is the so called "acoustic" charge of the carriers (taken as $1/\sqrt{2}$ times the deformation potential in a strain of unit energy density), τ is the intervalley scattering time, S is the ultrasonic power density, e is the electric charge of the carriers, λ is the ultrasonic wavelength and kT is the thermal energy. Knowing E_{ae} and S from these measurements allows calculation of $q^2 \tau$ in (50.1). This turns out to be about 1.5×10^{-44} cgs. units. The acoustic charge q, calculated from the measured piezoresistance effect, was used at 78° K to obtain a value of $\tau = 1.3 \times 10^{-10}$ sec.

Measurements of the acoustoelectric field were also made over a temperature range above liquid nitrogen temperature, and because in this temperature range the intervalley scattering time should vary as $e^{\Theta/T}$ where $\Theta = (\hbar\omega/k)$ is the phonon characteristic temperature, the product $E_{ae}T$ was plotted as a function of $1/T$. The result was a straight line yielding a slope leading to a value of $\Theta = 250°$ K in good agreement with the value obtained from piezoresistance effects.

Weinreich has also pointed out that since an ultrasonic wave carries a flux of momentum equal to $1/v$ times the energy flux, a loss in energy from the wave is equivalent to a proportional loss in momentum. This loss in momentum constitutes a constant force acting on the object absorbing the energy (radiation pressure). In the particular case in which the absorbers are free charge carriers the radiation pressure is exactly the acoustoelectric effect. The attenuation is then proportional to E_{ae}, the acoustoelectric field, by a factor that is independent of the detailed mechanism of the loss process. One can show that the attenuation in the amplitude of the stress wave is

$$\alpha = \frac{1}{2} e\, n\, v\, \frac{E_{ae}}{S} \tag{50.2}$$

where n is the density of carriers, v is the wave velocity and E_{ae} and S are as previously mentioned. Using the value of E_{ae}/S from WEINREICH and WHITE [*166*] one finds that the decrement

$$\delta = \alpha\, \lambda \approx \pi\, n\, 10^{-21}. \tag{50.3}$$

References.

[1] MASON, WARREN P.: Physical Acoustics and the Properties of Solids. New York: D. Van Nostrand Co. Inc. 1958.
[2] LOVE, A. E. H.: Theory of Elasticity. Cambridge: Cambridge University Press 1934.
[3] KOLSKY, H.: Stress Waves in Solids. Oxford: Clarendon Press 1953.
[4] PRAGER, W.: Introduction to Mechanics of Continua. Boston: Ginn and Company 1961.
[5] SOMMERFELD, A.: Mechanics of Deformable Bodies. New York: Academic Press Inc. 1950.
[6] BORGNIS, F. E.: Specific Directions of Longitudinal Wave Propagation in Anisotropic Media. Phys. Rev. **98**, 1000 (1955).
[7] LEVY, SHELDON, and ROHN TRUELL: Ultrasonic Attenuation in Magnetic Single Crystals. Rev. Mod. Phys. **25**, 140 (1953).
[8] YING, C. F., and ROHN TRUELL: Scattering of a Plane Longitudinal Wave by a Spherical Obstacle in an Isotropically Elastic Solid. J. Appl. Phys. **27**, 1086 (1956).
[9] WATERMAN, P. C.: Orientation Dependence of Elastic Waves in Single Crystals. Phys. Rev. **113**, 1240 (1959).
[10] LANDAU, L. D., and E. M. LIFSHITZ: Theory of Elasticity. Reading: Addison-Wesley Publishing Company 1959.
[11] MURNAGHAN, F. D.: Finite Deformations of an Elastic Solid. New York: John Wiley & Sons Inc. 1951.
[12] RIVLIN, RONALD: Some Topics in Finite Elasticity. Symposium on Structural Mechanics. London: Pergamon Press 1960. See also: The Strain Energy Function for Anisotropic Materials. Trans. Amer. Math. Soc. **88**, 175 (1958).
[13] BIRCH, F.: Finite Elastic Strain of Cubic Crystals. Phys. Rev. **71**, 809 (1947).
[14.1] HEARMON, R. F. S.: The Elastic Constants of Anisotropic Materials. II. Phil. Mag., Suppl. **5**, 323 (1956). Advances In Physics.
[14.2] NYE, J. F.: Physical Properties of Crystals. Oxford: Clarendon Press 1957.
[14.3] HUNTINGTON, H. B.: Elastic Constants of Crystals. Solid State Physics, Vol. 7, p. 213—351. New York: Academic Press 1958.
[15] SEEGER, ALFRED, and OTTO BUCK: Die experimentelle Ermittlung der elastischen Konstanten höherer Ordnung. Z. Naturforsch. **15**, 1056 (1960).
[16] BATEMAN, T., W. P. MASON and H. J. McSKIMIN: Third Order Elastic Moduli of Germanium. J. Appl. Phys. **32**, 928 (1961).
[17] Lord RAYLEIGH: Theory of Sound. London: MacMillan & Company Ltd. 1925.
[18] SEWELL, C. J. T.: Extinction of Sound in a Viscous Atmosphere by Small Cylinders and Spheres. Trans. Roy. Soc. Lond. A **20**, 239 (1910).
[19] HERZFELD, K. F.: The Scattering of Sound-waves by Small Elastic Spheres. Phil. Mag. **9**, 741 (1930).
[20] ANDERSON, V. C.: Sound Scattering from a Fluid Sphere. J. Acoust. Soc. Amer. **22**, 426 (1950).
[21] HART, R. W.: Sound Scattering of a Plane Wave From a Nonabsorbing Source. J. Acoust. Soc. Amer. **23**, 323 (1951).
[22] EPSTEIN, P. S., and R. R. CARHART: The Absorption of Sound in Suspensions and Emulsions. J. Acoust. Soc. Amer. **25**, 553 (1953).

[23] White, R.M.: Elastic Wave Scattering at a Cylindrical Discontinuty in a Solid. J. Acoust. Soc. Amer. **30**, 771 (1958).

[24] Spence, R.D., and S. Granger: The Scattering of Sound From a Prolate Spheroid. J. Acoust. Soc. Amer. **23**, 701 (1951).

[25] Mason, W.P., and H.J. McSkimin: Attenuation and Scattering of High Frequency Sound Waves in Metals and Glasses. J. Acoust. Soc. Amer. **19**, 464 (1947).

[26] Roth, W.: Scattering of Ultrasonic Radiation in Polycrystalline Metals. J. Appl. Phys. **19**, 901 (1948).

[27] Mason, W.P., and H.J. McSkimin: Energy Losses of Sound Waves in Metals Due to Scattering and Diffusion. J. Appl. Phys. **19**, 940 (1948).

[28] Roney, R.K.: The Influence of Metal Grain Structure on the Attenuation of an Ultrasonic Wave. Ph-D Thesis, California Institute of Technology 1950.

[29] Roderick, R.L., and Rohn Truell: On the Measurement of Ultrasonic Attenuation in Solids by the Pulse Technique. J. Appl. Phys. **23**, 267 (1952).

[30] Huntington, H.B.: Ultrasonic Delay Lines. J. Franklin Inst. **245**, I 1, II 101 (1948).

[31] Einspruch, Norman, E.J. Witterholt and Rohn Truell: Scattering of a Plane Transverse Wave by a Spherical Obstacle in an Elastic Medium. J. Appl. Phys. **31**, 806 (1960).

[32] Waterman, P.C., and Rohn Truell: Multiple Scattering of Waves. J. Math. Phys., July-August (1961).

[33] Foldy, L.L.: The Multiple Scattering of Waves. Phys. Rev. **67**, 107 (1945).

[34] Papadakis, E.P.: Ultrasonic Attenuation in Steel. J. Acoust. Soc. Amer. **32**, 1628 (1960).

[35] Truell, Rohn: Nature of Defects Arising from Fast Neutron Irradiation of Silicon Single Crystals. Phys. Rev. **116**, 890 (1959).

[36] Truell, Rohn, L.J. Teutonico and Paul W. Levy: Detection of Directional Neutron Damage in Silicon by Means of Ultrasonic Double Refraction Measurements. Phys. Rev. **105**, 1723 (1957).

[37] Zener, C.: Elasticity and Anelasticity of Metals. Chicago, Ill.: Chicago University Press 1948.

[38] Lücke, Kurt: Ultrasonic Attenuation Caused by Thermoelastic Heat Flow. J. Appl. Phys. **27**, 1433 (1956).

[39] Read, T.A.: The Internal Friction of Single Metal Crystals. Phys. Rev. **58**, 371 (1940). Read, T.A.: Internal Friction of Single Crystals of Copper and Zinc. Trans. Amer. Inst. Min. Metallurg. Engrs. **143**, 30 (1941).

[40] Koehler, J.S.: Imperfections in Nearly Perfect Crystals, p. 197. New York: John Wiley & Sons Inc. 1952.

[41] Granato, A., and K. Lücke: Theory of Mechanical Damping Due to Dislocations. J. Appl. Phys. **27**, 583 (1956).

[42] Nowick, A.S.: Internal Friction in Metals. Progr. Metal Phys. **4**, 1 (1954). Nowick, A.S.: Variation of Amplitude-Dependent Internal Friction in Single Crystals of Copper with Frequency and Temperature. Phys. Rev. **80**, 249 (1950). Nowick, A.S.: Internal Friction and Dynamic Modulus of Cold-Worked Metals. J. Appl. Phys. **25**, 1129 (1954). Nowick, A.S.: Creep and Recovery, p. 146. Amer. Soc. Metals 1956.

[43] Weertman, J.: Internal Friction of Metal Single Crystals. J. Appl. Phys. **26**, 202 (1955).

[44] Weertman, J., and E.I. Salkovitz: The Internal Friction of Dilute Alloys of Lead. Acta metallurg. **3**, 1 (1955).

[45] DeWit, G., and J.S. Koehler: Interaction of Dislocations with an Applied Stress in Anisotropic Crystals. Phys. Rev. **116**, 1113 (1959).

[46] Mason, W.P.: Phonon Viscosity and its Effect on Acoustic Wave Attenuation and Dislocation Motion. J. Acoust. Soc. Amer. **32**, 458 (1960).

[47] Leibfried, G.: Über den Einfluß thermisch angeregter Schallwellen auf die plastische Deformation. Z. Physik **127**, 344 (1950).

[48] Beshers, D.N.: Internal Friction of Copper and Copper Alloys. J. Appl. Phys. **30**, 252 (1959).

[49] Baker, George S.: Internal Friction in the Presence of a Static Stress. J. Appl. Phys. **28**, 734 (1957).

[50] Granato, Andrew, John DeKlerk and Rohn Truell: Dispersion of Elastic Waves in Sodium Chloride. Phys. Rev. **108**, 895 (1957).

[51] Gordon, R.B., and A.S. Nowick: The Pinning of Dislocations by X-Irradiation of Alkali Halide Crystals. Acta metallurg. **4**, 514 (1956).

[52] Bauer, Charles L., and Robert B. Gordon: Dislocation Damping Effects in Rock Salt. J. Appl. Phys. **31**, 945 (1960).

[53] TRUELL, ROHN: Ultrasonic Methods and Radiation Effects in Solids. J. Appl. Phys. 30, 1275 (1959).

[54] TRUELL, ROHN: Influence of Deformation and Temperature on the Cobalt Gamma Irradiation of Sodium Chloride. Evidence for Electrical Interaction Between Dislocations and Point Defects. J. Appl. Phys. 32, 1601 (1961).

[55] CHICK, BRUCE B., GEORGE P. ANDERSON and ROHN TRUELL: Ultrasonic Attenuation Unit and Its Use in Measuring Attenuation in Alkali Halides. J. Acoust. Soc. Amer. 32, 186 (1960).

[56] THOMPSON, DONALD O., and DAVID K. HOLMES: Effects of Neutron Irradiation Upon the YOUNG's Modulus and Internal Friction of Copper Single Crystals. J. Appl. Phys. 27, 713 (1956).

[57] ALERS, G. A., and D. O. THOMPSON: Dislocation Contributions to the Modulus and Damping in Copper at Megacycle Frequencies. J. Appl. Phys. 32, 283 (1961).

[58] EINSPRUCH, N. G., and F. WEST: Calorimetric Measurement of Ultrasonically Produced Strains in Solids. J. Acoust. Soc. Amer. 32, 1160 (1960).

[59] GRANATO, A. V., and R. STERN: Overdamped Resonance of Dislocations in Copper. Acta Metallurgica 10, 358 (1962).

[60] THOMPSON, D. O., and D. K. HOLMES: Dislocation Contribution to the Temperature Dependence of the Internal Friction and YOUNG's Modulus of Copper. J. Appl. Phys. 30, 525 (1959).

[61] HIKATA, AKIRA, BRUCE B. CHICK, CHARLES ELBAUM and ROHN TRUELL: Ultrasonic Attenuation and Velocity Data on Aluminum Single Crystals as a Function of Deformation and Orientation. Acta Metallurgica 10, 423 (1962).

[62] BORDONI, P. G.: Teoria della dissipazione elastica nei monocristalli secondo la meccanica quantistica; un nuovo effetto di rilassamento. Ric. Sci. 19, 851 (1949).

[63] BORDONI, P. G.: Elastic and Anelastic Behavior of Some Metals at Very Low Temperatures. J. Acoust. Soc. Amer. 26, 495 (1954).

[64] NIBLETT, D., and J. WILKS: The Internal Friction of Cold-Worked Copper at Low Temperatures. Phil. Mag. 2, 1427 (1957).

[65] PARE, V. K.: Thesis, Cornell University 1958.

[66] EINSPRUCH, N. G., and R. TRUELL: Megacycle Attenuation and Bordoni Peaks. Phys. Rev. 109, 652 (1958).

[67] THOMPSON, D. O., and D. K. HOLMES: Dislocation Contribution of the Temperature Dependence of the Internal Friction and YOUNG's Modulus of Copper. J. Appl. Phys. 30, 525 (1959).

[68] CASWELL, H. L.: Investigation of Low-Temperature Internal Friction. J. Appl. Phys. 29, 1210 (1958).

[69] BORDONI, P. G., M. NUOVO and L. VERDANI: Relaxation of Dislocations in Copper. Nuovo Cim. 14, 273 (1959).

[70] BRUNER, L. J.: Low-Temperature Internal Friction in Face-Centered and Body-Centered Cubic Metals. Phys. Rev. 118, 399 (1960).

[71] MASON, W. P.: Relaxations in the Attenuation of Single Crystals of Lead at Low Temperatures and Their Relation to Dislocation Theory. J. Acoust. Soc. Amer. 27, 643 (1955).

[72] SEEGER, A., H. DONTH and F. PFAFF: The Mechanism of Low Temperature Mechanical Relaxation in Deformed Crystals. Disc. Faraday Soc. 23, 19 (1957).

[73] SEEGER, A.: Theorie der Gitterfehlstellen. Handbuch der Physik, Bd. 7/1, S. 383. 1955.

[74] SEEGER, A.: On the Theory of the Low-Temperature Internal Friction Peak Observed in Metals. Phil. Mag. 1, 651 (1956).

[75] PEIERLS, R. E.: The Size of a Dislocation. Proc. Phys. Soc. Lond. 52, 34 (1940).

[76] NABARRO, F. R. N.: The Mathematical Theory of Stationary Dislocations. Adv. Physics 1, 269 (1952).

[77] DONTH, H.: Zur Theorie des Tieftemperaturmaximums der inneren Reibung von Metallen. Z. Physik 149, 111 (1957).

[78] LOTHE, J.: Aspects of the Theories of Dislocation Mobility and Internal Friction. Phys. Rev. 117, 704 (1960).

[79] MASON, W. P.: Conference on Internal Friction, Cornell University, July 1961. (To be published.)

[80] BRAILSFORD, A. D.: Abrupt-Kink Model of Dislocation Motion. Phys. Rev. 122, 778 (1961).

[81] AMELINCKX, S., J. VENNIK and G. REMAUT: Electrical Effects During Cyclic Stressing of Sodium Chloride. J. Phys. Chem. Solids 11, 170 (1959); 16, 158 (1960).

[82] DeWit, G., and J. S. Koehler: Interaction of Dislocations with an Applied Stress in Anisotropic Crystals. Phys. Rev. **116**, 1113 (1959).
Koehler, J. S., and G. DeWit: Influence of Elastic Anisotropy on the Dislocation Contribution to the Elastic Constants. Phys. Rev. **116**, 1121 (1959).

[83] Jacobson, E. H.: Generation of 9.2 kMc/s Ultrasonics in Quartz. Phys. Rev. Letters **2**, 249 (1959).

[84] Bömmel, H. E.: Ultrasonic Attenuation in Superconducting Lead. Phys. Rev. **96**, 220 (1954).

[85] Mackinnon, L.: The Absorption of 10 Mc/s Sound Pulses by a Superconducting Polycrystaline Tin Rod. Phys. Rev. **98**, 1181 (1955).

[86] Mackinnon, L.: The Absorption of 10 Mc/sec Sound in a Polycrystalline Magnesium Rod Between 1.5 and 4.2° K. Phys. Rev. **98**, 1210 (1955).

[87] Morse, R. W.: Ultrasonic Attenuation in Metals by Electron Relaxation. Phys. Rev. **97**, 1716 (1955).

[88] Mason, W. P.: Ultrasonic Attenuation Due to Lattice-Electron Interaction in Normal Conducting Metals. Phys. Rev. **97**, 557 (1955).

[89] Kittel, C.: An Electron Transfer Mechanism for Ultrasonic Attenuation in Metals. Acta mettallurg. **3**, 295 (1955).

[90] Bardeen, J., L. N. Cooper and J. R. Schrieffer: Theory of Superconductivity. Phys. Rev. **108**, 1175 (1957).

[91] Morse, R. W., and H. V. Bohm: Superconducting Energy Gap from Ultrasonic Attenuation Measurements. Phys. Rev. **108**, 1094 (1957).

[92] Cooper, L. N.: Specific Heat Measurements and the Energy Gap in Superconductors. Phys. Rev. Letters **3**, 17 (1959).

[93] Morse, R. W., T. Olsen and J. D. Gavenda: Evidence for Anisotropy of the Superconducting Energy Gap from Ultrasonic Attenuation. Phys. Rev. Letters **3**, 15 (1959).

[94] Morse, R. W.: Ultrasonic Attenuation in Metals at Low Temperatures. Progress in Cryogenics, vol. 1, p. 221. London 1959.

[95] Pippard, A. B.: Ultrasonic Attenuation in Metals. Phil. Mag. **46**, 104 (1955).

[96] Steinberg, M. S.: Ultrasonic Attenuation and Dispersion in Metals at Low Temperatures. Phys. Rev. **111**, 425 (1958).

[97] Schiff, L. I.: Quantum Mechanics, 1st. edit., p. 65. New York: McGraw-Hill 1949.

[98] Bardeen, J., and W. Shockley: Deformation Potentials and Mobilities in Non-Polar Crystals. Phys. Rev. **80**, 72 (1950).

[99] Wilson, A. H.: The Theory of Metals, 2nd edit., p. 256. Cambridge: Cambridge University Press 1954.

[100] Steinberg, M. S.: Magnetic Effects in the Attenuation of Transverse Acoustical Waves by Conduction Electron Transport. Phys. Rev. **110**, 772 (1958).

[101] Bömmel, H. E.: Ultrasonic Attenuation in Superconducting and Normal-Conducting Tin at Low Temperatures. Phys. Rev. **100**, 758 (1955).

[102] Pippard, A. B.: A Proposal for Determining the Fermi Surface by Magneto-Acoustical Resonance. Phil. Mag., Ser. VIII **2**, 1147 (1957).

[103] Morse, R. W., and J. D. Gavenda: Magnetic Oscillations of Ultrasonic Attenuation in a Copper Crystal at Low Temperature. Phys. Rev. Letters **2**, 250 (1959).

[104] Reneker, D. H.: Ultrasonic Attenuation in Bismuth at Low Temperatures. Phys. Rev. **115**, 303 (1959).

[105] Morse, R. W., A. Myers and C. T. Walker: Fermi Surfaces of Gold and Silver from Ultrasonic Attenuation. Phys. Rev. Letters **4**, 605 (1960).

[106] Kjeldas, T., and T. Holstein: Oscillatory Magneto-Acoustical Effects in Metals. Phys. Rev. Letters **2**, 340 (1959).

[107] Cohen, M. H., M. J. Harrison and W. J. Harrison: Magnetic Field Dependence of the Ultrasonic Attenuation in Metals. Phys. Rev. **117**, 937 (1960).

[108] Morse, R. W., and T. Olson: Magnetic Dependence of Ultrasonic Attenuation in Tin. Bull. Amer. Phys. Soc., Sec. II **4**, 167 (1959).

[109] DeKlerk, J.: Effect of a Magnetic Field on the Propagation of Sound Waves in a Ferromagnetic Material. Nature, Lond. **168**, 963 (1951).

[110] Bozorth, R. M., W. P. Mason and H. J. McSkimin: Frequency Dependence of Elastic Constants and Losses in Nickel. Bell Syst. Techn. J. **30**, 970 (1951).

[111] Rogers, T. F., and S. J. Johnson: Some Magneto-Acoustic Effects in Nickel. J. Appl. Phys. **21**, 1067 (1950).

[112] Levy, Sheldon, and Rohn Truell: Ultrasonic Attenuation in Magnetic Single Crystals. Rev. Mod. Phys. **25**, 140 (1953).

[113] DeKlerk, J.: Ultrasonic Wave Propagation in a Nickel Single Crystal. Proc. Phys. **73**, 337 (1959).

[114] Becker, R., and W. Döring: Ferromagnetismus, p. 102. Berlin: Springer 1939.

[115] LEVY, S.: Ultrasonic Attenuation in Magnetic Single Crystals. Ph.D. Thesis, Brown University 1952.

[116] VAN KRANENDONK, J., and J. H. VAN VLECK: Spin Waves. Rev. Mod. Phys. **30**, 1 (1958).

[117] KITTEL, C.: Interaction of Spin Waves and Ultrasonic Waves in Ferromagnetic Crystals. Phys. Rev. **110**, 836 (1958).

[118] AKHIESER, A. I., and L. A. SHISHKIN: On the Theory of Thermal Conductivity and Absorption of Sound in Ferromagnetic Dielectrics. J. Exp. Theor. Phys. USSR. **34**, 1267 (1958).

[119] KAGANOV, M. I., and Y. M. CHIKVASHVILI: A Theory of Absorption of Sound in Uni-axial Ferromagnetic Dielectrics. Soviet Phys. Solid State **3**, 200 (1961).

[120] BÖMMEL, H., and K. DRANSFELD: Excitation of Hypersonic Waves by Ferromagnetic Resonance. Phys. Rev. Letters **3**, 83 (1959).

[121] AKHIESER, A.: Theory of Relaxation Processes in Ferromagnetics at Low Temperatures. J. Phys. USSR. **10**, 217 (1946).

[122] KAGANOV, M. I., and V. M. TSUKEVNIK: Phenomenological Theory of Kinetic Processes in Ferromagnetic Dielectrics, Interaction of Spin Waves with Phonons. Soviet Phys. JETP **9**, 151 (1959); (translation) USSR. **36**, 224 (1959).

[123] KAGANOV, M. I., and V. M. TSUKEVNIK: Non-resonance Absorption of Oscillating Magnetic Field Energy In Ferromagnetic Dielectric. Soviet Phys. JETP **10**, 587 (1960); (translation) USSR. **37**, 823 (1959).

[124] BÖMMEL, H., and K. DRANSFELD: Attenuation of Hypersonic Waves in Quartz. Phys. Rev. Letters **2**, 298 (1959).

[125] BÖMMEL, H., and K. DRANSFELD: Excitation and Attenuation of Hypersonic Waves in Quartz. Phys. Rev. **117**, 1245 (1960).

[126] JACOBSEN, E. H.: Piezoelectric Production of Microwave Phonons. Phys. Rev. Letters **2**, 249 (1959).

[127] JACOBSEN, E. H.: Experiments with Phonons at Microwave Frequencies. Quantum Electronics (Symposium 1959), p. 468, Columbia Univ. Press.

[128] TRUELL, ROHN, BRUCE CHICK and THOMAS FITZGERALD: Radiation Induced Defects in Natural Quartz and Their Effect on Phonon-Phonon Scattering. Proc. Cornell Conference on Internal Friction 1961.

[129] DOBBS, E. ROLAND, BRUCE B. CHICK and ROHN TRUELL: Attenuation of Sound in a Germanium Crystal at High Frequencies and Low Temperatures. Phys. Rev. Letters **3**, 332 (1959).

[130] DOBBS, E. ROLAND: Attenuation of Sound in Germanium at Ultra-High Frequencies. Proc. VII Internat. Conference Low Temperature Physics, p. 291, Toronto 1960.

[131] VERMA, G. S., and S. K. JOSHI: Origin of Hypersonic Attenuation in Germanium at Low Temperatures. Phys. Rev. **121**, 396 (1961).

[132] AKHIESER, A.: On the Absorption of Sound in Solids. J. Phys. USSR. **1**, 277 (1939).

[133] LANDAU, L., and G. RUMER: On the Absorption of Sound in Solids. Phys. Z. Sowjet **11**, 18 (1937).

[134] POMERANCHUCK, I. J.: Sound Absorption in Dielectrics. J. Phys. USSR. **4**, 529 (1941).

[135] WOODRUFF, T. O., and H. EHRENREICH: Absorption of Sound in Insulators. Phys. Rev. **123**, 1553 (1961).

[136] ZIMAN, J. M.: Electrons and Phonons. Oxford: Clarendon Press 1960.

[137] PROCTOR, W. C., and W. H. TANTILLA: Saturation of Nuclear Electric Quadrupl Energy Levels by Ultrasonic Excitation. Phys. Rev. **98**, 1854 (1955).

[138] PROCTOR, W. C., and W. H. TANTILLA: Influence of Ultrasonic Energy on the Relaxation of Chlorine Nuclei in Sodium Chlorate. Phys. Rev. **101**, 1757 (1956).

[139] PROCTOR, W. G., and W. ROBINSON: Ultrasonic Excitation of Nuclear Magnetic Energy Levels of Na[23] in NaCl. Phys. Rev. **104**, 1344 (1956).

[140] STERNHEIMER, R. M.: Effect of the Atomic Core on Nuclear Quadruple Coupling. Phys. Rev. **105**, 158 (1957).

[141] YOSIDA, K., and T. MORIYA: The Effects of Convalency on the Nuclear Magnetic Resonance in Ionic Crystals. J. Phys. Soc. Japan **11**, 33 (1956).

[142] KRAUS, O., and W. H. TANTILLA: Nuclear Magnetization in the Presence of Ultrasonic Excitation. Phys. Rev. **109**, 1052 (1958).

[143] VERMA, G. S.: Ultrasonic Investigation of Nuclear Spin-Lattice Relaxation. Suppl. del Nuovo Cim., Ser. X **12**, No. 1, 41 (1959).

[144] JENNINGS, D. A., W. H. TANTILLA and O. KRAUS: Ultrasonically Induced Spin Transitions in Sodium Iodide. Phys. Rev. **109**, 1059 (1958).

[145] KITTEL, C.: Paramagnetic Centers as Detectors of Ultrasonic Radiation at Microwave Frequencies. Phys. Rev. Letters **1**, 5 (1958).

[146] Jacobsen, E.H., N.S. Shiren and E.B. Tucker: Effects of 9.2 kmc/sec Ultrasonics on Electron Spin Resonance in Quartz. Phys. Rev. Letter 3, 81 (1959).
[147] Shiren, N.S., and E.B. Tucker: Effects of 9.2 kmc/sec Ultrasonics on Electron Spin Resonances. Quantum Electronics, p. 485. New York: Columbia University Press 1960.
[148] Kittel, C.: Introduction to Solid State Physics, 2nd edit., p. 215. New York: John Wiley & Sons Inc. 1956.
[149] Transient and Steady State Response of Ultrasonic Piezoelectric Transducers. I.R.E. Convention Record Part 9, p. 61 (1956).
[150] Jacobsen, E.H.: Sources of Sound in Piezoelectric Crystals. J. Acoust. Soc. Amer. 32, 949 (1960).
[151] Rosenberg, H.M.: The Thermal Conductivity of Germanium and Silicon at Low Temperature. Proc. Phys. Soc. Lond. A 67, 837 (1954).
[152] Hill, R.W., and D.H. Parkinson: The Specific Heats of Germanium and Grey Tin at Low Temperatures. Phil. Mag. 43, 309 (1952).
[153] Fine, M.E.: Elastic Constants of Germanium Between 1.7° and 80° K. J. Appl. Phys. 26, 862 (1955).
[154] Kyame, J.J.: Wave Propagation in Piezoelectric Crystals. J. Acoust. Soc. Amer. 21, 159 (1949).
[155] Koga, I., M. Aruga and Y. Yoshinaka: Theory of Plane Elastic Waves in a Piezoelectric Crystalline Medium and Determination of Elastic and Piezoelectric Constants of Quartz. Phys. Rev. 109, 1467 (1958).
[156] Pailloux, H.: Piezoélectricité, Calcul des Vitesses de Propagation. J. Phys. Radium 19, 523 (1923).
[157] Hutson, A.R., and D.L. White: Elastic Wave Propagation in Piezoelectric Semiconductors. J. Appl. Phys. 33, 40 (1962).
[158] Gobrecht, H., and A. Bartschat: On the Effect of Activators and the Accumulation of Irradiation Energy on the Piezo-Electric and Elasto-Mechanical Characteristics of Cadmium Sulfide Single Crystals. (German). Z. Physik 153, 529 (1959).
[159] Nine, Harmon D.: Photosensitive Ultrasonic Attenuation in CdS. Phys. Rev. Letters 4, 359 (1960).
[160] Nine, Harmon D., and Rohn Truell: Photosensitive Ultrasonic Properties of Cadmium Sulphide. Phys. Rev. 123, 799 (1961).
[161] Hutson, A.R.: Piezoelectricity and Conductivity in ZnO and CdS. Phys. Rev. Letters 4, 505 (1960).
[162] Hutson, A.R., J.H. McFee and D.L. White: Ultrasonic Amplification in CdS. Phys. Rev. Letters 7, 237 (1961).
[163] Parmenter, R.H.: The Acousto-Electric Effect. Phys. Rev. 89, 990 (1953).
[164] Weinreich, Gabriel: Acoustodynamic Effects in Semiconductors. Phys. Rev. 104, 321 (1956).
[165] Weinreich, Gabriel, and H.G. White: Observation of the Acoustoelectric Effect. Phys. Rev. 106, 1104 (1957).
[166] Weinreich, Gabriel: Ultrasonic Attenuation by Free Carriers in Germanium. Phys. Rev. 107, 317 (1957).

The Effects of Intense Ultrasonics in Liquids.

By

B. E. NOLTINGK.

With 20 Figures.

1. Introduction. In the course of the development of ultrasonics, the transmission of acoustic waves through matter has been studied in two broad categories. These can be distinguished as (i) when the waves provide information about material which they do not significantly alter, and (ii) when the waves produce changes in the matter traversed, which changes are themselves of primary interest.

For (i) a relatively low intensity is normally adequate. It is the category (ii) that we are concerned with here, particularly when the medium is a liquid. It is to be expected that increasing intensity of the acoustic wave will increase the effects observed.

For a plane wave in a liquid, simple theory gives expressions for the peak values of the alternating components of pressure, displacement, velocity and acceleration as follows:

$$P_0 = \sqrt{2J \varrho v}, \qquad (1.1)$$

$$A_0 = \frac{1}{\omega} \sqrt{\frac{2J}{\varrho v}}, \qquad (1.2)$$

$$U_0 = \sqrt{\frac{2J}{\varrho v}}, \qquad (1.3)$$

$$B_0 = \omega \sqrt{\frac{2J}{\varrho v}}. \qquad (1.4)$$

Here J is the intensity and ω the angular frequency of the wave, while ϱ and v refer to the medium, being respectively density and velocity of sound therein. It can be seen immediately that, for a given intensity, pressure and velocity are independent of frequency, so that for effects which are governed by either of these two parameters the frequency of the acoustic vibration should be less important.

If the intensity J is measured in watts/cm² (which is convenient for relating to electrical transducers), ϱ in g/cm³ and v in cm/sec, then for P_0 in atmospheres peak there is a constant of proportionality.

$$P_0 = 4.4 \times 10^{-3} \sqrt{J \varrho v}. \qquad (1.5)$$

In many cases, the medium is water, for which $\varrho = 1$, $v = 1.46 \times 10^5$ so that

$$P_0 \text{ (water)} = 1.7 \sqrt{J} \text{ atm cm/watt}^{\frac{1}{2}}. \qquad (1.6)$$

Similarly, in cm/sec peak,

$$U_0 \text{ (water)} = 11.7 \sqrt{J}. \qquad (1.7)$$

17*

At a frequency of 1 Mc/sec

$$A_0 \text{ (water)} = 1.86 \times 10^{-6}\sqrt{J} \quad \text{in cm} \tag{1.8}$$

and

$$B_0 \text{ (water)} = 7.35 \times 10^7 \sqrt{J} \quad \text{in cm/sec}^2. \tag{1.9}$$

In most techniques, vibrations are introduced into the liquid from a solid. This may be either directly by contact with the transducer or through a coupling member, but in either case the intensity is limited by the strength of the solid, independent of the operating frequency. (Exceptions are the Pohlmann whistle[1] and electrodynamic excitation[2].)

The order of magnitude to which we are thus restricted in intensity is tens of watts/cm². Over very small volumes, this can be increased another order of magnitude by focussing devices, but the maximum practicable value of $\sqrt{J}$ will not be more than 10 to 100. It can therefore be seen that the primary acoustic pressures, velocities and particle amplitudes involved are quite modest. Exceptional effects are scarcely to be expected from them.

Particle accelerations, on the other hand, can be very large; 10^5 times that of gravity is attainable. It should also be noted that while the peak pressures expected are only a few atmospheres, pressure gradients—on account of the short wavelength of ultrasonics—can be appreciable, while the time differential of pressure can be very large. Unusual effects arising from intense ultrasonics are likely to be attributed to these parameters. The special phenomenon of cavitation can be thought of as resulting from the rapid variation with time of the acoustic pressure at a point, and, as will be explained later, cavitation is in fact responsible for most of the effects associated with intense ultrasonics in liquids.

There is, of course, no essential difference in general behaviour between audible vibrations and those of frequency too high to be heard by the human ear, i.e. ultrasonic. Some effects are smoothly frequency dependent, as we shall see, but it is often only from convenience that investigations have been made at super-audible frequencies, partly to avoid the discomfort not infrequently associated with intense audible sounds and sometimes partly because, the wavelength being shorter, the bulk of the apparatus can be less in experiments with ultrasonics.

2. Cavitation and de-gassing.
Although in every day experience a liquid cannot sustain any tensile stress, this is because it is normally able to change its shape and flow in such a direction as to relieve the stress. There are theoretical grounds for expecting that when a liquid is suitably restrained, so that failure must occur in the body of it rather than by gross change of shape, it should have a tensile strength of the order of 10^3 or 10^4 atmospheres. Measurements of tensile strength, however, — most of them have been made on water — have given figures some two orders of magnitude lower. The discrepancy is attributed to the presence of nuclei in all liquids; under the action of sufficient tensile stress, the liquid surface can recede from a nucleus, giving rise to a macroscopic strain and, if the stress is maintained, to failure.

The phenomenon of cavitation occurs when a tensile stress in a liquid is applied but not maintained indefinitely. If the volume element containing the nucleus is subjected first to tension and subsequently to compression, the nucleus will grow initially and then shrink. When this growth and shrinkage is large, it is called cavitation. Growth and shrinkage of a nucleus implies radial movement

[1] W. JANOVSKI and R. POHLMANN: Z. angew. Phys. **1**, 222 (1948).
[2] H. J. SEEMANN and H. STAATS: J. Acoust. Soc. Amer. **29**, 698 (1957).

of the liquid, and can lead to numerous marked effects. The kinetic energy of the moving liquid can be shown to be concentrated in a shell surrounding the nucleus. Thus during the shrinkage phase the energy abstracted from the gross pressure field is being concentrated in a continually shrinking volume; it is therefore not surprising that unusual results can follow from the very high local energy density.

The essential prerequisite to cavitation, namely a transition from tensile to compressive stress, can be brought about in two ways. The liquid may move as a whole from regions of negative pressure to regions of positive pressure. Alternatively it may remain fixed in space while the applied pressure is changed. The former is associated with hydraulic systems and can lead to cavitation in them; the latter is the basis for ultrasonic cavitation. While hydraulic cavitation has so far had greater technological importance, ultrasonically induced cavitation is receiving increased attention; even when the results are to be applied in hydraulics, ultrasonic excitation is often used in experiments because conditions can then be more easily controlled and varied.

Cavitation results from the action of a suitably varying pressure on a nucleus. Pressures can normally be measured and, apart from corrections for small-scale and high-speed effects, known. The other factor, the nucleus, is also important and is much less understood; it is possible that some apparently contradictory experimental results could be reconciled by considering differences in condition with regard to nuclei.

When intense ultrasonic vibration is applied to a liquid containing gas in solution, it is found that some of the gas is liberated. This is not merely a case of accelerating the attainment of equilibrium conditions: gas can be liberated from under-saturated as well as super-saturated solutions. LINDSTRÖM[1] has found that the gas content is reduced to about half its saturation value for a wide range of applied ultrasonic intensities. Such gas, of course, appears in the form of bubbles, and in any consideration of the effects of intense ultrasonics it is necessary to distinguish this de-gassing from cavitation. (Sometimes the terms "gaseous cavitation" as opposed to "vaporous cavitation" or "pure cavitation" have been used.) The early stages of de-gassing are closely related to the phenomenon of cavitation, but the former leads to large gas-filled bubbles, that are relatively permanent. They may be located at particular points in a standing wave field[2]. At times their movement becomes very rapid; presumably this occurs when they have reached a resonant size for some particular mode. It will, in fact, become apparent that the resonance of bubbles plays an important part in several aspects of intense ultrasonics. The sharp difference between the two effects, under simple observation, is that when the ultrasonic excitation is stopped cavitation voids disappear, while gas-filled bubbles persist at their full size. Often, both phenomena occur simultaneously in irradiated liquids.

To account for the de-gassing action, it is necessary to explain how, in the aggregate, dissolved gas can diffuse out of an under-saturated liquid into a vibrating bubble.

The effect has become known as "rectified diffusion" as named by BLAKE[3] who developed a theory to account for it. KURTZE[4] however, has pointed out a fallacy in the original calculations and suggests that the nett liberation of gas should be explained in terms of the liquid shell immediately surrounding the bubble,

[1] O. LINDSTRÖM: J. Acoust. Soc. Amer. 27, 654 (1955).
[2] L. BOHN: Acustica 7, 201 (1957).
[3] F. G. BLAKE: Acoust. Res. Lab. Harvard Univ. Tech. Rep. 12 (1949).
[4] G. KURTZE: Nachr. Akad. Wiss. Göttingen, math.-phys. Kl. 1 (1958).

which changes shape more rapidly than it changes its gas concentration. When a bubble undergoes oscillations of large amplitude, we shall see that it is at low pressure in an expanded state for a much larger fraction of the total time that it spends contracted; this of itself would lead to rectified diffusion of gas into the bubble. GÜTH has suggested[1] that the asymmetry expected in the pressure/radius relation for liquid surrounding a bubble should drive other, smaller bubbles to coalesce with the main one. However, these two mechanisms cannot provide the whole solution, since they are only operative after a bubble has reached a certain size.

3. Signs of cavitation.

As the vibration amplitude of a surface vibrating in a liquid is increased, an observer becomes aware of increasing activity in the neighbourhood of the surface. Bubbles appear, sometimes moving about rapidly, sometimes locked at a particular point. A characteristic "tearing" sound can often be heard. These effects are much more marked at the lower ultrasonic frequencies, say 20 kc/sec, becoming barely detectable to the unaided eye and ear even at hundreds of watts per cm² at frequencies above 1 Mc/sec. At low frequencies, it is generally simple to distinguish "gassy" bubbles from true cavitation. The latter sometimes appears in a local-

Fig. 1. Cavitation in water under a vibrator (Photo S. H. BOUTLE).

ised, roughly spherical, region, and is sometimes manifested as streamers or trees of curious, branched shape. Examples can be seen in Fig. 1 which is a photograph of tap water under a vibrator generating a few watts/cm² at 18 kc/sec.

The behaviour of a cavitating region is affected by many parameters. We may list the following:

(i) Frequency and amplitude of exciting ultrasonic vibration.

(ii) Duration of applied ultrasonics.

(iii) Standing-wave pattern, if any.

(iv) Liquid irradiated; material of liquid itself, dissolved gas content, nuclei contained.

(v) Overall geometry of system.

---

[1] W. GÜTH: Acustica 6, 526 (1956).

These between them will control such factors as the ability of bubbles to grow to resonant size, the effect of radiation pressure on an individual bubble, and the tendency of the liquid to flow on a macroscopic scale either from the "quartz wind" phenomenon or from convection.

With so many relevant variables, it is not surprising that different investigators have reported different observations. It would seem, however, that most descriptions are qualitatively in accordance with the following outline. Liquids contain nuclei. These may be relatively large gas bubbles capable of rising to the surface if given time, or may be smaller nuclei persisting indefinitely in the body of the liquid. When an element of liquid is subjected to an alternating pressure of given amplitude, nothing happens unless it contains a nucleus large enough to cause cavitation at that pressure. Therefore, there may be periods of waiting until a nucleus drifts into the appropriate region. When cavitation occurs, there is in general an accretion to the bubble of gas previously dissolved in the neighbouring liquid. Also, at the final collapse of a cavitating bubble it often breaks up into a number of bubbles which may become separated from each other and, until they re-dissolve, serve as nuclei for further cavitation. Detailed descriptions of the sequence of events in cavitating liquids have been given by BLAKE[1], WILLARD[2] and MOHR[3].

4. Cinematography of bubbles. A development from the simple visual observation of cavitation is to photograph the phenomenon, particularly with high speed techniques, in order to elucidate the life history of a bubble. Such photographic techniques had previously been applied to hydraulic cavitation, but recently two accounts have been given[4,5] of experiments with ultrasonics.

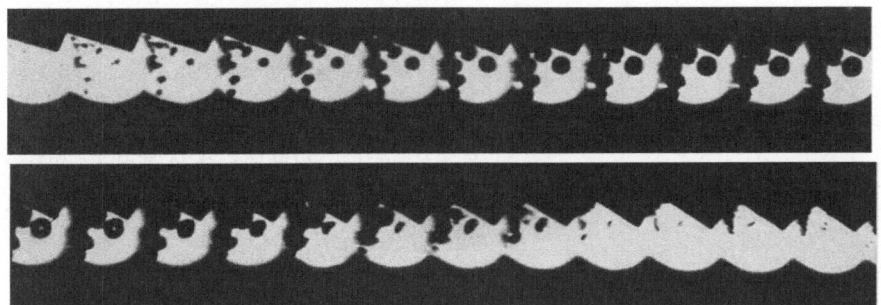

Fig. 2. Succession of frames in cinematograph of cavitation bubble (successive frames left to right, top row; then left to right, bottom row) (Photo A.T. ELLIS).

In order to obtain better phase-resolution for a given framing rate, and to have larger bubbles to study, the experiments have been made at low frequencies (strictly speaking, sonic rather than ultrasonic). With picture repetition rates up to a million per second, differences occurring in a small fraction of a period may be measured at acoustic frequencies of several kilocycles per second. Since photographs were always taken when many bubbles were present, the exact conditions holding for any one bubble could not be determined. However, some valuable qualitative results have been established. Thus Fig. 2, cine photos taken at 400000 frames/sec of degassed water cavitating in a 10 kc/sec acoustic

[1] F.G. BLAKE: Acoust. Res. Lab. Harvard Univ. Tech. Rep. 12 (1949).
[2] G.W. WILLARD: J. Acoust. Soc. Amer. 25, 669 (1953).
[3] W. MOHR: Acustica 7, 267 (1957).
[4] A.T. ELLIS: Cavitation in Hydrodynamics, p. 8. London: H.M.S.O. 1956.
[5] E. MUNDRY and W. GÜTH: Acustica 7, 241 (1957).

field, demonstrates that expanding bubbles remain relatively spherical while collapsing bubbles can become markedly unsymmetrical. It also illustrates—as is shown more conclusively in other examples—that a collapsing bubble can disappear virtually completely. By careful measurement of bubble dimensions from similar photographs, it is possible to plot the variation of radius with time. Fig. 3 shows the behaviour of a bubble having a "rest radius" R_r of 0.3 mm in a strong 2.5 kc/sec sound field; the "rest radius" is determined by still photography after the acoustic vibrations have been switched off. The amount of gas in the bubble corresponds to a larger value of R_0 than dealt with in the theoretical treatment of Sect. 6 but the behaviour agrees with what might be expected from that, a very low ratio for R_m/R_0 preventing true cavitation

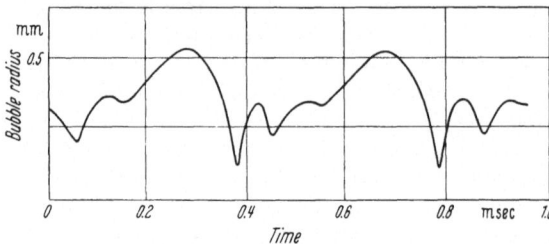

Fig. 3. Plot of mean radius against time for gas-filled bubble (E. MUNDRY and W. GÜTH).

(R_m is maximum and R_0 initial radius); the secondary maxima and minima show that the bubble's oscillations include a component at its own resonant frequency, which has been shock excited by the collapse. By contrast, Fig. 4 depicts the more violent collapse associated with a bubble having much smaller R_0 under similar circumstances.

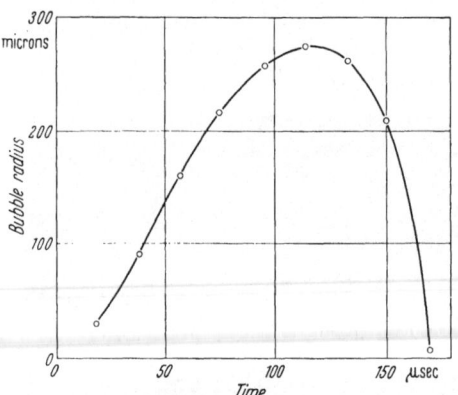

Fig. 4. Plot of mean radius against time of cavitation bubble (E. MUNDRY and W. GÜTH).

From these direct photographic observations, then, we receive general support for the outline we have given of the course of ultrasonic cavitation which will be developed more precisely later. There is also evidence that in practice it is seldom justifiable to think of a single, isolated cavitation bubble; it is much commoner for a number to occur in close proximity.

5. Cavitation noises. In the same way that cinematography is an extension of the simple visual observation of bubbles, so is the recording and analysis of cavitation sounds an obvious development from simply listening by ear. If a hydrophone—specially chosen to have a very high frequency response—is placed in the liquid that is irradiated with ultrasonics then it should pick up secondary sound which will give information about the cavitating bubbles.

ESCHE[1] has made a comprehensive study along these lines, covering a range of frequencies from 100 c/sec to 3.3 Mc/sec. For each exciting frequency, he analysed the cavitation sounds received into a complete spectrum giving the intensity at different frequencies. There was always a continuous background, on which were superposed "spectral lines" corresponding to the exciting frequency and its harmonics, also some of its subharmonics and their series of harmonics. The two types of sound can be explained as caused by non-linear oscillations of bubbles accompanied by sharp sounds originating at the sudden

[1] R. ESCHE: Acustica (Akustische Beihefte) **2**, 208 (1952).

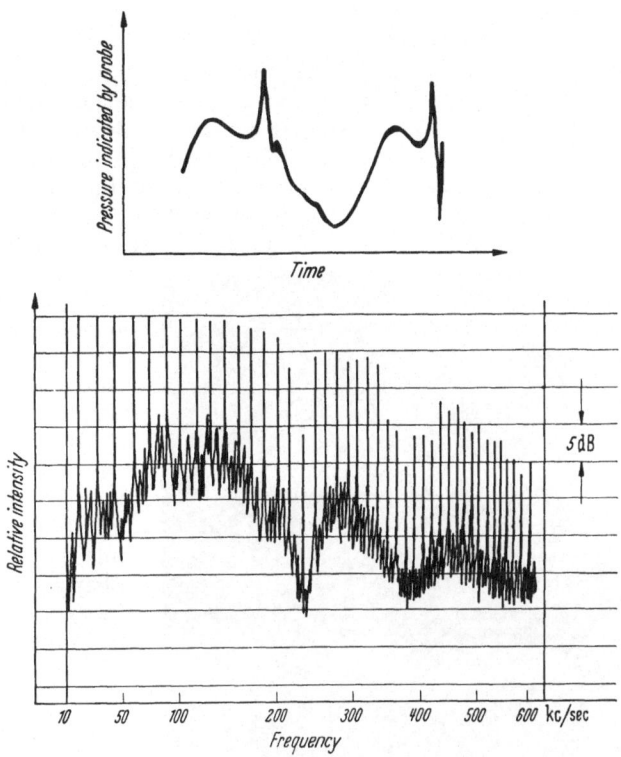

Fig. 5. Stress recorded by probe during acoustic cycle and corresponding sound spectrum (L. Bohn).

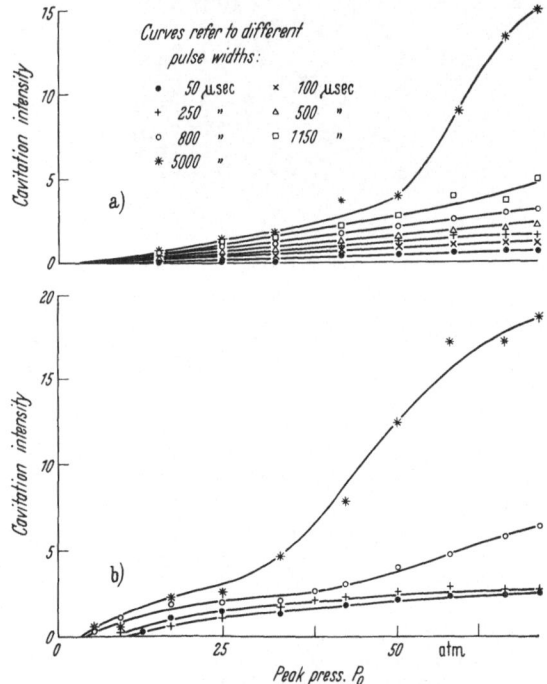

Fig. 6. Variation of intensity of cavitation noise with applied pressure (W. Mohr).

collapse of voids; GÜTH[1] has developed the theory of this in greater detail. ESCHE observed that above 10^4 c/sec the primary acoustic pressure needed to excite a just detectable cavitation sound rose more than proportionally to the frequency.

BOHN[2] continued more detailed studies along these lines. His hydrophone was a fine nickel wire, the stresses in which were detected by virtue of its magneto-striction. He was able to produce controlled bubbles at the tip of the wire by electrolysis. By working at a frequency as low as 15 kc/sec it was possible to record the stresses associated with a single acoustic cycle. An example of these is given in Fig. 5 together with the corresponding sound spectrum, obtained by integrating over many cycles. They refer to an acoustic intensity of 3 to 4 atm. The spectrum confirmes ESCHE'S findings. In the stress recording, a sharp peak caused by the cavitation collapse can be seen: since, even with the fine wire probe, the collpase only impinges on a very small fraction of the total area, this peak demonstrates the very high local stresses caused by cavitation.

Fig. 7. Schlieren photograph of shock wave arising from cavitation (E. MUNDRY and W. GÜTH).

MOHR[3] also used a hydrophone in the study of cavitation. He was not concerned with frequency analysis, but used a method of averaging the heights of the sound pulses received to provide an index of cavitation intensity. He noted the variation of this index with the primary applied pressure using pulses of 370 kc/sec ultrasonics; the results for different pulse widths are shown in Fig. 6 corresponding to pulse separations of (a) 10 and (b) 120 seconds.

Direct evidence for the existence of shock waves set up in liquids by the collapse of cavitating bubbles has been given by MUNDRY and GÜTH[4]. Fig. 7 is a photograph of theirs, obtained using schlieren optical techniques with their 2.5 kc/sec transducer. They confirmed that "gassy" bubbles could not produce shocks. With quite different apparatus, starting with a single large bubble, MELLEN[5] concluded from hydrophone measurements that a shock wave was

[1] W. GÜTH: Acustica 6, 532 (1956).
[2] L. BOHN: Acustica 7, 201 (1957).
[3] W. MOHR: Acustica 7, 267 (1957).
[4] E. MUNDRY and W. GÜTH: Acustica 7, 241 (1957).
[5] R. H. MELLEN: J. Acoust. Soc. Amer. 28, 447 (1956).

excited when the motion of the cavity wall reached supersonic velocities. ELLIS[1] has also had suggestions of very high stresses by applying cinematography to photo-elastic materials; pressures of the order of 5×10^4 atmospheres were indicated in solid bodies in the neighbourhood of collapsing voids.

6. Theoretical analysis of cavitation. It is of interest to attempt a mathematical treatment of cavitation. This was first undertaken by RAYLEIGH[2], who considered the subsequent behaviour of the liquid in the neighbourhood of a postulated spherical void, the ambient pressure being taken as constant. In hydraulic systems, of course, no general expression can be given for this pressure, but with ultrasonic cavitation it is an obvious approximation to take the ambient pressure variation as sinusoidal—though the very presence of the bubble studied, and of neighbouring bubbles, must prevent this from being exact.

On this basis, and with the additional assumptions that (i) the liquid is incompressible and (ii) the gas content of the cavitation bubble remains constant, a mathematical treatment of the motion of a spherical bubble has been developed[3].

Consider a bubble of initial radius R_0, containing gas at the equilibrium pressure $(P_A + 2S/R_0)$, P_A being the ambient pressure and S the liquid's surface tension.

When the bubble radius is R, the kinetic energy of the whole liquid, density ϱ, is

$$\int_R^\infty 2\pi \varrho\, r^2 \left(\frac{dr}{dt}\right)^2 dr = 2\pi \varrho\, R^3 \left(\frac{dR}{dt}\right)^2 \tag{6.1}$$

since the incompressibility of the liquid makes

$$r^2 \frac{dr}{dt} = \text{const} = R^2 \frac{dR}{dt}. \tag{6.2}$$

The kinetic energy can then be equated to the algebraic sum of the work done by the surface tension, gas pressure and liquid pressure at infinity, giving the energy equation:

$$\int_{R_0}^R \left\{ 4\pi R^2 \left[P_0 \sin\omega t - P_A + \left(P_A + \frac{2S}{R_0}\right)\frac{R_0^3}{R^3}\right] - 8\pi R S \right\} dR = 2\pi \varrho\, R^3 \left(\frac{dR}{dt}\right)^2 \tag{6.3}$$

for isothermal gas changes, if a sinusoidal pressure variation is applied at a great distance, making the instantaneous pressure

$$P = P_A - P_0 \sin\omega t. \tag{6.4}$$

Differentiating with respect to R, it becomes

$$2R\left[P_0 \sin\omega t - P_A + \left(P_A + \frac{2S}{R_0}\right)\frac{R_0^3}{R^3}\right] = 4S + 3\varrho R\left(\frac{dR}{dt}\right)^2 + 2\varrho R^2 \frac{d^2R}{dt^2}. \tag{6.5}$$

If the cavitating bubble is thought of as starting from a spherical solid particle, which it does not wet, and so containing no gas, the term $(P_A + 2S/R_0)\, R_0^3/R^3$ is omitted. In the cases that have been studied, this makes little difference to the overall behaviour.

General solutions are not possible, but specific examples have been evaluated. Fig. 8 shows a typical growth and collapse curve, the bubble radius R being

[1] A.T. ELLIS: Cavitation in Hydrodynamics, p. 8. London: H.M.S.O. 1956.
[2] Lord RAYLEIGH: Phil. Mag. **34**, 94 (1917).
[3] B.E. NOLTINGK and E.A. NEPPIRAS: Proc. Phys. Soc. Lond. B **63**, 674 (1950).

depicted along with the pressure P for the case $\varrho = 1$, $S = 80$ (water), $P_A = 10^6$, $P_0 = 4 \times 10^6$, $\omega = 3 \times 10^6$, $R_0 = 3.2 \times 10^{-4}$, all in CGS units. The initial value of dR/dt is taken as zero. The general resemblance to the experimental Fig. 4 is clear.

The point corresponding to the maximum void radius—R_m, occurring at time t_m—is of interest, since it determines the size of void which is to collapse and the remote liquid pressure causing the collapse (the collapse time being short enough so that we can roughly regard P as constant throughout it). R_m and t_m are plotted against different parameters in Figs. 9 to 11. The constant parameters are $\omega = 9 \times 10^4$, $P_0 = 4 \times 10^6$, and $R_0 = 3.2 \times 10^{-4}$. Fig. 10 shows that the phase at which t_m occurs does not vary with the frequency of the acoustic radiation, while R_m is inversely proportional to frequency. However, this and other conclusions must be applied guardedly, since they are only known to hold for certain values of the relevant parameters; in particular, as the acoustic frequency rises towards the resonant frequency of the bubble, cavitation stops altogether. Development from the work of Minnaert[1] shows that for small amplitudes this is given by

$$\varrho \, \omega^2 \, R_0^2 = 3\gamma \, (P_A + 2S/R_0) - 2S/R_0 \qquad (6.6)$$

γ being the ratio of specific heats for the gas in the bubbles. Kurtze[2] has considered the effects of large vibration amplitudes on resonant frequency, and has shown that the latter may be reduced very considerably.

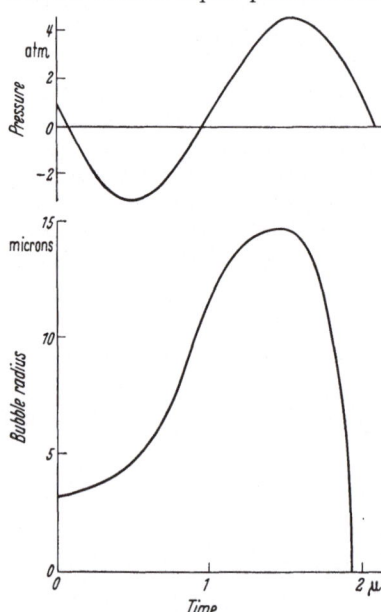

Fig. 8. Calculated variation of radius with time for cavitating bubble (B. E. Noltingk and E. A. Neppiras).

A closer examination of the basic equations shows that definite limits can be assigned to the conditions under which cavitation will occur[3]. These can be summarised in the statements that cavitation will not occur,

(i) For

$$R_T < \frac{4S}{3(P_0 - P_A)} \qquad (6.7)$$

(i.e. $P_0 < P_A + 4S/3\,R_T$) whatever the frequency, where $2SR_T^2 = 3P_A\,R_0^3 + 6\,S\,R_0^2$.

(ii) For

$$\omega^2 > \frac{3\gamma\,(P_A + 2\,S/R_0) - 2\,S/R_0}{\varrho\,R_0^2}$$

[i.e. R_0 greater than a root of the cubic $\varrho\omega^2 R_0^3 = 3\gamma\,P_A R_0 + 2S\,(3\gamma - 1)$] approximately, although variations in P_0 will also play some part, the formulae being more accurate the smaller P_0 is.

(iii) For $P_A \geq P_0$ approximately whatever ω and R_0 may be.

(iv) For P_A so small (or P_0 so large) that the cavitation bubble has no time to collapse while the pressure is still positive.

[1] M. Minnaert: Phil. Mag. **16**, 235 (1933).
[2] G. Kurtze: Nachr. Akad. Wiss. Göttingen, math.-phys. Kl. 1 (1958).
[3] E. A. Neppiras and B. E. Noltingk: Proc. Phys. Soc. Lond. B **64**, 1032 (1951).

R_0 is often a parameter that cannot easily be controlled, so it is of interest that, for any particular values of ω and P_A, (i) and (ii) can be combined to give

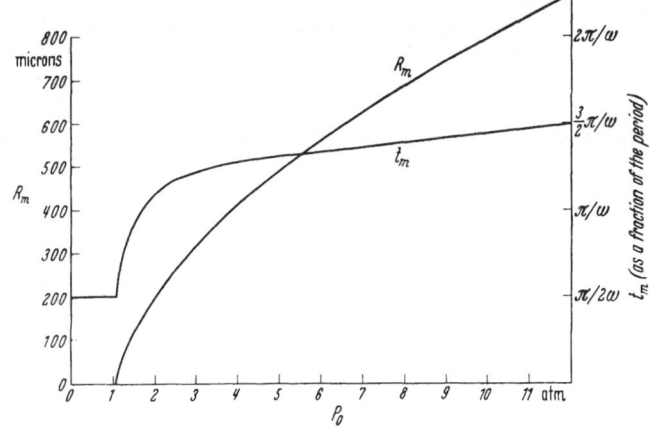

Fig. 9. Calculated variation with driving pressure of time and size of maximum bubble radius
(B. E. Noltingk and E. A. Neppiras).

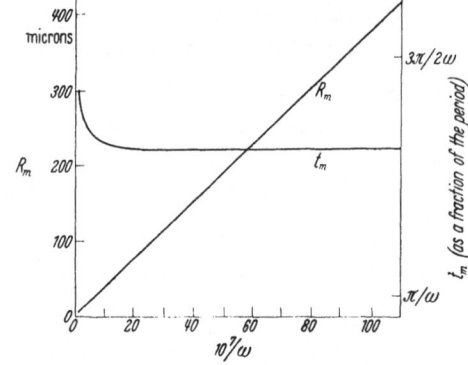

Fig. 10. Calculated variation with frequency of time and size of maximum bubble radius
(B. E. Noltingk and E. A. Neppiras).

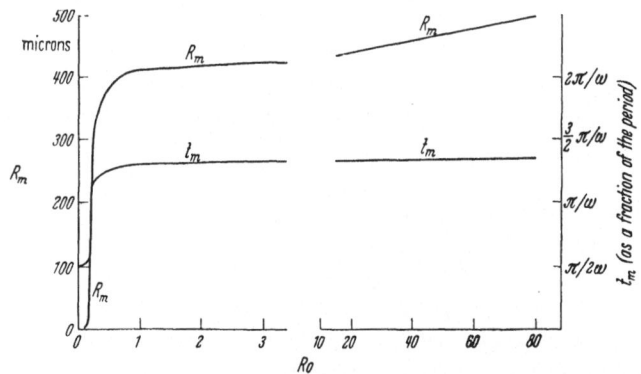

Fig. 11. Calculated variation with initial radius of time and size of maximum bubble radius
(B. E. Noltingk and E. A. Neppiras).

an absolute threshold for P_0 below which cavitation will not occur whatever the size of nuclei. This will be compared with experimental findings in a later section

Case (iv) implies that P_0 can actually become too large to allow cavitation. It may be doubted whether this will be experienced in practice, since an increase in the applied ultrasonic pressure will lead to an increase in the number and size of cavitating bubbles, and this will automatically reduce the effective P_0 for any one bubble. It is, however, a valid deduction that the violence of collapse of a single bubble cannot be increased indefinitely by increasing the applied ultrasonic pressure.

The treatment above is concerned with the growth phase of a cavitation bubble. Once the maximum size has been reached, collapse follows in a manner determined mainly by R_m, the corresponding value of P, and the pressure Q in the bubble at $t = t_m$.

None of the changes will be strictly adiabatic or strictly isothermal. Some idea of the conditions can be derived from the thermal time constant τ of a sphere

$$\tau = \frac{R^2 \varrho \sigma}{\pi^2 k} \tag{6.8}$$

(σ being specific heat, ϱ density and k thermal conductivity of the material). Events occurring on a time scale much less than τ can be thought of as adiabatic. For a sphere of air at S.T.P., $\tau = 0.4 R^2$ C.G.S. units, so that if, for instance, a bubble has a 10 micron radius (when it is at S.T.P.), times of less than a microsecond will not allow thermal equilibrium to be established. This suggests that it is not unreasonable to treat the slower growth phase as isothermal, and the more rapid collapse as adiabatic. The energy equation for collapse is then

$$\frac{3\varrho}{2} \left(\frac{dR}{dt}\right)^2 = P(Z - 1) - \frac{Q(Z^\gamma - Z)}{\gamma - 1} \tag{6.9}$$

where $Z = (R_m/R)^3$ and ϱ refers to the liquid. Surface tension can be ignored during collapse, and, since we are only concerned with orders of magnitude, P taken as constant.

From this it can be deduced that the minimum bubble size (before some rebound occurs) is given by

$$Z_{max}^{\gamma-1} = P(\gamma - 1)/Q. \tag{6.10}$$

The pressure in the gas and the innermost liquid shell then reaches its maximum value of

$$Q \left[\frac{P(\gamma - 1)}{Q}\right]^{\frac{\gamma}{\gamma-1}}.$$

The maximum radial velocity V_{max} is given by

$$V_{max}^2 = \frac{2 P(\gamma - 1)}{3\varrho\gamma} \left[\frac{P(\gamma - 1)}{Q\gamma}\right]^{\frac{1}{\gamma-1}} \tag{6.11}$$

occurring when the bubble has shrunk to a radius

$$R = \gamma^{\frac{1}{3(\gamma-1)}} R_{min}. \tag{6.12}$$

It may be noted, however, that computation of V_{max} for some realisable conditions yields values greater than the velocity of sound in the liquid. This of course invalidates the initial assumption that the liquid was incompressible for the final collapse phase.

The other initial assumption—that the contents of the bubble remain unchanged—may also be questioned. With more volatile liquids it is found that cavitation intensity is much less, and this is presumably to be attributed to the gas phase content of the bubble being increased by evaporation and lessening the severity of the collapse.

A more serious error in the simple treatment of the collapse is that spherical symmetry has been assumed, ignoring "Taylor instability" of accelerated liquid surfaces. BINNIE[1] has developed TAYLOR's original theory[2] for the special case of a cavitation bubble and has shown that there is a critical value of $4S/R^2$ for the inward acceleration $-d^2R/dt^2$. If this is exceeded, as would be predicted (for air in water) if $P/Q > 100$ and R_m more than a few microns, then the liquid surface will break up, finally collapsing round many randomly shaped bubbles. As we have seen, such a departure from spherical symmetry has in fact been observed. While, therefore, the theory cannot be expected to be valid for the most rapid collapses, it still appears to have considerable value in describing the earlier stages of a bubble's life and in indicating the effects of varying different parameters.

The assumption of isothermal expansion and adiabatic compression necessarily implies that high temperatures will be reached during collapse. The simple theory indicates a maximum temperature

$$T_{max} = T_0 \left(\frac{P}{3Q}\right)^{3(\gamma-1)} \tag{6.13}$$

where T_0 is the absolute temperature of the surrounding liquid. Inserting $\gamma = \frac{4}{3}$, $T_0 = 300°$ K and $P/Q = 200$, we have $T_{max} = 20000°$ K. In practice this must be considerably modified by non-spherical collapse and by imperfectly adiabatic conditions, but it appears reasonable to expect very high temperatures.

7. Cavitation thresholds.
In any attempt to develop the concept of "intensity of cavitation", we are faced with the problem that the expression is an indefinite one. Cavitation produces several different results, and if one state of cavitation is more effective than another with regard to one result we cannot be sure that it will also be more effective for another type of result. Uncertainties such as these have suggested that a more definite line of investigation is to study threshold conditions, noting the acoustic intensity required before the phenomenon which is being observed takes place at all.

The first point to establish, of course, is whether there is in fact any threshold, or whether any particular effect continues, only increasingly weakly, as the acoustic intensity is reduced to zero. That a definite threshold is at least sometimes to be expected is indicated by the theoretical curve in Fig. 9, where a sharp change in the curve can be seen at P_0 slightly greater than P_A. The satisfactory way of establishing a threshold is by making quantitative measurements over a range of acoustic intensities; qualitative estimates whether or not an effect occurs are really only justified when that effect never occurs in a small way. Some investigators have worked along these lines (though a surprisingly large number have not!), noting variations in behaviour at different input powers. Any effect for which cavitation is necessary can in principle be used to study thresholds, and a number of such results have been brought together for comparison in an earlier discussion on thresholds[3].

If optical observation is used, there is a danger of confusing true cavitation with the formation of permanent bubbles by de-gassing; the latter can take place at very low levels of the exciting vibration. Unless it has been prevented by working with a liquid that has a sufficiently small gas content, any conclusions as to

[1] A.M. BINNIE: Proc. Cambridge Phil. Soc. **49**, 151 (1953).

[2] G.I. TAYLOR: Proc. Roy. Soc. Lond., Ser. A **201**, 192 (1950).

[3] B.E. NOLTINGK: Cavitation in Hydrodynamics, x. London: H.M.S.O. 1956.

thresholds can have only doubtful value. When, in de-gassed liquids, cavitation is seen to occur in bursts, and it is ultimately the frequency of these that is measured, optical methods appear sound and have, as we shall see, given valuable results which have been interpreted in terms of the pre-existing nuclei in the liquid.

As mentioned in Sect. 5, Esche has made extensive measurements on the acoustic intensity needed to cause detectable sounds, finding very high intensities at the highest frequencies. However, Mohr's curve relating sound intensity to primary driving pressure (Fig. 6) shows that there is not a sharp threshold to this effect; it may be that below some input level no sound at all occurs, but the sound does not increase rapidly and steadily for greater inputs. Kurtze[1] has considered this subject further and suggested how Esche's results can be explained as a measurement of the acoustic pressure input needed to form a cavitation bubble of a certain minimum value (10^{-2} cm) for its maximum radius. The noise from the collapse of such a bubble is presumed to be the least that could be detected with the particular apparatus employed and it is this artificial threshold that has been measured rather than anything more directly related to the high radial collapse velocities which are the true characteristic of cavitation.

The other principle effects associated with cavitation and amenable to quantitative study are luminescence, chemical actions and bactericidal effects. Where these have been studied carefully thresholds have been found in the region of 1 to 10 atmospheres peak exciting pressure, even at frequencies of several megacycles per second. There appears, therefore, to be general agreement with the theoretical prediction of an absolute threshold of 1 atm below 10^6 c/sec rising slowly to 2 atm at 10^7 c/sec. This is an absolute lower limit; if there is a lack of suitable nuclei, higher acoustic pressures will be required, which may be particularly relevant to special cases such as that of bacteria. It should be emphasised that in many experimental arrangements it is difficult to determine the exciting intensity at all accurately—and in many it has never been attempted!

8. Nuclei. We have pointed out that liquids are found to cavitate much more readily than would be expected from simple considerations of cohesive forces. The theoretical treatment developed has been based on the idea of nuclei existing in the liquid; expanding bubbles grow round these nuclei. The theoretical treatment could be used to derive an exact relation between the size of a nucleus —expressed, say, as the radius R_0 of a corresponding spherical element—and the minimum acoustic pressure P_T for cavitation [$P_T = P_0 - P_A$ in the nomenclature of Eq. (6.4)]. Since, however, we are more interested in orders of magnitude than in exact solutions, we may note that at least for $R_0 < 10^{-3}$ cm in water at one atmosphere ($S = 80$, $P_A = 10^6$) Eq. (6.7) has a solution $R_T = 3 R_0$ within a factor of 2. We thus find that approximately

$$P_T = 0.5 \frac{S}{R_0}. \tag{8.1}$$

Hence a threshold of 10 atm corresponds to bubbles having initial diameters about 0.1 micron.

The cavitation action would follow according to theoretical analysis if the nuclei were simply bubbles of free gas; their rate of travel under Stokes' Law would be so low that they would remain suspended indefinitely. The anomaly to be explained is why the gas they contain does not go into solution immediately unless the liquid is supersaturated. It might be expected that incomplete wetting

---

[1] G. Kurtze: Nachr. Akad. Wiss. Göttingen, math.-phys. Kl. 1 (1958).

by the liquid of the walls of its container would allow cavitation to originate there, but in fact cavitation can also start in the body of the liquid. Approximate calculations of KURTZE's[1] give a lifetime τ_b for a bubble of

$$\tau_b = \frac{R_0^2}{3 D \alpha} \tag{8.2}$$

where D is the diffusion constant for gas in the bubble and α the ratio of the numbers of gas molecules per unit volume that are in solution in the liquid and free in the gas respectively. Taking $D = 2 \times 10^{-5}$ cm²/sec, $\alpha = 0.02$, this indicates that a bubble of 1 micron radius should dissolve in 0.01 sec.

Three solutions have been proposed to this problem of instability of nuclei: (i) FOX and HERZFELD[2] have suggested that surface active organic molecules may always be present in sufficient quantities to form a skin round a bubble: the skin is taken to have negligible tensile strength so that it will not hinder bubble expansion, but a finite compressive strength so that the gas inside the bubble can be at a lower pressure than would otherwise be possible. (ii) HARVEY[3] pointed out that solid particles in the liquid can play an important role. For, while the bond strength between a liquid and a solid which

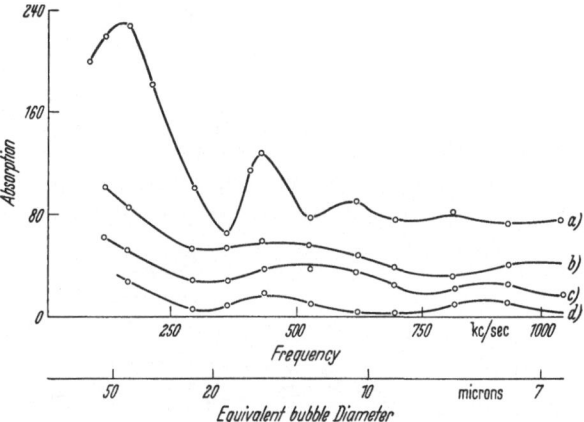

Fig. 12. Variation of ultrasonic absorption as water stands for different times (after E. G. RICHARDSON).

it wets is expected to be of the same order as the internal strength of the liquid, if the liquid does not wet a solid particle, the interface would provide the required weakness in the structure. (iii) If the solid has crevices in its surface which are filled with gas, the curvature of the gas-liquid interface can be markedly convex to the gas, so allowing stabilisation; KURTZE[1] has developed this concept.

It is a difficult matter to make independent investigations on the existence of such nuclei. Even bubbles that are some microns in size will be very difficult to detect optically, since a relatively large volume must be searched. NAAKE et al.[4] have attempted this. RICHARDSON[5] and STRASBERG[6] have investigated nuclei using ultrasonic absorption, relating their results to cavitation thresholds. Since the absorption is much stronger at the resonance frequency of a bubble, and this depends on its size [Eq. (6.6)] a measurement of the absorption at different frequencies gives information about the size distribution of nuclei. Thus Fig. 12 shows how the excess absorption of tap water (measured with a reverberation technique) decreases on standing; the four curves (a) to (d) correspond to waiting times of 0, 0.5, 1 and 20 hours respectively. The absorption

[1] G. KURTZE: Nachr. Akad. Wiss. Göttingen, math.-phys. Kl. 1 (1958).
[2] F. E. FOX and K. HERZFELD: J. Acoust. Soc. Amer. 26, 984 (1954).
[3] E. N. HARVEY: J. Amer. Chem. Soc. 67, 156 (1945).
[4] H. J. NAAKE, K. TAMM, P. DÄMMIG and H. W. HELBERG: Acustica 8, 65 (1958).
[5] E. G. RICHARDSON: Wear 2, 97 (1958).
[6] M. STRASBERG: Cavitation in Hydrodynamics, p. 6. London: H. M. S. O. 1956.

can be correlated with the volume of undissolved air in the water which was measured as 5, 4, 3 and 0 parts per thousand in the four cases. Calculations from Stokes' Law on the rate of rise agree with the order of magnitude observed, though it may be noted that curve (b), for instance, does not appear to approach (a) rather than (d) at the high frequency end, as might be expected. Such absorption measurements have not been made in absolute terms, only as an excess over that found when the liquid has stood for long periods. There is thus a limit to the sensitivity of the technique and Strasberg calculates the capabilities of his apparatus to be as shown in the table. The very high sensitivity in terms of

Table.

Range of bubble radius (microns)	3 to 9	7 to 20	11 to 33
Min. detectable number of bubbles in that range (per cm³)	0.7	0.03	0.005
Fraction of total volume occupied by bubbles	6×10^{-10}	3×10^{-10}	2×10^{-10}

percentage volume indicates that it is unlikely that measurements of compressibility—which have been suggested as an alternative technique—could yield comparable information. All the same, as the bubble size is reduced, it can be seen that the sensitivity in terms of the minimum numbers detectable grows rapidly worse. In fact, while ultrasonic absorption measurements are very valu-

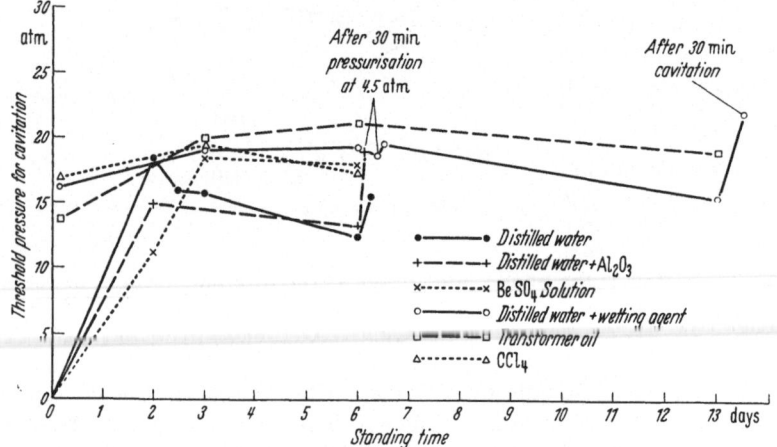

Fig. 13. Change of cavitation thresholds on standing (G. Kurtze).

able for studying the larger nuclei—which are generally associated with a relatively gross contamination—they are of little help for the postulated smaller nuclei with dimensions of a micron or less.

Ideas about the smallest nuclei have so far only been developed by considering the phenomenon of cavitation itself. Kurtze has done this at some length, producing cavitation in different liquids by means of a single, short pulse of negative pressure. This was shown to allow any bubble formed to reach a size detectable optically and has the advantage of preventing complications from bubbles growing in the course of successive pressure cycles, while the spherical shell transducer used (as by most recent workers) eliminated any peculiar effects from the walls of the container. The threshold negative pressures needed to cavitate various liquids after different waiting times are shown in Fig. 13. From these and other experiments, it could be concluded that when free access to air is possible all the liquids studied eventually reached similar threshold values, independent of surface tension, dissolved gas and water structure ($BeSO_4$ solution). This is in general agreement with RICHARDSON's thresholds at 442 kc/sec for ethyl ether (19.5 atm), pentane (13.7), acetone (15), hexane (10) and carbon tetrachloride (19), though he could not produce cavitation in acetaldehyde, isopentane, benzene or toluene even at 40 atmospheres. RICHARDSON detected cavitation optically, using liquids supersaturated with air. Since KURTZE's "waiting time" before high thresholds were reached is much more marked for liquids with higher surface tensions it is presumably associated with the need for wetting.

KURTZE "pressurised" samples only to a few atmospheres, when the effect was apparently a small one of hastening the wetting. RICHARDSON subjected samples to much higher pressures and found a sharp distinction in the effect on water and on the organic liquids tried. None of the latter had their thresholds affected by pressurising, whereas the water threshold increased linearly with the hydrostatic pressure that had been applied. The slope varies with air content of the water, but is apparently about $\frac{1}{3}$, 30 atm pressurising being needed to increase the threshold by 10 atm. It should be emphasised that RICHARDSON's work is mainly concerned with larger nuclei, KURTZE's with the small residuals.

KNAPP[1] has also done extensive work on the effect of pressurising water, though he used hydraulic techniques to measure tensile strength. He confirms the finding that pressurising increases strength initially, but his slope is only $\frac{1}{15}$, 200 atmospheres pressurising giving mean strengths of about 14 atmospheres, while increasing pressurising from 200 to 1500 atm has little additional effect. A considerable scatter can be attributed to the randomness of nuclei. Considerations of the frequency of breakdown indicate that on the average, even after pressurisation, there is 1 nucleus in 16 cm³ which cannot resist more than 32 atm and 1 in 100 cm³ which can only resist 1 atm.

From the complexity of all these findings, it would seem that a simple single mechanism can scarcely be adequate to explain all nuclei. Results are generally in agreement with the idea that the larger nuclei, responsible for low thresholds, are temporary bubbles of entrained gas. The form of the ultimate smaller nuclei is less certain. The organic skin model, apart from having little positive evidence in its favour, appears to be unlikely to occur in all the wide range of liquids that have been studied. It seems possible, however, that some such picture might account for the unusual behaviour of sea-water, which was found[2] to show an exceptionally large ultrasonic absorption in the region of 1 Mc/sec and a low cavitation threshold that did not increase on standing. Nuclei consisting of gas bubbles stabilised in cracks are in accordance with many observations and KURTZE has summarised the arguments in favour of them. It is a little difficult to reconcile such a model with the finding of (organic) liquids whose threshold does not rise after pressurising at high pressure or with increases in threshold that are much less than the pressurising. Again, the magnitude of the solid particles called for is a little unsatisfactory. If they are to contain crevices of the order 0.1 micron wide, then, unless the particle has a very strange shape, its own dimensions might be expected to be a hundredfold greater, i.e. 10 microns. Unless its specific gravity was nearly unity, it should then separate out more rapidly than is found to be the case. The possibilities of detecting such solid particles—visually or ultrasonically—would also be higher; there is already some slight positive evidence here, inasmuch as specks have been seen persisting

---

[1] R.T. KNAPP: Trans. Amer. Soc. Mech. Engrs. **80**, 1315 (1958).
[2] E. G. RICHARDSON: Wear **2**, 97 (1958).

afterwards at the point of collapse of cavitating bubbles. The third possibility —of solid particles with no gas attached, but unwetted by the liquid—does not meet the objections we have listed to the gas—in—crevice theory, but does not account for cavitation threshold being independent of surface tension. Finally, we may mention the idea of small gas bubbles being stabilised by electrostatic forces, a suggestion which has been put forward in general terms but not developed in any detail.

9. Erosion. The destructive action of cavitation in hydraulic systems was the practical effect which first drew attention to the phenomenon, and it has remained a subject of great interest to hydraulic engineers. As experiments have continued, there has seemed to be little basic difference between the damage caused by cavitation initiated in different ways, and a vibration method, because of its greater simplicity and ease of control, is becoming common for test purposes. It must be remembered, however, that the time scales vary between hydraulic and ultrasonic cavitation. In one experiment[1], it was found that only slight damage could be done at 400 kc/sec and none at 700 kc/sec; the intensity was not stated but appears to have been of the order of 100 W/cm².

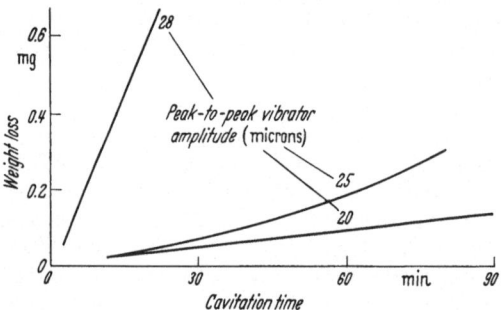

Fig. 14. Weight loss for specimens recorded at different amplitudes of vibration (after W. H. WHEELER).

The rate of erosion increases rapidly with the intensity of vibration. Fig. 14 illustrates this for annealed pure iron specimens vibrated at 8 kc/sec in $N/10$ KCl solution saturated with air[2].

An important issue is the extent to which cavitation damage is strictly erosion and how far it is due to corrosion. CALLIS[3] has summarised the evidence demonstrating the part that must be played by corrosion, particularly the finding of PETRACCHI[4] that damage could be prevented by polarising a specimen. WHEELER[2] confirmed the importance of electrolytic action, showing also the great reduction in cavitation attack when an inert liquid, toluene, was used as working fluid. He distinguished between matter removed from specimens as solid particles and that which went into solution, the former nearly always being more, sometimes by a factor of ten; however this does not give a clear distinction between corrosion and erosion, since a corrosive attack that produced deep crevices (as suggested by some micro-sections) could dislodge solid particles. More convincing evidence that some direct erosion occurs is provided by the fact that cavitating toluene does have some destructive action; similarly, RASMUSSEN[5] shows that non-metallic materials (Perspex and ebonite) can receive damage comparable to that in some metals.

BEBCHUK[6] conducted a range of experiments in which he measured the damage arising with different cavitating liquids. If weight loss is plotted against the vapour

[1] J. Z. LICHTMAN, D. H. KALLS, C. K. CHATTEN and E. P. COCHRAN: Trans. Amer. Soc. Mech. Engrs. **80**, 1325 (1958).
[2] W. H. WHEELER: Cavitation in Hydrodynamics, p. 21. London: H.M.S.O. 1956.
[3] G. T. CALLIS: Cavitation in Hydrodynamics, p. 18. London: H.M.S.O. 1956.
[4] G. PETRACCHI: Metallurg. ital. **41**, 1 (1949).
[5] R. E. H. RASMUSSEN: Cavitation in Hydrodynamics, p. 20. London: H.M.S.O. 1956.
[6] A. S. BEBCHUK: Soviet Phys. (Acoustics) **3**, 395 (1957).

pressure of the liquid for the temperature of the experiment, identical curves are found (peaking at vapour pressures of 100 mm Hg) for alcohol, acetone and trichlorethylene, while water is several times more destructive. However, the greater activity of water holds for ceramic and for gold specimens, as well as aluminium, zinc and tin, so that BEBCHUK maintains that it is not an oxidation effect.

From all these experiments, there can be little doubt that both erosion and corrosion occur. In some cases, the surface of a specimen has been seen to be indented after subjection to cavitation, without significant quantities of material being removed. WHEELER has suggested that a chemical action could result from the high temperature associated with bubble collapse (Sect. 6) but it seems improbable that significant amounts of solid are raised to high temperature. The more plausible suggestion is that mechanical damage to the film between metal and liquid allows chemical actions to proceed which the film would normally retard or prevent. An important factor then is the resistance of such a film to the "water-hammer" forces associated with the collapse of a cavitation bubble.

10. Dispersion and emulsification. Since we have seen that in ultrasonic cavitation there are agencies strong enough to erode solid surfaces, it is not surprising that material adhering to surfaces can also be removed. This holds for both solids and liquids. The effect has been known since the earliest experiments with intense ultrasonics; summaries of the situation a few years ago have been published[1,2]. Recent interest in the subject has been more in its possibilities as a practical technique—to clean surfaces from contaminants, or to produce emulsions—than in its scientific interest, so that there is a paucity of fundamental investigations.

OLAF[3] conducted a series of experiments at different frequencies, using an optical method to give quantitative results for the cleanliness of a glass surface. He concluded that cleaning only took place when cavitation occurred or when a bubble vibrated in resonance. However, his criterion for cavitation was based on acoustic measurements, as developed by ESCHE[4], and we have seen that this does not lead to a genuine cavitation threshold but rather a minimum value for maximum bubble size. OLAF's experiments therefore indicate that a certain violence of collapse must be exceeded for the dispersion of contaminants to take place; consequently a higher intensity is needed at higher frequencies. His own conclusion was that the optimum frequency depended on the particular geometry of the experimental arrangement used. For cleaning large surfaces, general experience supports the prediction that for the most effective action the frequency should be reduced indefinitely until practical considerations supervene. The observation of cleaning coming both from cavitation and from resonant bubbles is again made by MAKAROV and ROZENBERG[5] who distinguish occasional catastrophic collapses making major attacks on the surface and vibrating gas-filled bubbles which can be stabilised for long periods in cracks which they penetrate and scour.

OLAF found the trend of results to be the same for dispersing jeweller's rouge in water and for oil, grease or wax in trichlorethylene. He emphasises the importance of selecting the correct liquid. It would seem that differences here might

---

[1] L. BERGMANN: Der Ultraschall. Zürich 1954, Suppl. 1957.

[2] B.E. NOLTINGK and N.B. TERRY: Technical Acoustics II (E.G. RICHARDSON ed.). Amsterdam:: Elsevier 1957.

[3] J. OLAF: Acustica **7**, 253 (1957).

[4] R. ESCHE: Acustica (Akustische Beihefte) **2**, 208 (1952).

[5] L.O. MAKAROV and L.D. ROZENBERG: Soviet Phys. (Acoustics) **3**, 403 (1957).

account for considerable variations found in the effectiveness of ultrasonic treatment; there is bound to be a danger that the contaminant, after being liberated once from a surface, will re-adhere unless the medium has been chosen to allow it to remain suspended.

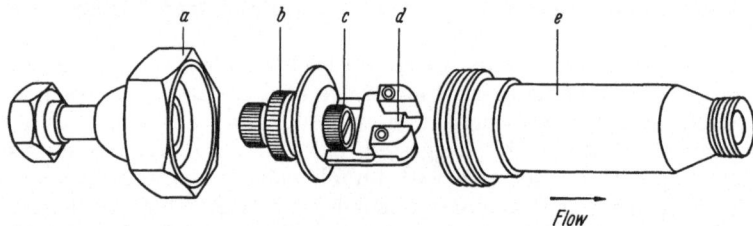

Fig. 15. Vibrating element of ultrasonic "liquid whistle" (J.S. SMITH). *a* base; *b* adjustable jet body; *c* jet insert; *d* vibrating blade; *e* resonant bell.

In some cases, ultrasonic vibrations have been used with the primary object of producing suspensions of fine solid particles. The action here must be similar to that causing cavitation erosion. Some summaries of the literature in this field have been made[1].

Surface chemistry is again of supreme importance in the emulsification of one liquid in another, which is of course analogous to dispersion and has also been brought about under the action of ultrasonics. JANOVSKI and POHLMANN[2] showed that the rate of emulsification was greatly increased when a "liquid whistle" type of generator was used instead of a magnetostrictor. SMITH[3] has enumerated many applications of such apparatus; the vibrating element used is shown in Fig. 15. It would appear that hydraulic factors—coarse pre-mixing and perhaps hydraulic cavitation—must play a large part in the performance of an emulsifying whistle, in addition to the ultrasonic action. Some cavitation is necessary for the ultrasonic emulsification of most liquids, but not when one of them is a metal. BONDY and SÖLLNER[4] with a conventional 250 kc/sec generator, investigated the dependence of the degree of emulsification on the time of treatment, finding a roughly asymptotic approach to the final concentration. It has been suggested[5] that an equilibrium is set up between emulsification by cavitation and a coagulation in the non-cavitating region of the ultrasonic field.

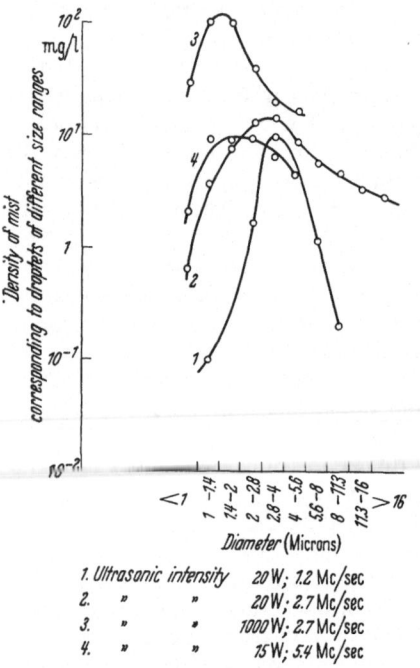

1. Ultrasonic intensity   20 W; 1.2 Mc/sec
2.   "   "   20 W; 2.7 Mc/sec
3.   "   •   1000 W; 2.7 Mc/sec
4.   "   "   15 W; 5.4 Mc/sec

Fig. 16. Droplet size in ultrasonic mists (K. BISA et al.).

When intense vibration occurs at a liquid-gas interface, droplets of liquid can be thrown off; they may be small enough to remain suspended in the gas

[1] L. BERGMANN: Der Ultraschall. Zürich 1954, Suppl. 1957.
[2] W. JANOVSKI and R. POHLMANN: Z. angew. Phys. **1**, 222 (1948).
[3] J.S. SMITH: Chemical and Process Engineering 54 (February 1959).
[4] C. BONDY and K. SÖLLNER: Trans. Faraday Soc. **32**, 556 (1936).
[5] H. CAMPBELL and C.A. LONG: Pharm. J. **163**, 127 (1949).

for some time as a mist or fog. This has been investigated by BISA, DIRNAGL and ESCHE[1] at frequencies above 1 Mc/sec, and by ANTONEVICH[2] at 20 kc/sec. The former used focussed, ceramic transducers, which gave mists of the composition indicated in Fig. 16 for water. It can be seen that the particle size falls off slightly with decreasing amplitude and considerably with decreasing frequency; the latter agrees with ANTONEVICH's findings that at 20 kc/sec the chief droplet sizes are 30 to 70 microns. While cavitation plays some part, it appears that the main action in the formation of ultrasonic fog follows from the excitation of surface waves. When these exceed a critical amplitude, a droplet can be detached from the surface, whereupon the various forces acting—from radiation pressure, surface particle acceleration, etc.—may be able to project it into suspension in the adjacent gas. The wavelength for surface waves is

$$\lambda_s = \left(\frac{2\pi S}{\varrho f^2}\right)^{\frac{1}{3}}, \tag{10.1}$$

S being surface tension and ϱ density of the liquid, while f is the exciting frequency. The predicted variation with f is as observed, and there is fair quantitative agreement for absolute sizes on the expectation that the droplet diameter will be about λ_s.

11. Luminescence. Under some conditions a faint luminescence is associated with the action of intense ultrasonics in liquids. The mechanism by which it is caused is not yet generally agreed. Many different liquids have been experimented on. For the most part, little correlation is found with luminescence resulting from ordinary chemical reactions, though solutions of luminol (3-aminophthalhydrazide), which is know to luminesce under other exciting influences, can give out a relatively strong light in ultrasonic fields. Where simpler materials, such as pure water, have been concerned, the phenomenon has been termed "sono-luminescence", to distinguish it from "sonic chemi-luminescence" which may have a more specifically chemical explanation. The light intensity concerned is generally so low that dark adaption of the eyes is needed before it can be appreciated, but in a few cases it has been claimed that luminescence was visible immediately.

BERGMANN[3] has summarised the earlier work, and more recently JARMAN[4] has reviewed the position. Even since then, however, more experimental work[5-8] has become available. Some details of the work appear contradictory, at any rate unless it is allowed that effects depend critically on small changes in frequency, state of liquid, etc.

Most of the earlier experiments were carried out at higher ultrasonic frequencies (1 Mc/sec) though CHAMBERS' extensive investigations[9] were at less than 10 kc/sec. GÜNTHER and colleagues[5] established a definite increase in intensity as the exciting frequency was increased. They, and others, observed that luminescence in standing waves occurred only in the neighbourhood of the nodes, i.e. where the pressure alternations were largest; this is to be expected, since it is recognised

[1] K. BISA, K. DIRNAGL and R. ESCHE: Siemens-Z. **28**, 341 (1954).

[2] J.N. ANTONEVICH: Trans. I.R.E. PGUE-7, 6 (Feb. 1959).

[3] L. BERGMANN: Der Ultraschall. Zürich 1954, Suppl. 1957.

[4] P. JARMAN: Sci. Progr. **46**, 632 (1958).

[5] P. GÜNTHER, W. ZEIL, U. GRISAR and E. HEIM: Z. Elektrochem. **61**, 188 (1957).

[6] D. SRINIVASAN and L.V. HOLROYD: Phys. Rev. **99**, 633 (1955).

[7] P. JARMAN: Proc. Phys. Soc. Lond. **73**, 628 (1959).

[8] E. MEYER and H. KUTTRUFF: Z. angew. Phys. **11**, 325 (1959).

[9] L.A. CHAMBERS: J. Chem. Phys. **5**, 290 (1937).

that cavitation necessarily accompanies ultrasonic luminescence and cavitation is associated with pressure changes rather than displacement. No precise experiments have been described to clarify the report[1] that at high frequencies travelling waves are much less effective than standing waves for producing luminescence, though Wagner[2] records that a ten second build-up time is needed after the ambient pressure has been changed.

Much of the variation with external conditions is in line with other observations on cavitation. Thus Jarman[3] notes that for consistent results the liquid sample must stand for two or three days; this agrees with Kurtze's investigations[4] into nuclei using pulses of negative pressure; and it may explain some earlier discrepancies (since the waiting time is thought to be associated with the movement of nuclei, much longer times might be needed for high viscosity liquids). Similarly the pronounced reduction in the intensity of luminescence at higher temperatures agrees with the concept of a higher vapour pressure allowing more liquid to evaporate into a cavitating bubble and so "cushion" its collapse. Jarman has extended the relationship to a number of liquids, finding a fair correlation between the light intensity and the ratio (Surface Tension)$^2 \div$ (Vapour Pressure). The addition of salts in solution to a liquid that is cavitating does not make a great difference to its luminescence, but the nature of dissolved gases is important. Thus Günther et al.[5] report a 200-fold increase in brightness when xenon is in solution in the water treated, compared with oxygen. Jarman records increased luminescence from the addition of carbon tetrachloride, but, as Srinivasan remarks of the work of Griffing and Sette[6], the likelihood of chemical reactions occurring prevents simple conclusions being drawn from this.

Some mention of the colour of the light from ultrasonic luminescence is made in earlier work, and Srinivasan and Holroyd and Günther and colleagues have made some precise spectral measurements. The former report a continuum equivalent to black-body radiation at about 10000° K. The latter observe also some spectral lines corresponding to matter in the liquid. The intensity of the continuum varies greatly from case to case, but its distribution across the spectrum changes little. Moreover, the spectral lines were suppressed when a source of intense γ-rays was brought up, while the continuum remained unchanged, suggesting basically different mechanisms for excitation.

It has been suggested[7] that chemical actions always play a part in ultrasonically excited luminescence, and we have indicated that it seems likely that in some cases the light observed may arise from energy that has been liberated chemically. Two recent experiments, however, give strong evidence against exclusively chemical origins for such luminescence. Srinivasan, in varying the type of gas dissolved in the liquid examined, was sometimes able to detect the chemical products of cavitation, but with some gases observed just as bright luminescence when no chemical products could be found. Again, Günther et al. measured the ultrasonic production of hydrogen peroxide at the same time as measuring the intensity of luminescence, both in pure water. There was little correlation between the two, the peroxide formed per unit of light emitted varying by a factor of 20:1 when oxygen was substituted for xenon as the solute.

[1] B.E. Noltingk and N.B. Terry: Technical Acoustics II (E.G. Richardson ed.). Amsterdam: Elsevier 1957.
[2] W.U. Wagner: Z. angew. Phys. 10, 445 (1958).
[3] P. Jarman: Proc. Phys. Soc. Lond. 73, 628 (1959).
[4] G. Kurtze: Nachr. Akad. Wiss. Göttingen, math.-phys. Kl. 1 (1958).
[5] P. Günther, W. Zeil, U. Grisar and E. Heim: Z. Elektochem. 61. 188 (1957).
[6] V. Griffing and D. Sette: J. Chem. Phys. 23, 503 (1955).
[7] V. Griffing: J. Chem. Phys. 20, 939 (1952).

An alternative proposal[1] is that ultrasonic luminescence is associated with an electrical discharge across low pressure gas in the cavitation void. This theory, in outline, is that a sudden rupture of the structure of the liquid may cause different parts of the surface of the void to be differently charged, because of the random distribution of the ions originally present. Theoretical development gives a formula for E the electric field across a lens-shaped void

$$E = \frac{4e}{r} \sqrt{Nd} \qquad\qquad (11.1)$$

where d is the separation of the faces of the void, which is taken to be circular with radius r. There are N ions per unit volume of liquid, each having charge e. Since this theory was originally propounded some years ago, but has recently been rather vaguely invoked to explain both chemical effects and luminescence produced by ultrasonics, it seems well to examine it more critically. It is not in accordance with current ideas on the tensile strength of liquids that it should be possible to break up their basic structure with stresses of a few atmospheres, as is required for the initial action, according to this theory. Nor is there any evidence of radical departures from spherical in the early stages of a bubble's growth. The recent work of KURTZE and of RICHARDSON (described in Sect. 8) appears to establish soundly that cavitation takes place on gaseous nuclei; it is very difficult to see how the movement of liquid outwards from such a nucleus could leave vastly different charges on opposite sides. Moreover, luminescence of comparable magnitude occurs in a wide variety of liquids, bearing little relation to their electrical properties: more accurate measurements by JARMAN[2] have contradicted the earlier report[3] of correlation between luminescence and dipole moment of the liquid irradiated. It seems, therefore, that if micro-discharges do occur inside cavitation bubbles some clearer explanation is desirable for the origin of the charges.

In Sect. 6 we have outlined the reasoning which leads to the expectation of very high temperatures in collapsing bubbles. It appears that such temperatures could well lead to incandescence in the gas phase, and it may be unnecessary to look further for the origin of much of the light observed.

It may be noted that on this explanation the luminescence should occur strictly at the instant of collapse of a bubble whereas discharges may arise earlier in the cycle. A detailed resolution of the time scale of events should therefore be very informative. WAGNER[4], JARMAN[2], and MEYER and KUTTRUFF[5,6] have attempted this and all report that the light comes in discrete pulses once per acoustic cycle—in contrast to the "chemi-luminescence" of luminol which lasts throughout the whole period. WAGNER and JARMAN, using standing waves at 260 and 16.5 kc/sec respectively, both related the phasing of the light flash to that of the liquid pressure measured with a hydrophone and reported it as occurring at or shortly after the minimum pressure for non-volatile liquids; JARMAN found a difference for the volatile liquids he used, for which flashes were recorded at or just before the maximum pressure. Neither identifies the instant of light emission with the collapse phase. We may compare the phasing of the cavitation pressure pulse shown in Fig. 5 (p. 265) which occurs at about the time that

[1] J. FRENKEL: Acta physicochem. URSS. **12**, 317 (1940).
[2] P. JARMAN: Proc. Phys. Soc. Lond. **73**, 628 (1959).
[3] L. A. CHAMBERS: J. Chem. Phys. **5**, 290 (1937).
[4] W. U. WAGNER: Z. angew. Phys. **10**, 445 (1958).
[5] E. MEYER and H. KUTTRUFF: Z. angew. Phys. **11**, 325 (1959).
[6] I am greatly indebted both to Mr. JARMAN and to Prof. MEYER for allowing me to see their results prior to publication.

would be expected for the collapse of a bubble. There is, however, an element of indirectness in this approach; the phase characteristics of the hydrophone and all the related circuits have to be established, which is not always easy at higher frequencies. Accordingly, MEYER and KUTTRUFF used an elegant technique at lower frequency. With a highly resonant (self-maintained) 2.4 kc/sec vibrator dipping into liquid, they found that the individual light flashes gave adequate output from a photo-cell so that a spark gap could be triggered. In this way, by using a variable delay, the bubbles themselves could be photographed at different phases relative to the pulse of luminescence, with no complication from an intermediate stage of pressure measurement. Their results showed clearly that light was emitted at the moment of collapse; if no delay was inserted, no bubbles were apparent. Further confirmation came from a study of position, in that bubbles were observed to move radially inwards, being formed nearer the periphery but always collapsing at the centre of the vibrating tip, and the luminescence was shown to originate precisely at the centre.

The conclusions appear to be that at the present time there is still some obscurity about the causes of ultrasonic luminescence, and it may well be that more than one mechanism contributes to the phenomenon. It seems probable, however, that a large part is played by the high temperatures reached through adiabatic heating as a bubble collapses.

12. Chemical effects. It has been established for many years that in certain circumstances, intense ultrasonics can have an effect on chemical reactions. Two classes may be distinguished; where the effect is to accelerate a known reaction, and where a direct chemical action, that would not otherwise have occurred, is brought about. The former may be attributable to a physical destruction of a protective film, as in the cavitation erosion of metals, or more generally to a "micro-stirring" of reagent which makes it unnecessary to depend on a slow diffusion process for a continuing supply of fresh material. NYBORG[1] has shown how this can be effective even at low ultrasonic intensities.

The more fundamental process of basically new reactions has received a good deal of attention in recent years. There is general agreement that cavitation is necessary. Little or no recent work appears to have been done at low ultrasonic frequencies, suggesting that, unlike many cavitation effects, chemical reaction rates do not fall off rapidly as the exciting ultrasonic frequency is raised. There are indications[2] that in the megacycle region an increase in frequency slightly reduces the effect. It has been claimed that in one case standing waves were needed before any action occurred but it is not clear whether this was simply to allow the formation of suitably sized nuclei under the conditions of the experiment.

In many cases, the phenomenon studied has been the formation of hydrogen peroxide in water: Some other reactions that have been observed could be explained as following upon this primary effect by simple chemical laws, but the breakdown of pyridine, for instance, examined by ZECHMEISTER and MAGOON[3], is difficult to account for in this way and is much more indicative of pyrolysis.

In studies on peroxide formation in water, it has been established that the oxidation action comes partly from water molecules that have been broken up: $H_2O \rightarrow H^{\cdot} + OH^{\cdot}$; $OH^{\cdot} + OH^{\cdot} \rightarrow H_2O_2$. When the water contains dissolved oxygen,

[1] W.L. NYBORG, R.K. GOULD, F.J. JACKSON and C.E. ADAMS: J. Acoust. Soc. Amer. **31**, 706 (1959).

[2] B.E. NOLTINGK and N.B. TERRY: Technical Acoustics II (E.G. RICHARDSON ed.). Amsterdam: Elsevier 1957.

[3] L. ZECHMEISTER and E.F. MAGOON: J. Amer. Chem. Soc. **78**, 2149 (1956).

that also takes part; isotope techniques have been used[1] to show that the two atoms from one oxygen molecule are incorporated into one peroxide molecule.

Several workers have remarked on the large changes that occur in reaction rates as the dissolved gas is varied[2-6], but the fact that different investigators differ widely in the results they report illustrates how complex the subject is. GUEGUEN et al.[7], demonstrate the importance of the concentration of the potassium iodide whose oxidation they study; presumably the larger number of oxidising agents produced when oxygen solutions are used instead of nitrogen ones demand a higher concentration of iodide to ensure that none of them are neutralised ineffectively. PARKE and TAYLOR observed that peroxide formation was much greater when less than 100% saturation content of oxygen was present. HENGLEIN found that a partial replacement of oxygen by other gases could greatly increase the yield as shown in Fig. 17. He explained this in terms of two effects:

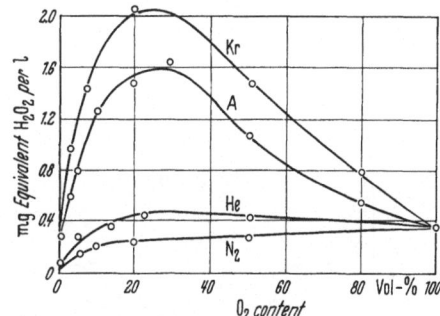

a primary break-up of water which could be much greater for the other gases: and a secondary interception of the radical formed, which had to occur before recombination of radicals if peroxide was to be formed, and which was more lik ly when dissolved oxygen was present. Analogous cases of reductions were observed with solutions of mixtures of argon and hydrogen.

It would seem that, although details remain obscure, results should generally be interpreted by consideration of these two steps. Active radicals must be

Fig. 17. Effect of the proportions of mixed dissolved gas on ultrasonically induced chemical actions (A. HENGLEIN).

produced in the course of cavitation; for water, these will be H' and OH', and for other substances, different radicals dependent on the bonds that can be broken. Radical production probably occurs in the vapour phase of a cavitation bubble; the precise mechanism is not yet established, but presumably has a close connection to the phenomenon of ultrasonic luminescence. The subsequent fate of the primary radical depends on what other matter is present, which thus affects the macroscopic chemical action observed. The parallel with radiation chemistry initiated by α-, β- and γ-rays has been pointed out several times[8,9], though pronounced differences have also been noted.

13. Effects in polymers. Solutions of long-chain high polymers can be modified when they are treated with intense ultrasonics. The phenomenon can quite readily be studied quantitatively, and many investigations have been made into different aspects. BERGMANN[10], GRASSIE[11] and JELLINEK[12] in their respective books

[1] M. DEL DUCA, E. YEAGER, M.O. DAVIES and F. HOVORKA: J. Acoust. Soc. Amer. **30**, 301 (1958).

[2] A. HENGLEIN: Naturwiss. **43**, 277 (1956).

[3] R.O. PRUDHOMME: J. Chim. phys. **54**, 332 (1957).

[4] M.E. FITZGERALD, V. GRIFFING and J. SULLIVAN: J. Chem. Phys. **25**, 926 (1956).

[5] A.V.M. PARKE and D. TAYLOR: J. Chem. Soc. 4442 (1956).

[6] O. LINDSTRÖM: J. Acoust. Soc. Amer. **27**, 654 (1955).

[7] H. GUEGUEN, P. RENAUD and N. SÉGARD: C. R. Acad. Sci., Paris **244**, 200 (1957).

[8] R.O. PRUDHOMME and P. GRABAR: J. Chim. phys. **46**, 323 (1949).

[9] N. MILLER: Trans. Faraday Soc. **46**, 546 (1950).

[10] L. BERGMANN: Der Ultraschall. Zürich 1954, Suppl. 1957.

[11] N. GRASSIE: Chemistry of high polymer degradation processes. London: Butterworth 1956.

[12] H.H.G. JELLINEK: Degradation of Vinyl Polymers. New York: Academic Press 1955.

have sections dealing with the subject up to the corresponding dates. Roberts, Yeager and Hovorka[1] review the literature as well as describing new experiments, while Mostafa's recent work[2] should also be mentioned.

In some cases, notably of naturally occurring high polymers, a spurious impression is given of degradation under the action of ultrasonics, because a temporary, thixotropic effect has brought about a reduction in viscosity of the solution. The viscosity, however, reverts to its original value after sufficient time, indicating that it was lowered by a dispersion of aggregates rather then by a breakage of chemical bonds. It is also, of course, necessary to guard against false values deriving from thermal effects.

When these are allowed for, a very definite effect of depolymerisation is observed. As the course of degradation is followed, curves are found such as those shown in Fig. 18, which refer to 25 ml samples of a 1% wt./vol. polystyrene solution in benzene irradiated at 0.75 Mc/sec. It is clear that a minimum molecular weight is approached, beyond which no degradation occurs.

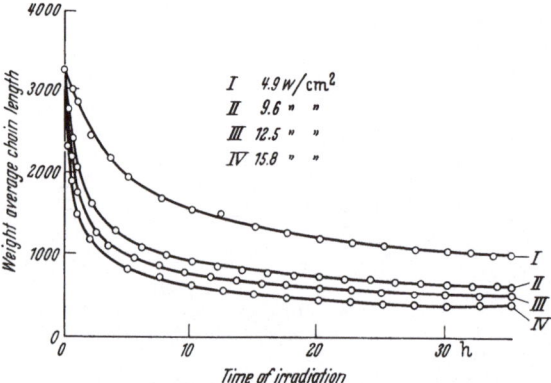

Fig. 18. Degradation of polystyrene under the action of ultrasonics (M.A.K. Mostafa).

Various attempts have been made to develop mathematical theories from this fact, which would allow the time variation and the size distribution of the solution to be predicted. These have been based on the assumption that for chain lengths greater than the limiting value there is a certain probability of breaking any chain link, which depends on the instantaneous acoustic conditions; the probability falls to zero for less than the critical chain length. It has been possible to reach a fair agreement with experimental results, but in view of the arbitrary constants available for adjustment, and the admittedly inexact assumptions made, it cannot be said that the conclusions are very convincing.

In any case, such calculations are not concerned with the primary problem of the mechanism of depolymerisation. Earlier suggestions had been that a viscous drag along the length of a molecule could lead to rupture—explaining immediately the minimum length effect. However, Roberts et al.[1] have analysed this possibility more carefully and shown that forces which might result are orders of magnitude too small to break chemical bonds. It is now agreed that the major part of ultrasonic depolymerisation is associated with the formation of bubbles in the liquid treated. Since it is known that polymers can be degraded by the stresses arising in mechanical mastication equipment, it is not altogether surprising that stresses associated with cavitation should also prove effective. A number of results, however, have shown that, even when true cavitation is prevented by increasing the ambient pressure beyond the peak alternating

[1] W. Roberts, E. Yeager and F. Hovorka: Tech. Rept. 18, Ultrasonic Research Lab. Western Reserve Univ. Cleveland, Ohio. 1957.
[2] M.A.K. Mostafa: J. Polymer Sci. 28, 499 and 519 (1958).

pressure, degradation still occurs as evidenced by viscosity changes. This can be seen, for instance, in Fig. 19 from the work of ROBERTS, YEAGER and HOVORKA which refers to 10 minute runs at 800 kc/sec in 0.5% solutions of polystyrene in toluene, saturated with nitrogen gas. Their data on variations with ultrasonic intensity at 1 atm ambient pressure are reported in Fig. 20. They show thresholds in the neighbourhood of 1 atm but not precisely at it, while MOSTAFA's threshold was about 3 atm. In view of the uncertainty of absolute intensity calibrations, the evidence from thresholds is not very conclusive either way. A further complication is that thresholds may be expected to vary with the degree of polymerisation of the material treated, since the critical minimum chain length has been shown to depend on intensity.

If depolymerisation is to be attributed to larger vibrating bubbles, moving symmetrically, rather than to collapsing voids, it is difficult to see how adequate stresses can be set up. It seems possible that more violent action could be associated with the break-up of larger bub-

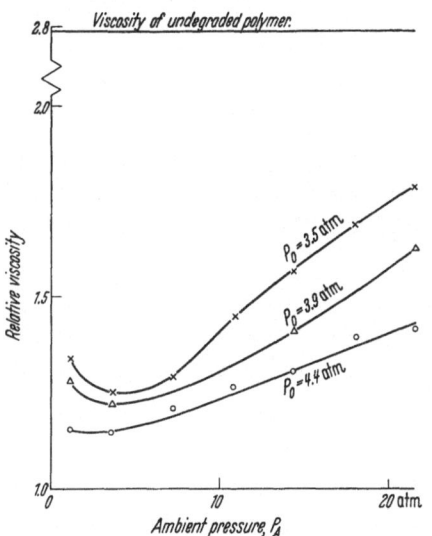

Fig. 19. Degradation at different ambient pressures (W. ROBERTS et al.).

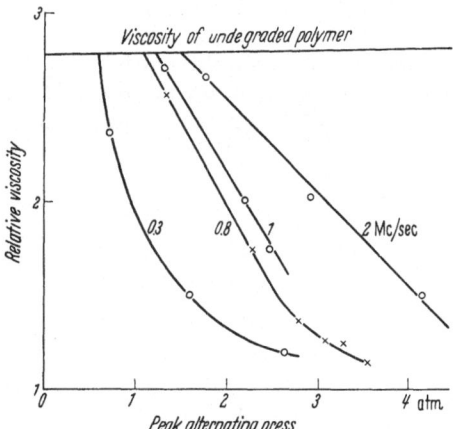

Fig. 20. Effect of intensity on degradation at different frequencies (W. ROBERTS et al.).

bles into smaller ones—presumably under the action of surface waves—which ROBERTS et al. have also remarked, as well as BOHN[1].

Fig. 20 indicates some, but not a large change in degradation effect with frequency. Elsewhere[2], even less variation over a wider frequency range has been reported.

A complex dependence on the concentration of solutions has been found. In some cases a marked variation has been observed according to the nature of the gas dissolved in the polymer solution, but here results appear contradictory.

Final conclusions over the mechanism of depolymerisation cannot yet be drawn. It appears probable that both true cavitation and strongly vibrating bubbles play their parts. It is possible that secondary reactions are sometimes important from substances such as hydrogen peroxide produced as described in the previous

[1] L. BOHN: Acustica 7, 201 (1957).
[2] G. SCHMID and W. POPPE: Z. Elektrochem. 53, 28 (1949).

chapter. For the few cases in which ultrasonically induced polymerisations have been reported, such secondary chemical actions seem the most likely explanation.

14. Effects in metals. A specialised case of the working of intense ultrasonics in liquids occurs when the liquid is a molten metal. There are particular difficulties in this field (apart from metals with very low melting points) because of the need to conduct the vibrations into a high temperature region. The effect of irradiating metal melts with ultrasonics as they solidify has, however, been studied and significant changes have been demonstrated, although in such a subject it is difficult to express results quantitatively.

Hiedemann[1] has reviewed experiments undertaken more than a few years ago. Few comprehensive investigations have been carried out more recently. The principle changes brought about by ultrasonic treatment are grain refinement and reduction in gas content. These are sometimes found to be associated with improved mechanical properties, but no very striking increase in strength has been reported. The great majority of work has been done with metals such as aluminium in order to avoid the highest temperatures. Frequencies used have ranged from 50 c/sec to 4.5 Mc/sec with apparently little difference between their effects. It has been suggested that the reduction of grain size is due to the breakdown of crystallites after they have been formed, but it is now widely accepted that vibration promotes the primary formation of nuclei at a larger number of different points. This might be expected to be a cavitation phenomenon, but there is no convincing experimental evidence either way.

Degassing presumably occurs along similar lines in molten metals and in other liquids. Another effect that has been observed is the emulsification of immiscible molten metals, resulting, on freezing, in a dispersion of one in the other which is difficult or impossible to achieve by other means.

Addendum.

Since this article was written, a number of papers have appeared dealing with the subjects treated. They are, in general, the sort of developments that might have been expected and do not seem to invalidate any of the main conclusions. For completeness, however, a summary is included here.

Hsieh and Plesset[2] have developed a more complete theory for the phenomenon of "rectified diffusion" discussed in Sect. 2; this appears to account more satisfactorily for observed effects.

An interesting approach to cavitation thresholds (Sect. 7) is outlined in a short note by Mikhailov and Shutilov[3]. They show that, if liquid in a closed vessel is subjected to ultrasonic radiation, a sudden increase in volume occurs when the intensity exceeds a critical value which they identify with the threshold for cavitation.

Developments in the field of dispersion (Sect. 10) have been mainly concerned with empirical technological applications. Rozenberg and Eknadiosyants [4], however, have used high speed cinematography in an extensive investigation into ultrasonic fog formation. They state that their results still leave it undecided how far fog formation results from surface waves and how far it is caused by cavitation.

[1] E. Hiedemann: J. Acoust. Soc. Amer. **26**, 831 (1954).
[2] D.-Y. Hsieh and M. S. Plesset: J. Acoust. Soc. Amer. **33**, 206 (1961).
[3] I. G. Mikhailov and V. A. Shutilov: Soviet Phys. (Acoustics) **5**, 385 (1960).
[4] L. D. Rozenberg and O. K. Eknadiosyants: Soviet Phys. (Acoustics) **6**, 369 (1961).

Several recent papers[1-3] have a bearing on the subject of luminescence dealt with in Sect. 11. These can be said to contain further evidence in favour of associating luminescence with the conditions occurring at the collapse of a bubble; JARMAN[4] has suggested the modification of attributing luminescence to the shock wave set up at collapse rather than directly to an adiabatic temperature rise.

GOOBERMAN and LAMB[5] have developed a theoretical treatment for ultrasonic depolymerisation (Sect. 13). As earlier workers, the mechanism they assume is viscous drag, but with the important difference that the velocity gradient needed to cause it is taken to arise from cavitation shock waves rather than from primary vibrations. Predictions from the theory are found to be in agreement with experiments. EL'PINER[6] has reported the interesting observation that molecular weight can increase or decrease according as the polymer solution irradiated is saturated with hydrogen or with oxygen; this conforms with the idea that secondary chemical actions must be taken into account.

[1] P. GÜNTHER, E. HEIM u. G. EICHKORN: Z. angew. Phys. **11**, 274 (1959).
[2] D. SRINIVASAN and L. V. HOLROYD: J. Appl. Phys. **32**, 446 (1961).
[3] R. Q. MACLEAY and L. V. HOLROYD: J. Appl. Phys. **32**, 449 (1961).
[4] P. JARMAN: J. Acoust. Soc. Amer. **32**, 1459 (1960).
[5] G. GOOBERMAN and J. LAMB: J. Polymer Sci. **42**, 25, 35 (1960).
[6] I. E. EL'PINER: Soviet Phys. (Acoustics) **6**, 399 (1961).

Sachverzeichnis.

(Deutsch-Englisch.)

Bei gleicher Schreibweise in beiden Sprachen sind die Stichwörter nur einmal aufgeführt.

Subject Index.

(English-German.)

Where English and German spelling of a word is identical the German version is omitted.

Dielectric constants of piezoelectric materials, *Dielektrizitätskonstanten piezoelektrischer Stoffe* 47.

Diffraction of light by an ultrasonic wave, *Beugung des Lichtes an einer Ultraschallwelle* 124—125.

— of sound by a Rayleigh disc, *Beugung des Schalls an einer Rayleighschen Scheibe* 59.

Diffusion of kinks in dislocations, *Diffusion von Versetzungssprüngen* 208.

Dipole nature of a jet-edge source, *Dipolnatur einer Strahl-Schneide-Quelle* 28.

Dipole source, *Dipol-Quelle* 1, 3, 13.

Directivity of condenser microphone, *Richtungsabhängigkeit des Kondensatormikrophons* 43, 44.

Directivity index, *Bündelungsindex* 2.

Directivity of ribbon microphones, *Richtungsabhängigkeit des Bandmikrophons* 57.

Directivity of sound radiation from a gun, *Richtungsverteilung der Schallstrahlung eines Geschützes* 33.

Directivity of an ultrasound generator, *Richtungsverteilung einer Ultraschallquelle* 103 to 105.

Dislocation damping of sound waves, *Versetzungsdämpfung von Schallwellen* 177 to 208.

Dislocation density, *Versetzungsdichte* 184.

Dislocation multiplication, *Versetzungsvervielfachung* 198.

Dislocation strain, *Versetzungsformänderung* 178.

Displacement gradients, *Verrückungsgradienten* 154.

Displacement vector, *Verschiebungsvektor* 154.

Dissipation of ultrasonic energy, *Dissipation der Ultraschallenergie* 162.

Distortion of amplitudes in direct loudspeakers, *Verzerrung der Amplituden in Direktlautsprechern* 12.

— by second harmonic, *durch Oberschwingungen* 8, 40.

Distribution of radiated intensity for special sources, *Verteilung der abgestrahlten Intensität für spezielle Quellen* 104—105.

Domain walls, interaction with stress waves, *Wände magnetischer Bezirke, Wechselwirkung mit elastischen Wellen* 223, 227.

Doppler effect for diffracted light, *Doppler-Effekt für gebeugtes Licht* 133—136.

Doublet source see dipole source, *Doppelquelle s. Dipolquelle.*

Dynamic microphone, *dynamisches Mikrophon* 37.

Ear, detection of pressure amplitudes, *Ohr, Nachweis von Druckamplituden* 35.

Edge hit by a jet, *Kante, die von einem Strahl angeblasen wird* 97.

Efficiency changes of electrostatic loudspeakers, *Wirkungsgradänderungen bei elektrostatischen Lautsprechern* 17.

Efficiency of an electrodynamic loudspeaker, *Wirkungsgrad eines elektrodynamischen Lautsprechers* 10.

— of the Hartmann generator, *des Hartmannschen Generators* 31.

— of a horn loudspeaker, *eines Hornlautsprechers* 20.

— of a siren, *einer Sirene* 25.

— of sound generation by cannons, *der Schallerzeugung durch Kanonen* 34.

— of St. Clair generator, *des St. Clairschen Schallgenerators* 24.

Elastic after-effect, *elastische Nachwirkung* 173.

Elastic coefficients, *elastische Koeffizienten* 154.

Elastic energy, *elastische Energie* 158 to 159.

Elastic energy flux, *elastischer Energiefluß* 158, 159.

Elastic potential, *elastisches Potential* 159.

— —, existence, *Existenz* 154.

Elastic sphere in an elastic medium, *elastische Kugel in einem elastischen Medium* 166.

Electrical detection interferometer for ultrasound, *elektrisches Nachweisinterferometer für Ultraschall* 142—143.

Electrical discharge in the cavitation void, *elektrische Entladung im Kavitationshohlraum* 281.

Electrical impedance, *elektrische Impedanz* 5.

Electrical quadrupole tensor, *elektrischer Quadrupoltensor* 236.

Electrodynamic loudspeaker, *elektrodynamischer Lautsprecher* 9—12.

Electrodynamic pistonphone, *elektrodynamisches Kolbenmikrophon* 63—64.

Electrodynamic pressure microphone, *elektrodynamisches Druckmikrophon* 36—38.

Electromagnetic transducers, *elektromagnetische Ultraschallerzeuger* 92—94.

Electron mean free path comparable with ultrasonic wavelength, *freie Weglänge der Elektronen vergleichbar mit Wellenlänge der Ultraschallwellen* 208, 210.

Electrostatic actuator of BALLANTINE, *elektrostatischer Aktuator von Ballantine* 64 to 66.

Electrostatic loudspeaker, *elektrostatischer Lautsprecher* 15—17.

Electrostatic probe, *elektrostatische Ultraschallsonde* 116.

Emission of a phonon, *Emission eines Schallquants* 216.

Emulsification by ultrasonics, *Emulsionsbildung durch Ultraschall* 277, 278.

Enclosures of electrodynamic loudspeakers, *Gehäuse für elektrodynamische Lautsprecher* 13—15.

Energy, elastic, *elastische Energie* 158—159.

— of kink formation in dislocations, *der Sprungbildung in Versetzungen* 205—206.

Energy barrier for a dislocation, *Energieschwelle für eine Versetzung* 189.

Plastic foils in electrostatic loudspeakers, *Plastikfolien in elektrischen Lautsprechern* 17.

Plastic lenses, *Plastiklinsen* 114.

Plate, circular, directivity, *kreisförmige Platte, Richtungsverteilung der abgestrahlten Intensität* 104.

—, rectangular, directivity, *rechteckige Platte, Richtungsverteilung der abgestrahlten Intensität* 104.

Plexiglass lens, *Plexiglaslinse* 114.

Pohlmann whistle, *Pohlmannsche Pfeife* 98, 260, 278.

Point defects, *Punktfehlordnungen* 188, 189, 193.

Polymerization by intense ultrasonics, *Polymerisierung durch intensiven Ultraschall* 285.

Polymers, treatment by intense ultrasonics, *Polymere, Behandlung mit intensivem Ultraschall* 283—285.

Potential energy of elastic strain, *potentielle Energie einer elastischen Verformung* 159.

Power radiated by an electrodynamic loudspeaker, *Leistung eines elektrodynamischen Lautsprechers* 10.

— — from piezoelectric quartz, *von einem Piezoquarz abgestrahlte Leistung* 81.

Pressure calibration by reciprocity method, *Druckeichung nach der Reziprozitätsmethode* 68—73.

Pressure ratio, *Druckverhältnis* 52.

Pressure sensitivity of barium titanate, *Druckempfindlichkeit von Bariumtitanat* 48.

Pressure in a sound wave, *Druck in einer Schallwelle* 115.

Probe microphone, *Sondenmikrophon* 50—56.

Propagation constants of ultrasound, determination, *Ausbreitungskonstanten des Ultraschalls, Bestimmung* 142—152.

Propagation direction, *Fortpflanzungsrichtung* 156.

— —, misorientation, *Orientierungsfehler* 160.

Propagation factor, *Wellenzahl* 162.

Propagation of stress waves, *Ausbreitung elastischer Wellen* 153, 156—159.

— of stress waves in piezoelectric crystals, *elastischer Wellen in piezoelektrischen Kristallen* 244—246.

Propagation velocity, *Fortpflanzungsgeschwindigkeit* 156.

Pulsating sphere, radiation impedance, *pulsierende Kugel, Strahlungsimpedanz* 1.

Pulse method, *Impulsmethode* 150.

Pure cavitation see cavitation, *reine Kavitation s. Kavitation.*

Push-pull electrostatic transducer, *elektrostatischer Kipp-Schallerzeuger* 8, 9, 16.

Quadrupole interaction of an atomic nucleus, *Quadrupolwechselwirkung eines Atomkerns* 236.

Quadrupole source, *Quadrupol-Quelle* 1, 3.

Quartz bar, excitation of hypersound, *Quarzstab, Erregung von Hyperschall* 101.

Quartz, electric constants, temperature dependence, *Quarz, elektrische Konstanten, Temperaturabhängigkeit* 47.

— as piezoelectric transducer, *als piezoelektrischer Schallerzeuger* 76, 77.

—, temperature dependence of ultrasonics, *Temperaturabhängigkeit der Ultraschallwellen* 230, 231.

Q-value of dissipation, *Q-Wert der Dissipation* 163, 181.

Radiation field of ultrasound source, *abgestrahltes Feld einer Ultraschallquelle* 102 to 114.

Radiation impedance of a circular piston, *Strahlungsimpedanz eines Kolbens von Kreisquerschnitt* 2.

— — of a pulsating sphere, *einer pulsierenden Kugel* 1.

Radiation induced defects, *strahlungserzeugte Fehlstellen* 191.

Radiation pressure, *Strahlungsdruck* 115, 116 to 123.

Radiometer for ultrasound waves, *Radiometer für Ultraschallwellen* 121—122.

Radius of cavitation bubbles, *Radius von Kavitationsblasen* 264, 267—268, 274.

RAMAN and NATH's formula, *Raman-Nathsche Formel* 128—131.

Rayleigh disc, *Rayleighsche Scheibe* 59 to 61.

Rayleigh scattering of stress waves, *Rayleighsche Streuung elastischer Wellen* 165, 166, 170.

Rayleigh wave, *Rayleighsche Oberflächenwelle* 76.

Reciprocity parameter, *Reziprozitätsparameter* 71.

Reciprocity relations for electromechanical transducers, *Reziprozitätsbeziehungen für elektromechanische Schallwandler* 66—68.

Rectangular plate, directivity, *Rechteckplatte, Richtungsverteilung der abgestrahlten Intensität* 104.

Rectified diffusion, *rektifizierte Diffusion* 261, 286.

Reflection producing standing waves, *Reflexion zur Erzeugung stehender Wellen* 111.

— of ultrasound waves, *von Ultraschallwellen* 106—108, 110.

Refraction index of light in an ultrasonic wave, *Brechungsindex des Lichtes in einer Ultraschallwelle* 125.

Refraction of ultrasound waves, *Brechung von Ultraschallwellen* 106—108, 110.

Relaxation leading to Bordoni peaks, *Relaxation als Ursache Bordonischer Maxima* 203.

— of phonon distribution, *der Schallquantenverteilung* 234.

Relaxation time, *Relaxationszeit* 174.